An Introduction to Enzyme and Coenzyme Chemistry

TIM BUGG
Department of Chemistry
University of Southampton

b

**Blackwell
Science**

© 1997 by
Blackwell Science Ltd
Editorial Offices:
Osney Mead, Oxford OX2 0EL
25 John Street, London WC1N 2BL
23 Ainslie Place, Edinburgh EH3 6AJ
350 Main Street, Malden
   MA 02148 5018, USA
54 University Street, Carlton
   Victoria 3053, Australia
10, rue Casimir Delavigne
   75006 Paris, France

Other Editorial Offices:
Blackwell Wissenschafts-Verlag GmbH
Kurfürstendamm 57
10707 Berlin, Germany

Blackwell Science KK
MG Kodenmacho Building
7–10 Kodenmacho Nihombashi
Chuo-ku, Tokyo 104, Japan

First published 1997
Reprinted 2000

Set by Semantic Graphics, Singapore
Printed and bound in Great Britain
at the Alden Press Ltd, Oxford and
Northampton

The Blackwell Science logo is a
trade mark of Blackwell Science Ltd
registered at the United Kingdom
Trade Marks Registry

DISTRIBUTORS

Marston Book Services Ltd
PO Box 269
Abingdon
Oxford OX14 4YN
(Orders: Tel: 01235 465500
          Fax: 01235 465555)

USA
Blackwell Science, Inc.
Commerce Place
350 Main Street
Malden, MA 02148 5018
(Orders: Tel: 800 759 6102
               781 388 8250
         Fax: 781 388 8255)

Canada
Login Brothers Book Company
324 Saulteaux Cresent
Winnipeg, Manitoba R3J 3T2
(Orders: Tel: 204 224-4068)

Australia
Blackwell Science Pty Ltd
54 University Street
Carlton, Victoria 3053
(Orders: Tel: 3 9347 0300
         Fax: 3 9347 5001)

A catalogue record for this title
is available from the British Library

ISBN 0-86542-793-3

Library of Congress
Cataloging-in-publication Data

Bugg, Tim.
   An Introduction to enzyme and
   coenzyme chemistry / Tim Bugg.
         p.   cm.
   Includes bibliographical references
   and index.
   ISBN 0-86542-793-3
   1. Enzymes.   2. Coenzymes.
   I. Title.
QP601.B955   1997
547.7'58—dc20   96-26495
                                      CIP

For further information on
Blackwell Science, visit our website:
www.blackwell-science.com

AN INTRODUCTION TO
ENZYME AND
COENZYME CHEMISTRY

Contents

Colour plates fall between pp. 152 and 153
Problems and further reading are included at the end of each chapter

Preface

The motivation for writing this book came from my own frustration at having to use a large number of sources to find material for second- and third-year undergraduate teaching about enzyme and coenzyme mechanisms. Whilst there are many biochemical texts which describe enzyme action, few contain sufficient chemistry to satisfy the appetite of an organic chemist. I thought, therefore, that a book containing a description of enzyme chemistry from an organic chemistry perspective would be of interest to a number of chemistry students, and perhaps also be of interest to biochemists and medical students who wanted a more 'molecular' treatment of enzyme catalysis.

The book is not intended to be a comprehensive treatment of enzyme chemistry, but to give an introduction to the principles of enzyme action, and to introduce the major classes of enzymatic reactions and the current ideas regarding enzyme mechanisms. Some of the material has been gleaned from my undergraduate and postgraduate lecture courses, but much of the material in Chapters 5–10 was collated with the help of my research group, with whom I have explored the various classes of enzyme mechanisms over the last 2–3 years. My thanks go to them for helping to find this material, and in some cases for translating research papers into plain English!

I think it is important to have a good mental picture of what enzymes and active sites look like, so I resolved to illustrate several examples in the book with colour pictures of their X-ray crystal structures. In preparing these images I would like to thank Dr Jonathan Essex for his time and assistance, and Dr Jim Knox for some nice pictures of D-alanine: D-alanine ligase.

Finally, I would like to thank several people. I would like to thank Dr Leslie Johnson for his incisive proofreading of the script, and my research students for proofreading and encouragement (especially John Pollard for the photograph of his enzyme purification). I would like to thank my former mentors Drs Chris Abell and Chris Walsh for inspiration: I now realize what a monumental effort Chris Walsh's excellent book on enzymatic reaction mechanisms must have been! I would like to thank my parents, and thanks to the members of Team Volleyball News Solent for putting it all into perspective. Thanks to the latter for a great season last year in Division 2 of the National Volleyball League, and a 'learning experience' this year in Division 1!

Tim Bugg
Southampton
April 1996

1   From Jack Beans to Designer Genes

1.1   Introduction

Enzymes are giant macromolecules which catalyse biochemical reactions. They are remarkable in many ways. Their three-dimensional structures are highly complex, yet they are formed by spontaneous folding of a linear polypeptide chain. Their catalytic properties are far more impressive than synthetic catalysts which operate under more extreme conditions. Each enzyme catalyses a single chemical reaction on a particular chemical substrate with very high enantioselectivity and enantiospecificity at rates which approach 'catalytic perfection'. Living cells are capable of carrying out a huge repertoire of enzyme-catalysed chemical reactions, some of which have little or no precedent in organic chemistry. In this book I shall seek to explain from the perspective of organic chemistry what enzymes are, how they work and how they catalyse many of the major classes of enzymatic reactions.

1.2   Discovery of enzymes

Historically, biological catalysis has been used by mankind for thousands of years, ever since fermentation was discovered as a process for brewing and bread-making in ancient Egypt. It was not until the 19th century AD, however, that scientists addressed the question of whether the entity responsible for processes such as fermentation was a living species or a chemical substance. In 1897 Eduard Buchner published the observation that extracts of yeast containing no living cells were able to carry out the fermentation of sugar to alcohol and carbon dioxide. He proposed that a species called 'zymase' found in yeast cells was responsible for fermentation. The biochemical pathway involved in fermentation was subsequently elucidated by Embden and Meyerhof—the first pathway to be discovered. Finally, in 1926 the crystallization of the enzyme urease (Fig. 1.1) from Jack beans proved beyond doubt that biological catalysis was carried out by a chemical substance.

Fig. 1.1 Reaction catalysed by the enzyme urease.

$$H_2N\!-\!\overset{\overset{\displaystyle O}{\|}}{C}\!-\!NH_2 + H_2O \xrightarrow{\text{Jack bean urease}} CO_2 + 2\,NH_3$$

The recognition that biological catalysis is mediated by enzymes heralded the growth of biochemistry as a subject, and the elucidation of the metabolic pathways catalysed by enzymes. Each reaction taking place on a biochemical pathway is catalysed by a specific enzyme. Without enzyme catalysis the uncatalysed chemical process would be too slow to sustain life. Enzymes catalyse reactions involved in all facets of cellular life: metabolism (the production of cellular building blocks and energy from food sources); biosynthesis (how cells make new molecules); detoxification (the breakdown of toxic foreign chemicals); and information storage (the processing of deoxyribonucleic acids (DNAs)).

In any given cell there are present several thousand different enzymes, each catalysing its specific reaction. How does a cell know when it needs a particular enzyme? The production of enzymes, as we shall see in Chapter 2, is controlled by a cell's DNA, both in terms of the specific structure of a particular enzyme and the amount which is produced. Thus, different cells in the same organism have the ability to produce different types of enzymes and to produce them in differing amounts according to the cell's requirements.

Since the majority of the biochemical reactions involved in cellular life are common to all organisms, a given enzyme will usually be found in many or all organisms, albeit in different forms and amounts. By looking closely at the structures of enzymes from different organisms which catalyse the same reaction, we can in many cases see similarities between them. These similarities are due to the evolution and differentiation of species by natural selection. So, by examining closely the similarities and differences of an enzyme from different species we can trace the course of molecular evolution, as well as learning about the structure and function of the enzyme itself.

Recent developments in biochemistry, molecular biology and X-ray crystallography now allow a far more detailed insight into how enzymes work at a molecular level. We now have the ability to determine the amino acid sequence of enzymes with relative ease, whilst the technology for solving the three-dimensional structure of enzymes is developing apace. We also have the ability to analyse their three-dimensional structures using molecular modelling and then to change the enzyme structure rationally using site-directed mutagenesis. We are now starting to enter the realms of enzyme engineering, where by rational design we can modify the genes encoding specific enzymes, creating the 'designer genes' of the title. These modified enzymes could in future perhaps be used to catalyse new types of chemical reactions, or via gene therapy to correct genetic defects in cellular enzymes which would otherwise lead to human diseases.

1.3   The discovery of coenzymes

At the same time as the discovery of enzymes in the late 19th and early 20th centuries, a class of biologically important small molecules was being discovered which had remarkable properties to cure certain dietary disorders. These molecules were christened the vitamins, a corruption of the phrase 'vital amines' used to describe their dietary importance (several of the first-discovered vitamins were amines, but this is not true of all the vitamins). The vitamins were later found to have very important cellular roles, shown in Table 1.1.

The first demonstration of the importance of vitamins in the human diet took place in 1753. A Scottish naval physician, James Lind, found that the disease scurvy, prevalent amongst mariners at that time, could be avoided by deliberately including green vegetables or citrus fruits in the sailors' diets. This discovery was used by Captain James Cook to maintain the good health of his crew during his voyages of exploration in 1768–76. The active ingredient was elucidated much later as vitamin C, ascorbic acid.

The first vitamin to be identified as a chemical substance was thiamine, lack of which causes the limb paralysis beriberi. This nutritional deficiency was first identified in the Japanese Navy in the late 19th century. The

Table 1.1 The vitamins.

Vitamin	Chemical name	Deficiency disease	Biochemical function	Coenzyme chemistry
A	Retinol	Night blindness	Visual pigments	—
B_1	Thiamine	Beriberi	Coenzyme (TPP)	Decarboxylation of α-keto acids
B_2	Riboflavin	Skin lesions	Coenzyme (FAD, FMN)	$1e^-/2e^-$ redox chemistry
Niacin	Nicotinamide	Pellagra	Coenzyme (NAD)	Redox chemistry
B_6	Pyridoxal	Convulsions	Coenzyme (PLP)	Reactions of α-amino acids
B_{12}	Cobalamine	Pernicious anaemia	Coenzyme	Radical re-arrangements
C	Ascorbic acid	Scurvy	Coenzyme, anti-oxidant	Redox agent (collagen formation)
D	Calciferols	Rickets	Calcium homeostasis	—
E	Tocopherols	Newborn haemolytic anaemia	Anti-oxidant	—
H	Biotin	Skin lesions	Coenzyme	Carboxylation
K	Phylloquinone	Bleeding disorders	Coenzyme, anti-oxidant	Carboxylation of glutamyl peptides
	Folic acid	Megaloblastic anaemia	Coenzyme (tetrahydrofolate)	1-carbon transfers
	Pantothenic acid	Burning foot syndrome	Coenzyme (CoA, phosphopantotheine)	Acyl transfer

CoA, coenzyme A; FAD, flavin adenine dinucleotide; FMN, flavin mononucleotide; NAD, nicotinamide adenine dinucleotide; PLP, pyridoxal-5'-phosphate; TPP, thiamine pyrophosphate.

incidence of beriberi in sailors was connected with their diet of polished rice by Admiral Takaki, who eliminated the ailment in 1885 by improvement of the sailors' diet. Subsequent investigations by Eijkman identified a substance present in rice husks able to cure beriberi. This vitamin was subsequently shown to be an essential 'cofactor' in cellular decarboxylation reactions, as we shall see in Chapter 7.

Over a number of years the family of vitamins shown in Table 1.1 was identified and their chemical structures elucidated. Some, like vitamin C, have simple structures, whilst others, like vitamin B_{12}, have very complex structures. It has taken much longer to elucidate the molecular details of their biochemical mode of action. Many of the vitamins are converted in animal cells into coenzymes: small organic cofactors which are used by certain types of enzyme in order to carry out particular classes of reaction. Table 1.1 gives a list of the coenzymes that we are going to encounter in this book.

1.4   The commercial importance of enzymes in biosynthesis and biotechnology

Many plants and micro-organisms contain natural products which possess potent biological activities. The isolation of these natural products has led to the discovery of many biologically active compounds such as quinine, morphine and penicillin (Fig. 1.2) which have been fundamental to the development of modern medicine.

The process of natural product discovery continues today, with the recent identification of important compounds such as cyclosporin A, a potent immunosuppressant which has dramatically reduced the rejection rate in organ transplant operations; and taxol, an extremely potent anticancer drug isolated from yew bark (Fig. 1.3).

Many of these natural products are structurally so complex that it is not feasible to synthesize them in the laboratory at an affordable price. Nature, however, is able to biosynthesize these molecules with apparent ease using enzyme-catalysed biosynthetic pathways. Hence, there is considerable interest in elucidating the biosynthetic pathways for important natural products

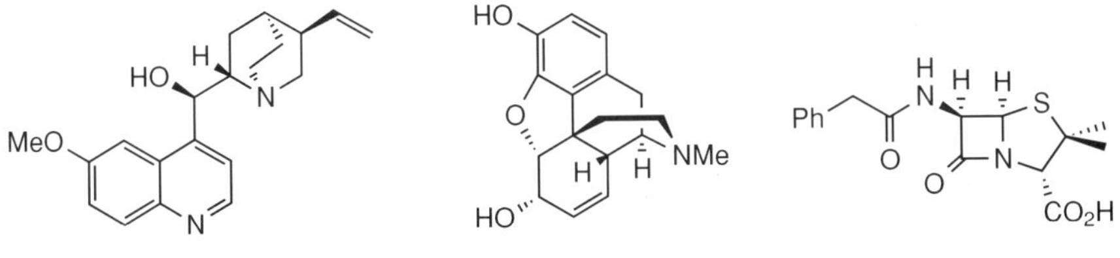

Quinine   Morphine   Penicillin G

Fig. 1.2 Structures of quinine, morphine and penicillin G.

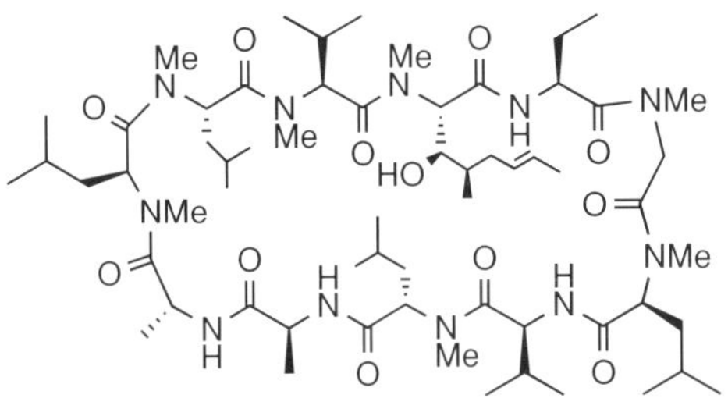

Fig. 1.3 Structures of cyclosporin A and taxol.

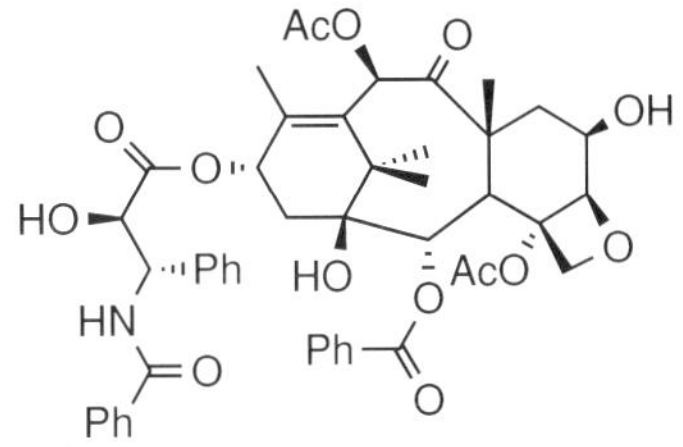

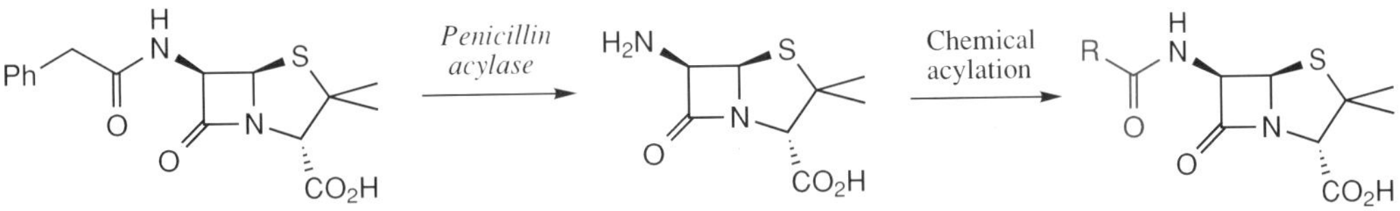

Fig. 1.4 Industrial production of a semi-synthetic penicillin using penicillin acylase.

and using the enzymes to produce natural products *in vitro*. One example of this is the industrial production of semi-synthetic penicillins using a naturally occurring enzyme, penicillin acylase (Fig. 1.4). Penicillin G, which is obtained from growing *Penicillium* mould, has certain clinical disadvantages; enzymatic deacylation and chemical re-acylation give a whole range of 'semi-synthetic' penicillins which are clinically more useful.

The use of enzyme catalysis for commercial applications is an exciting area of the biotechnology industry. One important application that we shall encounter is the use of enzymes in asymmetric organic synthesis. Since enzymes are highly efficient catalysts that work under mild conditions and are enantiospecific, they can in many cases be used on a practical scale to resolve racemic mixtures of chemicals into their optically active components. This is becoming increasingly important in drug synthesis, since one enantiomer of a drug usually has very different biological properties from the other. The unwanted enantiomer might have detrimental side-effects, as in the case of thalidomide, where one enantiomer of the drug was useful in relieving morning sickness in pregnant women, but the other enantiomer caused serious deformities in the newborn child when the racemic drug was administered.

1.5   The importance of enzymes as targets for drug discovery

If there is an *essential* enzyme found uniquely in a certain class of organisms or cell type, then a selective *inhibitor* of that enzyme could be used for

Table 1.2 Commercial applications of enzyme inhibitors.

Antibacterial agents	**Penicillins** and **cephalosporins** inactivate the *transpeptidase* enzyme which normally makes cross-links in the bacterial cell wall (peptidoglycan), leading to weakened cell walls and eventual cell lysis. **Streptomycin** and **kanamycin** inhibit protein synthesis on bacterial ribosomes, whereas mammalian ribosomes are less affected.
Antifungal agents	**Ketoconazole** inhibits *lanosterol 14α-demethylase*, an enzyme involved in the biosynthesis of an essential steroid component of fungal cell membranes. **Nikkomycin** inhibits *chitin synthase*, an enzyme involved in making the chitin cell walls of fungi.
Antiviral agents	**AZT** inhibits the *reverse transcriptase* enzyme required by the human immunodeficiency virus (HIV) in order to replicate its own RNA.
Insecticides	Organophosphorus compounds such as **dimethoate** derive their lethal activity from the inhibition of the insect enzyme *acetylcholinesterase* involved in the transmission of nerve impulses.
Herbicides	**Glyphosate** inhibits the enzyme *EPSP synthase* which is involved in the biosynthesis of the essential amino acids phenylalanine, tyrosine and tryptophan (see Chapter 8, Section 8.5).

AZT, 3′-azido,3′-deoxy-thymidine; EPSP, 5-enolpyruvyl-shikimate-3-phosphate.

selective toxicity against that organism or cell type. Similarly, if there is a significant difference between a particular enzyme found in bacteria as compared with the same enzyme in humans, then a selective inhibitor could be developed for the bacterial enzyme. If this inhibitor did not inhibit the human enzyme, then it could be used as an antibacterial agent. Thus, *enzyme inhibition is a basis for drug discovery.*

This principle has been used for the development of a range of pharmaceutical and agrochemical agents (Table 1.2); we shall see examples of important enzyme targets later in the book. In many cases resistance to these agents has emerged due to mutation in the structures of the enzyme targets. This has provided a further incentive to study the three-dimensional structures of enzyme targets, and has led to the development of powerful molecular modelling software for analysis of enzyme structure and *de novo design* of enzyme inhibitors.

The next two chapters are 'theory' chapters on enzyme structure and enzyme catalysis, followed by a 'practical' chapter on methods used to study enzymatic reactions. Chapters 5–10 cover each of the major classes of enzymatic reactions, noting each of the coenzymes used for enzymatic reactions. Finally, there is a brief introduction in Chapter 11 to other types of biological catalysis. In cases where discussion is brief the interested reader will find references to further reading at the end of each chapter.

2   All Enzymes are Proteins

2.1   Introduction

Enzymes are giant molecules. Their molecular weight varies from 5000 to
5 000 000 Da, with typical values in the range 20 000–100 000 Da. At first
sight this size suggests a bewildering complexity of structure, yet we shall see
that enzymes are structurally assembled in a small number of steps in a fairly
simple way.

Enzymes belong to a larger biochemical family of macromolecules known
as proteins. The common feature of proteins is that they are polypeptides:
their structure is made up of a linear sequence of α-amino acid building
blocks joined together by amide linkages. This linear polypeptide chain then
folds to give a unique three-dimensional structure.

2.2   The structures of the L-α-amino acids

Proteins are composed of a family of 20 α-amino acid structural units whose
general structure is shown in Fig. 2.1. α-Amino acids are chiral molecules:
that is, their mirror image is not superimposable upon the original molecule.

Each α-amino acid can be found as either the L- or D-isomer depending
on the configuration at the α-carbon atom (except for glycine where R = H).
All proteins are composed only of L-amino acids; consequently enzymes are
inherently chiral molecules — an important point. D-Amino acids are rare in
biological systems, although they are found in a number of products and
notably in the peptidoglycan layer of bacterial cell walls (see Chapter 9).

The α-amino acids used to make up proteins number only 20, whose
structures are shown in Fig. 2.2. The differences between these 20 lie in the
nature of the side chain R. The simplest amino acids are glycine (abbreviated
Gly or simply G), which has no side chain, and alanine (Ala or A), whose
side chain is a methyl group. A number of side chains are hydrophobic
('water-hating') in character: for example, the thioether of methionine (Met);

general structure of
an L-α-amino acid

general structure of
a D-α-amino acid

Fig. 2.1 General structure of L and
D-amino acids.

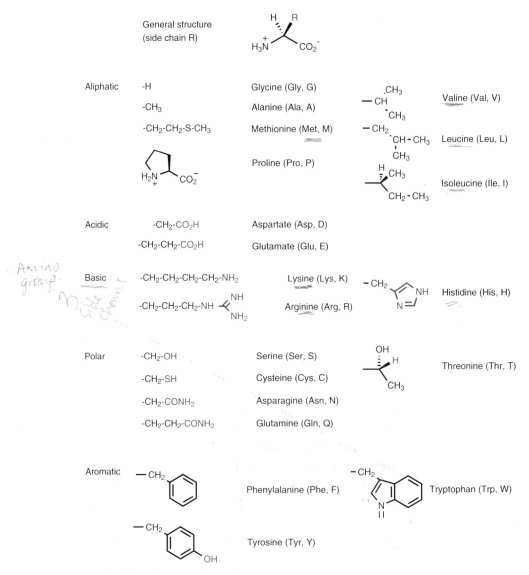

Fig. 2.2 The side chains of the 20 α-amino acids found in proteins. Whole amino acid structure shown for proline. Functionally important groups highlighted in red.

the branched aliphatic side chains of valine (Val), leucine (Leu) and isoleucine (Ile); and the aromatic side chains of phenylalanine (Phe) and tryptophan (Trp). The remainder of the amino acid side chains are hydrophilic ('water-loving') in character. Aspartic acid (Asp) and glutamic acid (Glu) contain carboxylic acid side chains, and their corresponding primary amides are found as asparagine (Asn) and glutamine (Gln). There are three basic side chains consisting of the ε-amino group of lysine (Lys), the guanidine group of arginine (Arg) and the imidazole ring of histidine (His). The polar nucleophilic side chains that will assume a key role in enzyme catalysis are the primary hydroxyl of serine (Ser), the secondary hydroxyl of threonine (Thr), the phenolic hydroxyl group of tyrosine (Tyr) and the thiol group of cysteine (Cys).

The nature of the side chain confers certain physical and chemical properties upon the corresponding amino acid, and upon the polypeptide chain in which it is located. The amino acid side chains are therefore of considerable structural importance and, as we shall see in Chapter 3, they play key roles in the catalytic function of enzymes.

2.3   The primary structure of polypeptides

To form the polypeptide chain found in proteins each amino acid is linked to the next via an amide bond, forming a linear sequence of 100–1000 amino acids — this is the primary structure of the protein. A portion of the amino-terminal (or N-terminal) end of a polypeptide is shown in Fig. 2.3, together with the abbreviated representations for this peptide sequence.

The sequence of amino acids in the polypeptide chain is all-important. It contains all the information to confer both the three-dimensional structure of proteins in general and the catalytic activity of enzymes in particular. How is this amino acid sequence controlled? It is specified by the nucleotide sequence of the corresponding *gene*, the piece of deoxyribonucleic acid (DNA) which encodes for that particular protein in that particular organism. To give an idea of how this is achieved, I will give a simplified account of how a polypeptide sequence is derived from a gene sequence. For a more detailed description the reader is referred to several excellent biochemical texts.

Genes are composed of deoxyribonucleotides containing four hetero-cyclic bases: adenine, guanine, cytosine and thymine arranged in a specific linear sequence. To give some idea of size, a typical gene might consist of

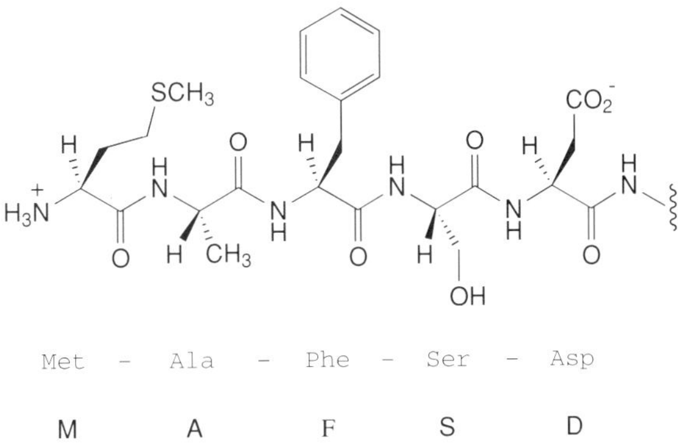

Fig. 2.3 A portion of the N-terminal end of a linear polypeptide chain.

Met – Ala – Phe – Ser – Asp

M   A   F   S   D

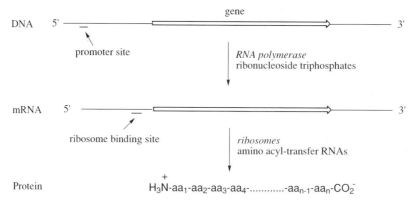

Fig. 2.4 Pathway for protein biosynthesis.

a sequence of 1000 nucleotide bases encoding the information for the synthesis of a protein of approximately 330 amino acids, whose molecular weight would be 35–40 kDa.

How is the sequence encoded? First, the deoxyribonucleotide sequence of the DNA strand is transcribed into messenger ribonucleic acid (mRNA) composed of the corresponding ribonucleotides, which contain the bases adenine, guanine, cytosine and uracil (found in place of thymine). The RNA strand is then translated into protein by the biosynthetic machinery known as ribosomes, as shown in Fig. 2.4. The RNA sequence is translated into protein in sets of three nucleotide bases, one set of three bases being known as a 'triplet codon'. Each codon encodes a single amino acid. The code defining which amino acid is derived from which triplet codon is the 'universal genetic code', shown in Fig. 2.5. This code is followed by the protein biosynthetic machinery of nearly all cells, with the exception of mitochondria and some protozoa.

AAA Lys	ACA Thr	AGA Arg	AUA Ile
AAG Lys	ACG Thr	AGG Arg	AUG Met
AAC Asn	ACC Thr	AGC Ser	AUC Ile
AAU Asn	ACU Thr	AGU Ser	AUU Ile
CAA Gln	CCA Pro	CGA Arg	CUA Leu
CAG Gln	CCG Pro	CGG Arg	CUG Leu
CAC His	CCC Pro	CGC Arg	CUC Leu
CAU His	CCU Pro	CGU Arg	CUU Leu
GAA Glu	GCA Ala	GGA Gly	GUA Val
GAG Glu	GCG Ala	GGG Gly	GUG Val
GAC Asp	GCC Ala	GGC Gly	GUC Val
GAU Asp	GCU Ala	GGU Gly	GUU Val
UAA Stop	UCA Ser	UGA Stop	UUA Leu
UAG Stop	UCG Ser	UGG Trp	UUG Leu
UAC Tyr	UCC Ser	UGC Cys	UUC Phe
UAU Tyr	UCU Ser	UGU Cys	UUU Phe

Fig. 2.5 The universal genetic code.

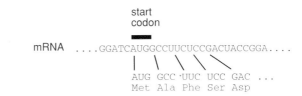

Fig. 2.6 Translation of
mRNA to proteins.

As an example we shall consider in Fig. 2.6 the N-terminal peptide
sequence Met–Ala–Phe–Ser–Asp illustrated in Fig. 2.3. The first amino acid
at the N-terminus of each protein is always methionine, whose triplet codon
is AUG. The next triplet GCC encodes alanine; UUC encodes phenylalanine; UCC encodes serine; and GAC encodes aspartate. Translation then
continues in triplets until one of three 'stop' codons is reached; at this point
protein translation stops. Note that for most amino acids there is more than
one possible codon: thus, if UUC is changed to UUU, phenylalanine is still
encoded, but if changed to UCC then serine is encoded.

In this way the nucleotide sequence of the gene is translated into the
amino acid sequence of the encoded protein. An important practical consequence is that the amino acid sequence of an enzyme can be determined by
nucleotide sequencing of the corresponding gene, which is now usually the
most convenient way to determine a protein sequence.

2.4   Alignment of amino acid sequences

Most biochemical reactions are found in more than one organism, in some
cases in all living cells. If the enzymes which catalyse the same reaction are
purified from different organisms and their amino acid sequences are
determined, then we often see similarity between two sequences. The degree
of similarity is usually highest in enzymes from organisms which have
evolved recently on an evolutionary timescale. The implication of such an
observation is that the two enzymes have evolved divergently from a
common ancestor.

Over a long period of time, changes in the DNA sequence of a gene can
occur by random mutation or by several types of rare mistakes in DNA
replication. Many of these mutations will lead to a change in the encoded
protein sequence in such a way that the mutant protein is inactive. These
mutations are likely to be lethal to the cell and are hence not passed down to
the next generation. However, mutations which result in minor modifications
to non-essential residues in an enzyme will have little effect on the activity of
the enzyme, and will therefore be passed on to the next generation.

So, if we look at an alignment of amino acid sequences of 'related'
enzymes from different organisms, we would expect that catalytically

```
Alignment of N-terminal 15 amino acids of four sequences in 3-letter codes:

                             1           5            10             15
E. coli MhpB                Met His Ala Tyr Leu His Cys Leu Ser His Ser Pro Leu Val Gly
A. eutrophus MpcI           Met Pro Ile Gln Leu Glu Cys Leu Ser His Thr Pro Leu His Gly
P. paucimobilis LigB    Met Met Arg Val Thr Thr Gly Ile Thr Ser Ser His Ile Pro Ala Leu Gly
E. coli HpcB            Met Met Lys Leu Ala Leu Ala Ala Lys Ile Thr His Val Pro Ser Met Tyr
                                                                     +   *         *
```

```
Alignment of N-terminal 60 amino acids of two sequences in 1-letter codes:

                    1         11        21        31        41        51
E. coli MhpB        MHAYLHCLSH SPLVGYVDPA QEVLDEVNGV IASARERIAA FSPELVVLFA PDHYNGFFYD
A. eutrophus MpcI   MPIQLECLSH TPLHGYVDPA PEVVAEVERV QAAARDRVRA FDPELVVVFA PDHFNGFFYD
                    *  **** +** ******  **+ **  *  *+**+*+ * * *****+** ***+******
```

*, identically conserved residue   +, functionally conserved residue

Fig. 2.7 Examples of amino acid sequence alignments.

important amino acid residues would be invariant or 'conserved' in all species. In this way seque. :e alignments can provide clues for identifying important amino acid residues in the absence of an X-ray crystal structure. For example, in Fig. 2.7 there is an alignment of the N-terminal portion of the amino acid sequence of 2,3-dihydroxyphenylpropionate 1,2-dioxygenase (MhpB) from *Escherichia coli* with 'related' dioxygenase enzymes from *Alcaligenes eutrophus* (MpcI) and *Pseudomonas* (LigB) and another *E. coli* enzyme HpcB. Clearly there are a small number of conserved residues (indicated by a *) which are very important for activity, and a further set of residues for which similar amino acid side chains are found (e.g. hydroxyl-containing serine and threonine, indicated with a +).

Furthermore, sequence similarity is sometimes observed between different enzymes which catalyse similar reactions or use the same cofactor, giving rise to 'sequence motifs' found in a family of enzymes. We shall meet some examples of sequence motifs in the course of the book.

2.5   Secondary structures found in proteins

When the linear polypeptide sequence of the protein is formed inside cells by ribosomes, a remarkable thing happens: the polypeptide chain folds to form the three-dimensional structure of the protein. All the more remarkable is that from a sequence of 100–1000 amino acids a *unique* stable three-dimensional structure is formed. It has been calculated that if the protein folding process were to sample each of the available conformations then it would take longer than the entire history of the universe—yet, in practice it takes a few seconds! In some cases the protein folds spontaneously, but in many cases the folding process requires 'helper' proteins known as chaperones. Factors that seem to be important in the folding process are:

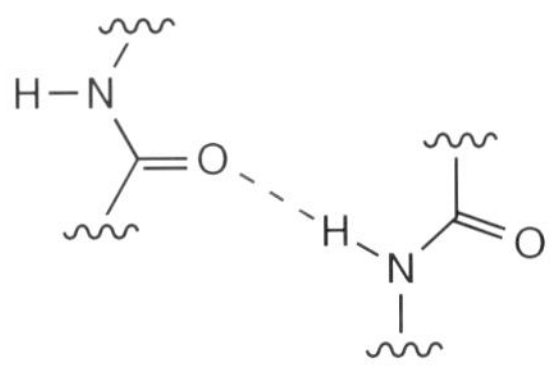

Fig. 2.8 A hydrogen bond formed between a carbonyl oxygen lone pair and the amide N–H of a neighbouring peptide strand.

1   packing of hydrophobic amino acid side chains and exclusion of solvent water;
2   formation of specific non-covalent interactions;
3   formation of secondary structures.

Secondary structure is the term given to local regions (10–20 amino acids) of stable, ordered three-dimensional structures held together by hydrogen-bonding, i.e. non-covalent bonding between acidic hydrogens (O–H, N–H) and lone pairs of electrons, as shown in Fig. 2.8.

There are at least three stable forms of secondary structure commonly observed in proteins: the α-helix, the β-sheet and the β-turn. The α-helix is a helical structure formed by a single polypeptide chain in which hydrogen bonds are formed between the carbonyl oxygen of one amide linkage and the N–H of the linkage four residues ahead in the chain, as shown in Fig. 2.9.

In this structure each amide linkage forms two specific hydrogen-bonds, making it a very stable structural unit. All of the amino acid side chains point outwards from the pitch of the helix. Consequently, amino acid side chains which are four residues apart in the primary sequence will end up close in space. Interactions between such side chains can lead to further favourable interactions within the helix, or with other secondary structures.

The β-sheet is a structure formed by two or more linear polypeptide strands, held together by a series of interstrand hydrogen bonds. There are two types of β-sheet structures: (i) parallel β-sheets, in which the peptide strands both proceed in the same amino-to-carboxyl direction; and (ii) antiparallel, in which the peptide strands proceed in opposite directions. Both types are illustrated in Fig. 2.10.

The β-turn is a structure often formed at the end of a β-sheet which leads

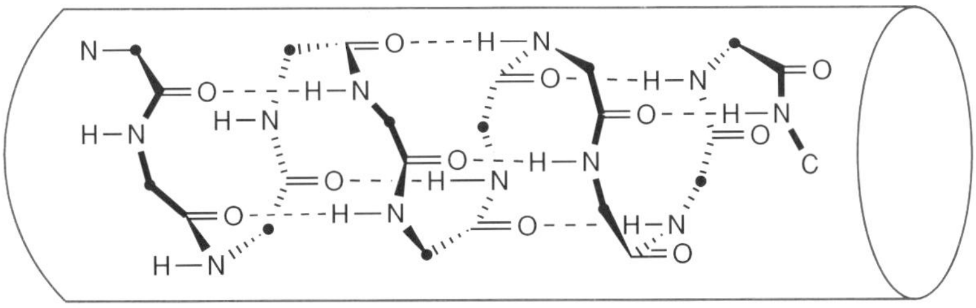

Fig. 2.9 Structure of an α-helix. Positions of amino acid α-carbons are indicated with dots.

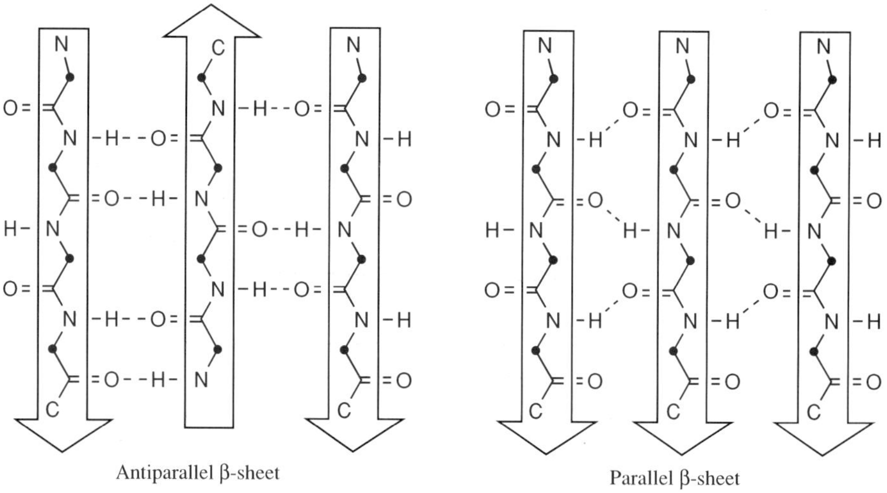

Antiparallel β-sheet   Parallel β-sheet

Fig. 2.10 Structure of β-sheets. Positions of amino acid α-carbons are indicated with dots. N and C indicate the amino- and carboxyl-termini of the peptide chains.

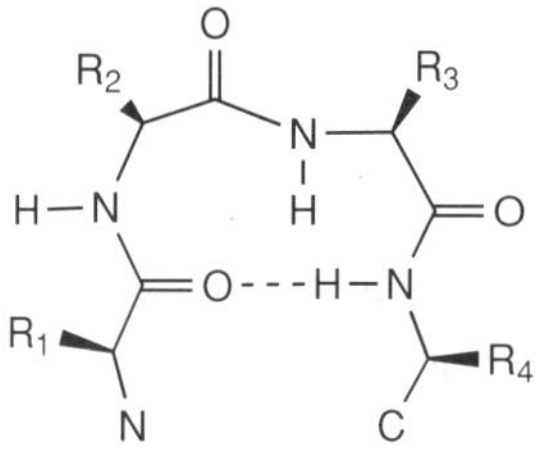

Fig. 2.11 Structure of a β-turn.

to a 180° turn in the direction of the peptide chain. An example of a β-turn is shown in Fig. 2.11, where the role of hydrogen-bonding in stabilizing such structures can be seen.

2.6   The folded tertiary structure of proteins

The three-dimensional structure of protein subunits, known as the tertiary structures, arises from packing together elements of secondary structure to form a stable global conformation, which in the case of enzymes is catalytically active. The packing of secondary structural units usually involves burying hydrophobic amino acid side chains on the inside of the protein and positioning hydrophilic amino acid side chains on the surface.

Although, in theory, the number of possible protein tertiary structures is virtually infinite, in practice proteins are often made up of common structural motifs, from which the protein structure can be categorized. Common families of protein structure are:

1 α-helical proteins;

2   α/β-structures;

3   antiparallel β-structures.

Members of each class are illustrated in Plate 2.1 (facing p. 152). The α-helical proteins are made up only of α-helices which pack onto one another to form the tertiary structure. Many of the haem-containing cytochromes which act as electron carriers (see Chapter 6) are four-helix 'bundles', illustrated Plate 2.1(a,b) in the case of cytochrome b_{562}. The family of globin oxygen carriers, including haemoglobin, consist of a more complex α-helical tertiary structure. α/β-Structures consist of regular arrays of β-sheet–α-helix–*parallel*-β-sheet structures. The redox flavoprotein flavodoxin contains five such parallel β-sheets, forming a twisted β-sheet surface interwoven with α-helices, as shown in Plate 2.1(c). Antiparallel β-structures consist of regular arrays of β-sheet–β-turn–*antiparallel* β-sheet. For example, the metallo-enzyme superoxide dismutase contains a small barrel of antiparallel β-sheets, as shown in Plate 2.1(d).

Protein tertiary structures are conveniently represented by 'topology diagrams' shown in Fig. 2.12, in which α-helices are represented by barrels or rectangles and β-sheets represented by ribbon arrows, the direction of the arrow indicating the direction of the peptide chain.

Frequently, proteins consist of a number of 'domains', each of which contains a region of secondary structure. Sometimes a particular domain has a specific function, such as binding a substrate or cofactor. Larger proteins often consist of more than one tertiary structure, which fit together to form the active 'quaternary' structure. In some cases a number of identical

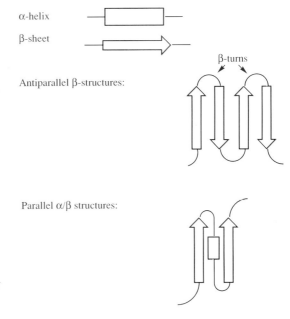

α-helix

β-sheet

Antiparallel β-structures:

β-turns

Parallel α/β structures:

Fig. 2.12 Representations of secondary structures in topology diagrams.

subunits can bind together to form a homodimer (two identical subunits), trimer or tetramer, or in other cases non-identical subunits fit together to form highly complex quaternary structures. One familiar example is the mammalian oxygen transport protein haemoglobin, which consists of a tetramer of two pairs of identical subunits.

How are protein tertiary structures determined experimentally? The most common method for solving three-dimensional structures of proteins is to use X-ray crystallography, which involves crystallization of the protein, and analysis of the diffraction pattern obtained from X-ray irradiation of the crystal. The first enzyme structure to be solved by this method was lysozyme in 1965, since which time several hundred crystal structures have been solved. Recent advances in nuclear magnetic resonance (NMR) spectroscopy have reached the point where the three-dimensional structures of small proteins (< 15 kDa) in solution can be solved using multi-dimensional NMR techniques.

2.7   Enzyme structure and function

All enzymes are proteins, but not all proteins are enzymes, the difference being that enzymes possess catalytic activity. The part of the enzyme tertiary structure that is responsible for the catalytic activity is called the 'active site' of the enzyme, and often makes up only 10–20% of the total volume of the enzyme. This is where the enzyme chemistry takes place.

The active site is usually a hydrophilic cleft or cavity containing an array of amino acid side chains which bind the substrate and carry out the enzymatic reaction, as shown in Fig. 2.13a. In some cases the enzyme active

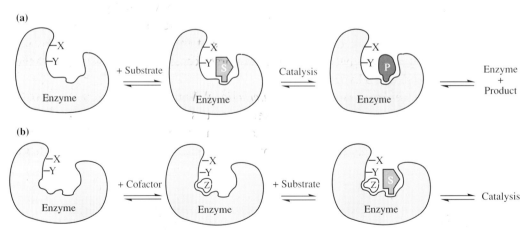

Fig. 2.13 Schematic figure of (a) enzyme plus substrate or (b) enzyme plus substrate plus cofactor.

site also binds one or more cofactors which assist in the catalysis of particular types of enzymatic reactions, as shown in Fig. 2.13b.

How does the enzyme bind the substrate? One of the hallmarks of enzyme catalysis is its high substrate selectively, which is due to a series of highly specific non-covalent enzyme–substrate binding interactions. Since the active site is chiral, it is naturally able to bind one enantiomer of the substrate over the other, just as a hand fits a glove. There are four types of enzyme–substrate interactions used by enzymes, as follows:

1 *Electrostatic interactions.* Substrates containing ionizable functional groups which are charged in aqueous solution at or near pH 7 are often bound via electrostatic interactions to oppositely charged amino acid side chains at the enzyme active site. Thus, for example, carboxylic acids (pK_a 4–5) are found as the negatively charged carboxylate anion at pH 7, and are often bound to positively charged side chains such as the protonated ε-amino side chain of a lysine or the protonated guanidine side chain of arginine, shown in Fig. 2.14.

Similarly, positively charged substrate groups can be bound electrostatically to negatively charged amino acid side chains of aspartate and glutamate. Energetically speaking, the binding energy of a typical electrostatic interaction is in the range 25–50 kJ mol^{-1}, the strength of the electrostatic interaction varying with $1/r^2$, where r is the distance between the two charges.

2 *Hydrogen-bonding.* Hydrogen bonds can be formed between a hydrogen-bond donor containing a lone pair of electrons and a hydrogen-bond acceptor containing an acidic hydrogen. These interactions are widely used for binding polar substrate functional groups. The strength of hydrogen bonds depends upon the chemical nature and the geometrical alignment of the interacting groups. Studies of enzymes in which hydrogen-bonding groups have been specifically mutated have revealed that hydrogen bonds between uncharged donors/acceptors are of energy 2.0–7.5 kJ mol^{-1}, whilst hydrogen bonds between charged donors/acceptors are much stronger, in the range 12.5–25 kJ mol^{-1}.

3 *Non-polar (Van der Waals interactions).* Van der Waals interactions arise from interatomic contacts between the substrate and the active site. Since the shape of the active site is usually highly complementary to the shape of the substrate, the sum of the enzyme–substrate Van der Waals interactions

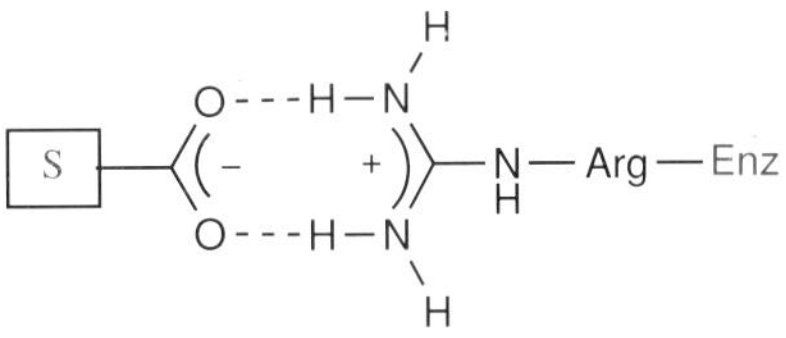

Fig. 2.14 Electrostatic enzyme–substrate interaction of carboxylate bound by lysine/ arginine.

can be quite substantial (50–100 kJ mol^{-1}), even though each individual interaction is quite weak (6–8 kJ mol^{-1}). Since the strength of these interactions varies with $1/r^6$ they are only significant at short range (2–4 Å), so a very good 'fit' of the substrate into the active site is required in order to realize binding energy in this way.

4 *Hydrophobic interactions.* If the substrate contains a hydrophobic group or surface, then favourable binding interactions can be realized if this is bound in a hydrophobic part of the enzyme active site. These hydrophobic interactions can be visualized in terms of the tendency for hydrophobic organic molecules to aggregate and extract into a non-polar solvent rather than remain in aqueous solution. These processes of aggregation and extraction are energetically favourable due to the maximization of inter-water hydrogen-bonding networks which are otherwise disrupted by the hydrophobic molecules, as shown in Fig. 2.15.

There are many examples of hydrophobic 'pockets' or surfaces in enzyme active sites which interact favourably with hydrophobic groups or surfaces in the substrates and hence exclude water from the two hydrophobic surfaces. As mentioned above, these hydrophobic interactions may be very important for maintaining protein tertiary structure, and as we shall see below they are central to the behaviour of biological membranes.

Having bound the substrate, the enzyme then proceeds to catalyse its specific chemical reaction using active site catalytic groups, and finally releases the product back into solution. Enzyme catalysis will be discussed in the next chapter; however, before finishing the discussion of enzyme structure three special classes of enzyme structural types will be introduced.

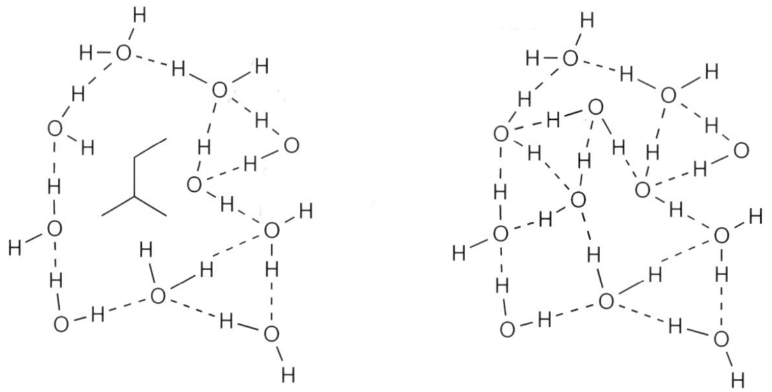

Hydrophobic molecule in water

Additional water–water hydrogen bonds possible if hydrophobic molecule is excluded from water

Fig. 2.15 Schematic illustration of the basis for the hydrophobic interaction.

Table 2.1 Metallo-enzymes.

Metal	Types of enzyme	Role of metal	Redox active?
Mg	Kinases, phosphatases, phosphodiesterases	Binding of phosphates/ polyphosphates	×
Zn	Metalloproteases, dehydrogenases	Lewis acid carbonyl activation	×
Fe	Oxygenases (P_{450}, non-haem)	Binding and activation of oxygen	✓
	[FeS] Clusters	Electron transport, hydratases	
Cu	Oxygenases	Activation of oxygen	✓
Mn	Hydratases	Lewis acid?	✓
Co	Vitamin B_{12} coenzyme	homolysis of Co–carbon bond	✓
Mo	Nitrogenase	Component of Mo/Fe cluster	✓

Co, cobalt; Cu, copper; Fe, iron; FeS, iron sulphide; Mg, magnesium; Mn, manganese; Mo, molybdenum; Zn, zinc.

2.8   Metallo-enzymes

Although the polypeptide backbone of proteins is made up only of the 20 common L-amino acids, many proteins bind one or more metal ions. Enzymes that bind metal ions are known as metallo-enzymes: in these enzymes the metal cofactor is usually found at the active site of the enzyme, where it may have either a structural or catalytic role.

A brief summary of the more common metal ions is given in Table 2.1. magnesium ions are probably the most common metal ion cofactor: they are found in many enzymes which utilize phosphate or pyrophosphate substrates, since magnesium ions effectively chelate polyphosphates (Fig. 2.16).

Zinc ions are used structurally to maintain tertiary structure, for example in the 'zinc finger' DNA-binding proteins by co-ordination with the thiolate side chains of four cysteine residues, as shown in Fig. 2.17a. In contrast, zinc is also used in a number of enzymes as a Lewis acid to co-ordinate carbonyl groups present in the substrate and hence activate them towards nucleophilic attack, as shown in Fig. 2.17b.

The other common role of metal ions is as redox reagents. Since none of the 20 common amino acids is able to perform any useful catalytic redox

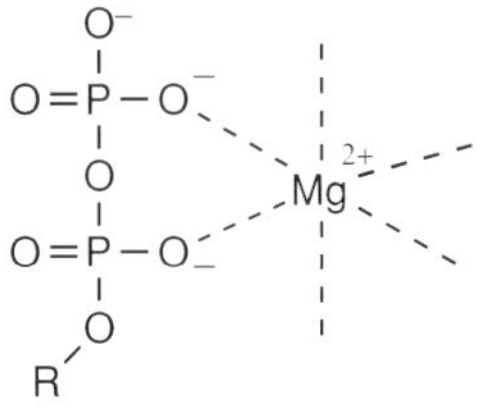

Fig. 2.16 Magnesium chelation of polyphosphates.

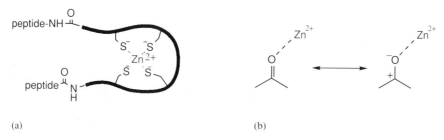

Fig. 2.17 (a) Zinc–cysteine co-ordination. (b) Zinc acting as a Lewis acid.

chemistry, it is not surprising that many redox enzymes employ redox-active metal ions. We shall meet a number of examples of these redox-active metallo-enzymes in Chapter 6. For a more detailed discussion of the role of metal ions in biological systems the reader is referred to several excellent texts in bio-inorganic chemistry.

2.9   Membrane-associated enzymes

Although the majority of enzymes are freely soluble in water and exist in the aqueous cytoplasm of living cells, there is a substantial class of enzymes which are associated with the biological membranes which encompass all cells. Biological membranes are made up of a lipid bilayer composed of phospholipid molecules containing a polar head group and a hydrophobic fatty acid tail. The phospholipid molecules aggregate spontaneously to form a stable bilayer in which the hydrophilic head groups are exposed to solvent water and the hydrophobic tails are packed together in a hydrophobic interior.

Enzymes which are associated with biological membranes fall into two classes, as illustrated in Fig. 2.18:
1 extrinsic membrane proteins which are bound loosely to the surface of the membrane, often by a non-specific hydrophobic interaction, or in some cases by a non-peptide membrane 'anchor' which is covalently attached to the protein;
2 intrinsic or integral membrane proteins which are buried in the membrane bilayer.

Why should some enzymes be membrane associated? Many biological processes involve passage of either a molecule or a 'signal' across biological membranes, and these processes are often mediated by membrane proteins. These membrane processes have very important cellular functions such as cell–cell signalling, response to external stimuli, transport of essential nutrients and export of cellular products. In many cases these membrane proteins have an associated catalytic activity and are therefore enzymes.

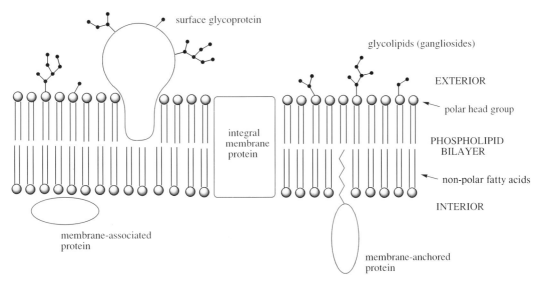

Fig. 2.18 Classes of membrane protein. Black dots represent monosaccharide units attached to glycolipids and glycoproteins.

Intrinsic membrane proteins which completely span the membrane bilayer often possess multiple transmembrane α-helices containing exclusively hydrophobic or non-polar amino acid side chains which interact favourably with the hydrophobic environment of the lipid bilayer. A schematic model of one such protein is shown in Fig. 2.19.

2.10   Glycoproteins

A significant number of proteins found in animal and plant cells contain an additional structural feature attached covalently to the polypeptide backbone of the protein: they are glycosylated by attachment of carbohydrates.

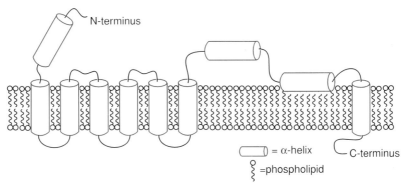

Fig. 2.19 Predicted structure of an integral membrane protein phospho-*N*-acetylmuramic acid-pentapeptide translocase from *Escherichia coli*.

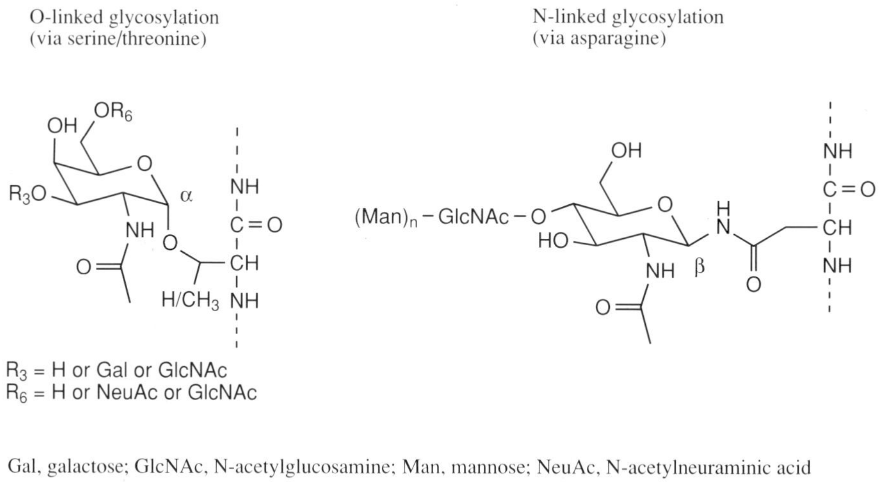
O-linked glycosylation
(via serine/threonine)
N-linked glycosylation
(via asparagine)
OR6
OH
O
R3O
α
NH
NH
C=O
O
CH
O
H/CH3
NH
R3 = H or Gal or GlcNAc
R6 = H or NeuAc or GlcNAc
OH
NH
C=O
(Man)n – GlcNAc – O
O
H
N
CH
HO
NH
β
O
NH
O
Gal, galactose; GlcNAc, N-acetylglucosamine; Man, mannose; NeuAc, N-acetylneuraminic acid

Fig. 2.20 O- and N-linked glycosylation.

Glycoproteins are usually membrane proteins residing in the cytoplasmic membrane of the cell, such that the sugar residues attached to the protein are located on the exterior of the cell membrane. Since these glycoproteins are exposed to the external environment of the cell, they are often important for cell–cell recognition processes. In this respect they act as a kind of 'bar-code' for the type of cell on which they are residing. This function has been exploited in a sinister fashion, as a means of recognition and entry into mammalian cells, by viruses such as influenza virus and human immunodeficiency virus (HIV).

The carbohydrate residues are attached in one of two ways shown in Fig. 2.20: (i) either to the hydroxyl group of a serine or threonine residue (O-linked glycosylation); or (ii) to the primary amide nitrogen of an asparagine residue (N-linked glycosylation).

The level of glycosylation can be very substantial: in some cases up to 50% of the molecular weight of a glycoprotein can be made up of the attached carbohydrate residues. The pattern of glycosylation can also be highly complex, for example highly branched mannose-containing oligosaccharides of 15–40 residues are often found. The sugar attachments are generally not involved in the active site catalysis, but are usually required for full activity of the protein.

Problems

1 Which of the amino acid side chains found in proteins would be (a) positively charged or (b) negatively charged at pH 4,7 and 10, respectively (see Figs 3.9 and 3.12)?

2 The amide bonds found in polypeptides all adopt a *trans* conformation in which the N–H bond is transcoplanar with the C=O. Why? Certain peptides containing proline have been found to contain *cis*-amide bonds involving the amine group of proline. Explain.

3 The following segment of RNA sequence is found in the middle of a gene, but the correct reading frame is not known. What amino acid sequences would be encoded from each of the three reading frame? Comment on which is the most likely of the three.

5′–ACGGCUGAAAACUUCGCACCAAGUCGAUAG–3′

4 You have just succeeded in purifying a new enzyme, and you have obtained an N-terminal sequence for the protein, which reads Met–Ala–Leu–Ser–His–Asp–Trp–Phe–Arg–Val. How many possible nucleotide sequences might encode this amino acid sequence? If you want to design a 12-base oligonucleotide 'primer' with a high chance of matching the nucleotide sequence of the gene as well as possible, what primer sequence would you suggest?

5 α-Helices in proteins have a 'pitch' of approximately 3.6 amino acid residues. In order to visualize the side chain interactions in α-helices, the structure of the helix is often represented as a 'helical wheel'. This representation is constructed by viewing along the length of the helix from the N-terminal end, with the amino acid side chains protruding from the central barrel of the helix, as shown below.

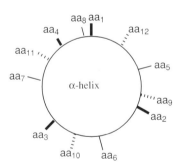

Draw helical wheels for the following synthetic peptides, which were designed to form α-helices with specific functions. Suggest what that function might be.

(a) Gly–Glu–Leu–Glu–Glu–Leu–Leu–Lys–Lys–Leu–Lys–Glu–Leu–Leu–Lys–Gly.

(b) Leu–Ala–Lys–Leu–Leu–Lys–Ala–Leu–Ala–Lys–Leu–Leu–Lys–Lys.

Inspired by the above examples, suggest a synthetic peptide which would fold into an α-helix containing aspartate, histidine and serine side chains in a line along one face of the helix.

Further reading

Protein structure

Branden, C. & Tooze, J. (1991) *Introduction to Protein Structure.* Garland, New York.
Chothia, C. (1984) Principles that determine the structure of proteins. *Ann Rev Biochem*, **53** 537–72.
Creighton, T.E. (1993) *Proteins—Structures and Molecular Properties.* Freeman. New York.
Schulz, G.E. & Schirmer, R.H. (1979) *Principles of Protein Structure.* Springer-Verlag, New York.

Protein folding

Jaenicke, R. (1991) *Biochemistry*, **30**, 3147–61.
Matthews, C.R. (1993) *Ann Rev Biochem*, **62**, 653–83.
Rossman, M.G. & Argos, P. (1981) *Ann Rev Biochem*, **50**, 497–532.

Protein evolution

Bajaj, M. & Blundell, T. (1984) Evolution and the tertiary structure of proteins. *Ann Rev Biophys Bioeng*, **13**, 453–92.
Doolittle, R.F. (1979) Protein evolution. In: *The Proteins* (eds H. Neurath & R.L. Hill), Vol. 4, pp 1–118. Academic Press, New York.

Metalloproteins

Bertini, I., Gray, H.B., Lippard, S.J. & Valentine, J.S. (1994) *Bio-inorganic Chemistry.* University Science Books, Mill Valley, California.

Biological membranes

Findlay, J.B.C. & Evans, W.H. (1987) *Biological Membranes—a Practical Approach.* IRL Press, Oxford.
Gennis, R.G. (1989) *Biomembranes: Molecular Structure and Function.* Springer-Verlag, New York.

Glycoproteins

Lee, Y.C. & Lee, R.T. (1995) *Acc Chem Res* **28**, 321–7.
Rademacher, T.W., Parekh, R.B. & Dwek, R.A. (1988) *Ann Rev Biochem*, **57**, 785–838.

3　Enzymes are Wonderful Catalysts

3.1　Introduction

The function of enzymes is to catalyse biochemical reactions. Each enzyme has evolved over millions of years to catalyse one particular reaction, so it is perhaps not surprising to find that they are extremely good catalysts when compared with synthetic catalysts.

The hallmarks of enzyme catalysis are: speed, selectivity and specificity. Enzymes are capable of catalysing reactions at rates well in excess of a million-fold faster than the uncatalysed reaction, typical ratios of k_{cat}/k_{uncat} being 10^6–10^{14}. Figure 3.1 shows an illustration of the speed of enzyme-catalysed glycoside hydrolysis. The rate of acid-catalysed glycoside hydro-lysis is accelerated 10^3-fold by intramolecular acid catalysis, but enzyme-catalysed hydrolysis is 10^4-fold faster still—some 10^7 faster than the uncatalysed reaction carried out at pH 1.

Enzymes are highly selective in the reactions that they catalyse. Since

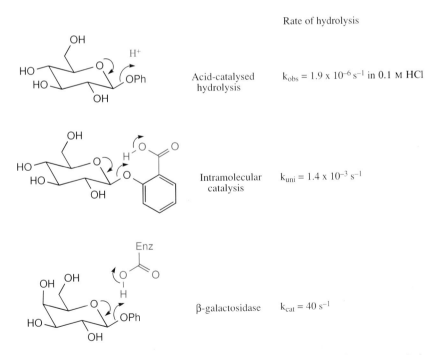

Fig. 3.1 Rate acceleration of glycoside hydrolysis by intramolecular and enzyme catalysis.

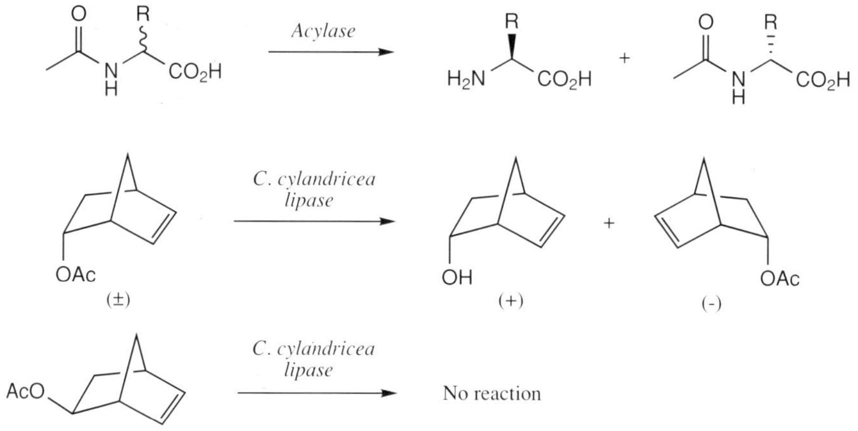

Fig. 3.2 Stereoselectivity in enzymatic hydrolysis reactions.

they bind their substrates via a series of selective enzyme–substrate binding interactions at a chiral active site, they are able to distinguish the most subtle changes in substrate structure, and are able to distinguish between regioisomers and between enantiomers, as shown in Fig. 3.2. Finally, enzymes carry out their reactions with near faultless precision: they are able to select a unique site of action within the substrate, and carry out the enzymatic reaction stereospecifically, as illustrated in Fig. 3.3.

In this chapter we shall examine the factors that contribute to the remarkable rate acceleration achieved in enzyme-catalysed reactions. Examples of enzyme stereospecificity will be discussed in Chapter 4. It is worth at this point distinguishing between *selectivity*, which is the ability of the *enzyme* to select a certain substrate or functional group out of many; and

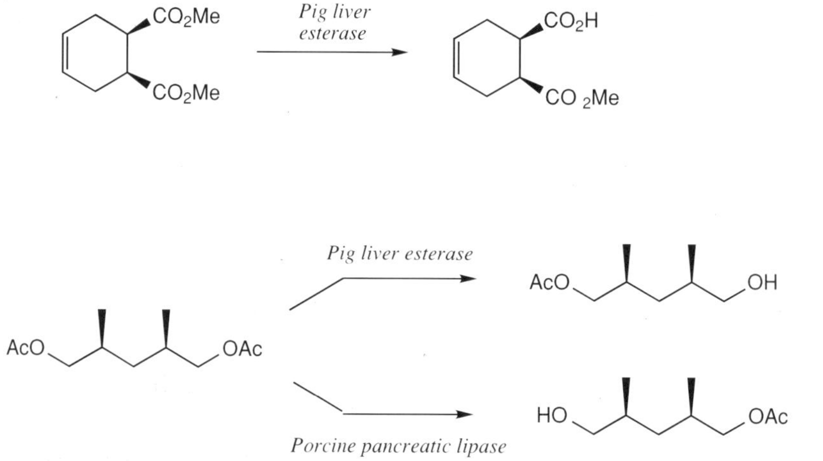

Fig. 3.3 Stereospecificity in enzymatic hydrolysis reactions.

specificity, which is a property of the *reaction* catalysed by the enzyme, being the production of a single regio- and stereoisomer of the product. Both are properties which are highly prized in synthetic reactions used in organic chemistry: enzymes are able to do both.

3.2   A thermodynamic model of catalysis

A catalyst may be defined as a species which accelerates the rate of a chemical reaction whilst itself remaining unchanged at the end of the reaction. In thermodynamic terms, catalysis of a chemical reaction is achieved by reducing the *activation energy* for that reaction, the activation energy being the difference in free energy between the reagent(s) and the transition state for the reaction. This reduction in activation energy can be achieved either by stabilization (and hence reduction in free energy) of the transition state by the catalyst, or by the catalyst finding some other lower energy pathway for the reaction.

Figure 3.4 illustrates the free energy profile of a typical acid-catalysed chemical reaction which converts a substrate S to a product P. In this case an intermediate chemical species SH^+ is formed upon protonation of S. If the conversion of SH^+ to PH^+ is 'easier' than the conversion of S to P, then the activation energy for the reaction will be reduced and hence the reaction will go faster. It is important at this point to define the difference between an intermediate and a transition state: an intermediate is a stable (or semi-stable) chemical species formed during the reaction and is therefore a *local energy minimum*, whereas a transition state is by definition a *local energy maximum*.

The rate of a chemical reaction (k) is related to the activation energy of the reaction (E_{act}) by the following equation (where T is the reaction temperature (in degrees k), R is the universal gas constant $8.314\ Jk^{-1}\ mol^{-1}$, and A is a reaction constant.):

$$k = A e^{(-E_{act}/RT)}$$

Therefore, the rate acceleration of the catalysed reaction versus the uncatalysed reaction can simply be calculated:

$$k_{cat}/k_{uncat} = e^{(E_{uncat} - E_{cat}/RT)}$$

If, for example, a catalyst can provide $10\ kJ\ mol^{-1}$ of transition stabilization energy for a reaction at 25°C a 55-fold rate acceleration will result, whereas a $20\ kJ\ mol^{-1}$ stabilization will give a 3000-fold acceleration and $40\ kJ\ mol^{-1}$ stabilization a 10^7-fold acceleration! A consequence of the exponential relationship between activation energy and reaction rate is that a little extra transition state stabilization goes a long way!

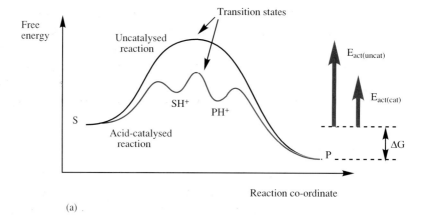

(a)

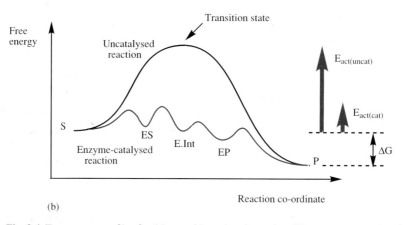

(b)

Fig. 3.4 Free energy profiles for (a) an acid-catalysed reaction (b) an enzyme-catalysed reaction of substrate S to product P.

An enzyme-catalysed reaction can be analysed thermodynamically in the same way as the acid-catalysed example, but is slightly more complicated. As explained in Chapter 2, enzymes function by binding their substrate reversibly at their active site, and then proceeding to catalyse the biochemical reaction using the active site amino acid side chains. Often, enzyme-catalysed reactions are multi-step sequences involving one or more intermediates, as illustrated in Fig. 3.4. An enzyme–substrate intermediate, ES, is formed upon binding of substrate, which is then converted to the enzyme–product complex, EP, either directly or via one or more further intermediates.

In both catalysed reactions shown in Fig. 3.4 the overriding consideration as far as rate acceleration is concerned is transition state stabilization. Just as in non-enzymatic reactions there is acid–base and nucleophilic catalysis taking place at enzyme active sites. However, the secret to the extraordinary power of enzyme catalysis lies in the fact that the reaction is taking place as the substrate is bound to the enzyme active site. So, what was in the non-enzymatic case an intermolecular reaction has effectively become an intramolecular reaction. The rate enhancements obtained from these types of *proximity effects* can be illustrated by intramolecular reactions in organic chemistry, which is where we shall begin the discussion.

3.3   Proximity effects

There are many examples of organic reactions that are intramolecular: that is, they involve two or more functional groups within the same molecule, rather than functional groups in different molecules. Intramolecular reactions generally proceed much more rapidly and under much milder reaction conditions than their intermolecular counterparts, which makes sense since the two reacting groups are already 'in close proximity' to one another. But, can we measure these types of effects quantitatively, and can we rationalize them?

A useful concept in quantitating these proximity effects is that of effective concentration. In order to define the effective concentration of a participating group (nucleophile, base, etc.), we compare the rate of the intramolecular reaction with the rate of the corresponding intermolecular reaction where the reagent and the participating group are present in separate molecules. The effective concentration of the participating group is defined as the concentration of reagent present in the intermolecular reaction required to give the same rate as the intramolecular reaction.

I will illustrate this using data for the rates of hydrolysis of a series of phenyl esters in aqueous solution at pH 7, given in Fig. 3.5. The reference reaction in this case is the hydrolysis of phenyl acetate catalysed by sodium acetate at the same pH. Introduction of a carboxylate group into the same molecule as the ester leads to an enhancement of the rate of ester hydrolysis, which for phenyl succinate (Fig. 3.5 (**3**)) is 23 000-fold faster than phenyl acetate (Fig. 3.5 (**1**)). This remarkable rate acceleration is achieved because the neighbouring carboxylate group can attack the ester to form a cyclic anhydride intermediate, shown in Fig. 3.6. This intermediate is more reactive than the original ester group and so hydrolyses rapidly.

Note that the rate acceleration is largest when a five-membered anhydride is formed, since five-membered ring formation is kinetically favoured over six-membered ring formation, which in turn is greatly favoured over four- and seven-membered ring formation. The effective concentration can be worked out by comparing the rates of these intramolecular reactions with the rates of the intermolecular reaction between phenyl acetate and sodium acetate in water. For phenyl succinate an effective concentration of 4000 M is found, so the hydrolysis of phenyl succinate proceeds much faster than if

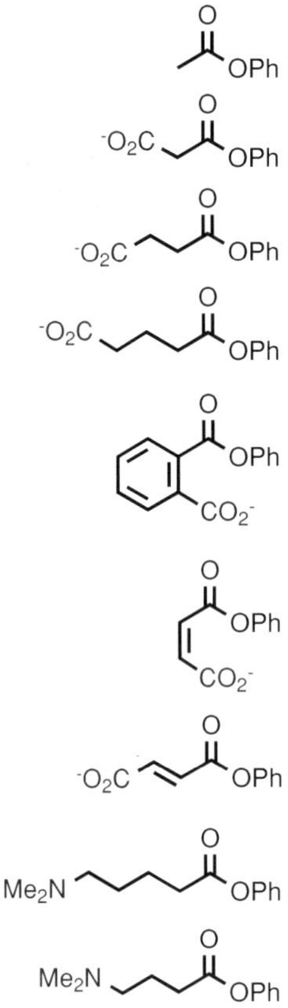

Fig. 3.5 Intramolecular catalysis of ester hydrolysis. Et_3N, triethylamine; NaOAc, sodium acetate.

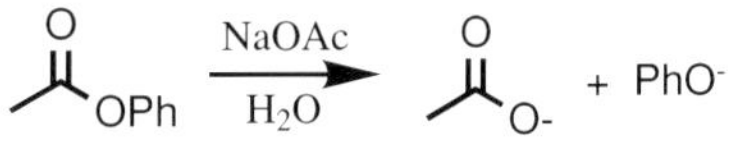

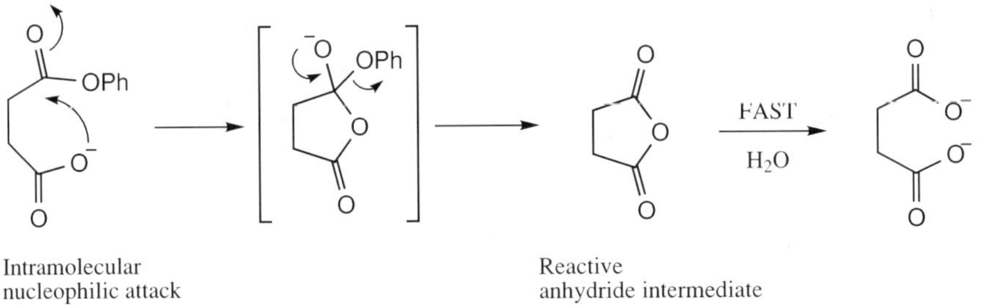

Intramolecular
nucleophilic attack

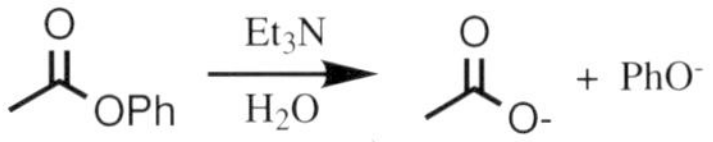

Reactive
anhydride intermediate

Fig. 3.6 Intramolecular mechanism for hydrolysis of phenyl succinate (3).

phenyl acetate was surrounded completely by acetate ions! Here we start to see the catalytic potential of these proximity effects.

In the same series of phenyl esters, if the possible ring size of five is maintained, but a *cis* double bond is placed in between the reacting groups, the observed rates of hydrolysis are even faster. Phenyl phthalate (see Fig. 3.5 (**5**)) has an effective concentration of acetate ions of 2×10^5 M, whilst phenyl maleate (see Fig. 3.5 (**6**)) has an astonishing effective concentration of 10^{10} M! Yet, the same molecule containing a *trans* double bond has no rate acceleration at all. So, it is clear that by holding the reactive groups rigidly in close proximity to one another remarkable rate acceleration can be achieved. Why is the hydrolysis of phenyl maleate, in which a five-membered anhydride is formed, so much faster than the hydrolysis of phenyl succinate, in which an apparently similar five-membered anhydride is formed? The answer is that in phenyl maleate the reactive groups are held in the right orientation to react, as shown in Fig. 3.7, so the *probability* of the desired reaction is increased.

In thermodynamic terms, the restriction of the double bond in the case of phenyl maleate has removed rotational degrees of freedom, so that in going to the transition state for the intramolecular reaction less degrees of freedom are lost, which means that the reaction is *entropically* more favourable. If you think of entropy as a measure of order in the system, then in the case of phenyl maleate the molecule is already ordered in the right way with respect to the reacting groups.

This entropic advantage due to the ordering of reactive groups lies at the heart of intramolecular proximity effects, and is a major factor in the catalytic power of enzyme catalysis. The binding of substrates and cofactors at an enzyme active site of defined three-dimensional structure brings the reagents into close proximity to one another and to the enzyme active site functional groups. This increases the probability of correct positioning for reaction to take place, so it speeds up the reaction. An important factor in this analysis is that the enzyme structure is already held rigidly (or fairly rigidly at least) in the correct conformation for binding and catalysis, so that no entropy is lost in proceeding from the enzyme–substrate complex to the transition state of the reaction. The synthesis of host–guest systems which mimic enzymes by binding substrates non-covalently has also provided examples of rate acceleration by proximity effects (see Chapter 11).

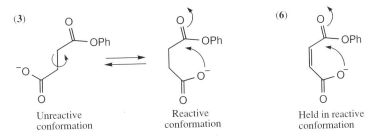

Fig. 3.7 Intramolecular catalysis in phenyl succinate (**3**) versus phenyl maleate (**6**).

3.4   The importance of transition state stabilization

All catalysts operate by reducing the activation energy of the reaction, by stabilizing the transition state for the reaction. Enzymes do the same, but the situation is somewhat more complicated since there are usually several transition states in an enzymatic reaction. We have already seen how an enzyme binds its substrate reversibly at the enzyme active site. One might imagine that if an enzyme were to bind its substrate very tightly that this would lead to transition state stabilization. This, however, is not the case; in fact, it is counter productive for an enzyme to bind its substrate(s) very tightly!

Figure 3.8 shows free energy curves for a hypothetical enzyme-catalysed reaction proceeding via a single rate-determining transition state. Suppose that we can somehow alter the enzyme E so that it binds the substrate S or the transition state more tightly. In each case the starting free energy (of E + S) is the same. In the presence of high substrate concentrations the enzyme will in practice be fully saturated with substrate, so the activation energy for the reaction will be governed by the energy difference between the ES complex and the transition state. In Fig. 3.8b the enzyme is able to bind

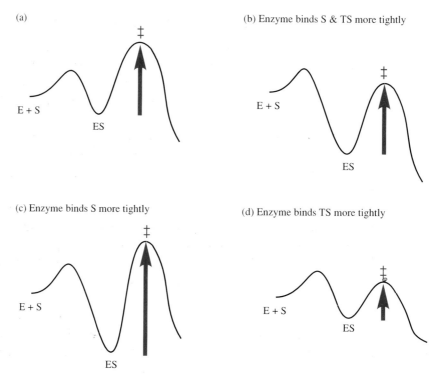

Fig. 3.8 The importance of transition state stabilization—a thought experiment.

both the substrate and the transition state more tightly (and hence lower their free energy equally). This, however, leads to no change in the activation energy, and hence no rate acceleration. In Fig. 3.8c the enzyme binds only the substrate more tightly; this generates a 'thermodynamic pit' for the ES complex and hence increases the activation energy, so the reaction is slower! However, if, as in Fig. 3.8d the enzyme can selectively bind the transition state, then it can reduce the activation energy and hence speed up the reaction.

The conclusion of this thought experiment is that in order to achieve optimal catalysis, enzymes should selectively bind the transition state, rather than the substrate. Hence, it is not advantageous for enzymes to bind their substrates too tightly. This is evident when one looks at substrate binding constants for enzymes (these will be discussed in more detail in Chapter 4, Section 4.3). Typical Michaelis constant (K_m) values for enzymes are in the milli- to micromolar range (10^{-3}–10^{-6} M), whereas dissociation constants for binding proteins and antibodies whose function is to bind small molecules tightly are in the nano- to picomolar range (10^{-9}–10^{-12} M). We shall meet several examples of specific transition state stabilizing interactions later in the book.

3.5   Acid/base catalysis in enzymatic reactions

Acid and base catalysis is involved in all enzymatic processes involving proton transfer, so in practice there are very few enzymes that do not have acidic or basic catalytic groups at their active sites. However, unlike organic reactions which can be carried out under a very wide range of pH conditions to suit the reaction, enzymes have a strict limitation that they must operate at physiological pH, in the range 5–9. Given this restriction, and the fairly small range of amino acid side chains available for participation in acid/base chemistry (shown in Fig. 3.9), a remarkably diverse range of acid/base chemistry is achieved.

General acid catalysis takes place when the substrate is protonated by a catalytic residue, which in turn gives up a proton, as shown in Fig. 3.10. The active site acidic group must therefore be protonated at physiological pH but its pK_a must be just above (i.e. in the range 7–10). If the pK_a of a side chain was in excess of 10 then it would become thermodynamically unfavourable to transfer a proton.

General base catalysis takes place either when the substrate is deprotonated, or when water is deprotonated prior to attack on the substrate, as shown in Fig. 3.11. Enzyme active site bases must therefore be deprotonated at physiological pH but have pK_a values just below. Typical pK_a ranges for amino acid side chains in enzyme active sites are shown in Fig. 3.12. They

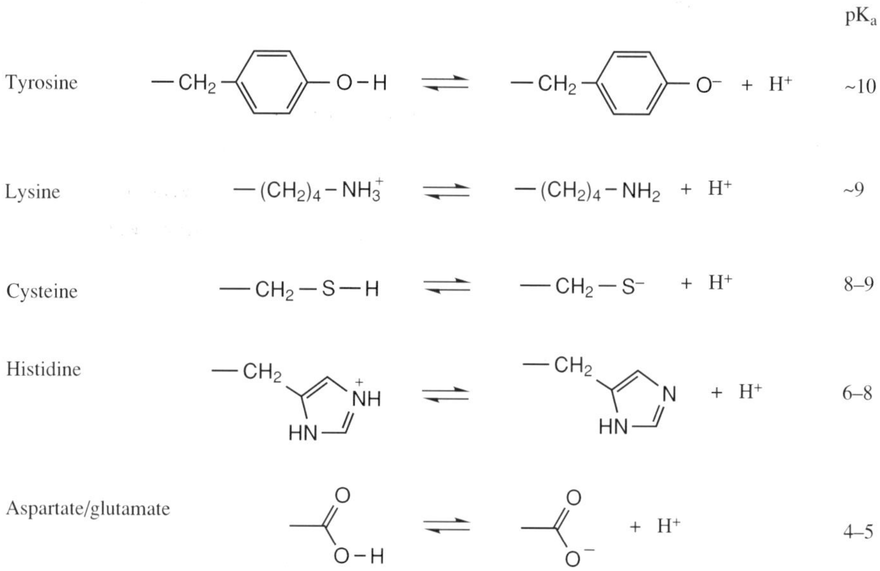

pK$_a$

Tyrosine $-CH_2$⟨benzene ring⟩$-O-H$ ⇌ $-CH_2$⟨benzene ring⟩$-O^-$ $+$ H^+ ~10

Lysine $-(CH_2)_4-NH_3^+$ ⇌ $-(CH_2)_4-NH_2$ $+$ H^+ ~9

Cysteine $-CH_2-S-H$ ⇌ $-CH_2-S^-$ $+$ H^+ 8–9

Histidine $-CH_2$ ⟨imidazole NH$^+$⟩ ⇌ $-CH_2$ ⟨imidazole N⟩ $+$ H^+ 6–8

Aspartate/glutamate   ⟨carboxyl O-H⟩ ⇌ ⟨carboxylate O$^-$⟩ $+$ H^+ 4–5

Fig. 3.9 Amino acid side chains used for acid/base catalysis.

Fig. 3.10 General acid catalysis.

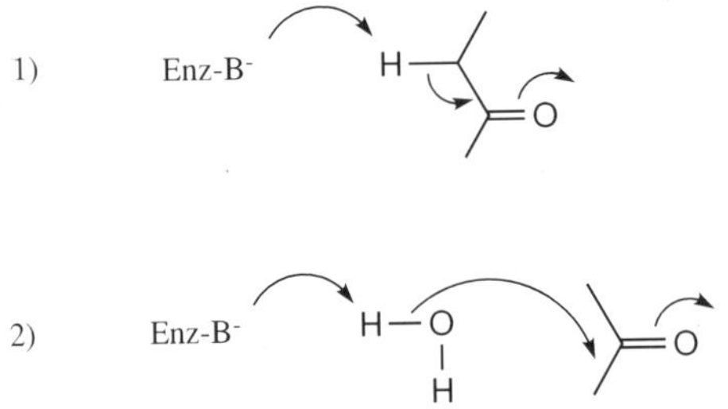

1)   Enz-B$^-$ ⟨H with carbonyl⟩

2)   Enz-B$^-$ ⟨H—O—H with carbonyl⟩

Fig. 3.11 General base catalysis.

can be measured by analysis of enzymatic reaction rate versus pH, as described in Chapter 4, Section 4.7.

Although the pK$_a$ values given in Fig. 3.12 are the typical values found in proteins, in some cases the pK$_a$ values of active site acidic and basic groups can be strongly influenced by their micro-environment. Thus, for example, the enzyme acetoacetate decarboxylase contains an active site lysine residue when forms an imine linkage with its substrate—its pK$_a$ value was found to be 5.9, much less than the expected value. When an active site peptide was obtained containing the catalytic lysine, it was found to be adjacent to another lysine residue. So, it is likely that the proximity of another positively

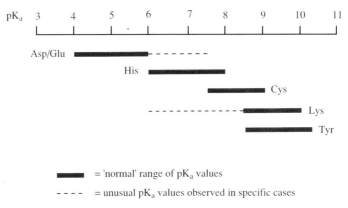

Fig. 3.12 Range of pK$_a$ values observed for amino acid side chains in enzyme active sites. Asp, aspartic acid; Cys, cysteine; Glu, glutamic acid; His, histidine; Lys, lysine; Tyr, tyrosine.

charged residue would make the protonated form thermodynamically less favourable, and hence reduce the pK$_a$. The same effect can be observed in the pK$_a$ values of ethylenediamine (Fig. 3.13), where the pK$_a$ for the mono-protonated form is 10.7 as usual, but the pK$_a$ for the doubly protonated form is 7.5.

Similarly, if a charged group was involved in a salt bridge with an oppositely charged residue, its pK$_a$ would be altered, or if it was in a hydrophobic region of the active site, which would destabilize the charged form of the group. So, for example, we shall see later protonated aspartic acid and glutamic acid residues acting as acidic groups in some enzymes. Finally, it is worth noting that histidine, whose side chain contains an imidazole ring of pK$_a$ 6–8, can act either as an acidic or a basic residue, depending on its particular local pK$_a$, making it a versatile reagent for enzymatic acid/base chemistry.

Acid/base catalysis by enzymes is made that much more effective by the optimal positioning of the active site acid/base groups in close proximity to the substrate, generating a high effective concentration of the enzyme reagent. This can be illustrated in the case of glycoside hydrolysis using the

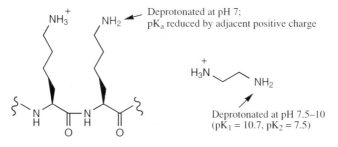

Fig. 3.13 Abnormally low lysine pK$_a$ in acetoacetate decarboxylase.

data from Fig. 3.1. The mechanism of the non-enzymatic reaction involves protonation of the glycosidic group by external acid to form a good leaving group, followed by formation of an oxonium intermediate. Glycoside hydrolysis can be accelerated dramatically by positioning an acidic group in close proximity to the glycosidic leaving group, as shown in Fig. 3.1. Enzymes which catalyse glycoside hydrolysis (see Chapter 5, Section 5.7) also employ an acidic catalytic group to protonate the glycosidic leaving group—but, the enzymatic reaction is some 30 000-fold faster than even the intramolecular reaction, suggesting that the enzyme is able to stabilize further the transition state for their reaction.

Enzymes also have the ability to carry out bifunctional catalysis: protonation of the substrate at the same time as deprotonation in another part of the molecule. An example of bifunctional catalysis is the enzyme ketosteroid isomerase, in which aspartate (Asp)-38 acts as an active site base and tyrosine (Tyr)-14 as an active site acidic group (Fig. 3.14). A dienol intermediate is formed via a concerted step involving simultaneous deprotonation of the substrate by Asp-38 and protonation of the substrate carbonyl by Tyr-14. In the second step the tyrosinate group acts as a base, and the substrate is re-protonated by the protonated Asp-38.

Bifunctional catalysis is thought to make possible the deprotonation of substrates with apparently high pK_a values. Thus, in the above example

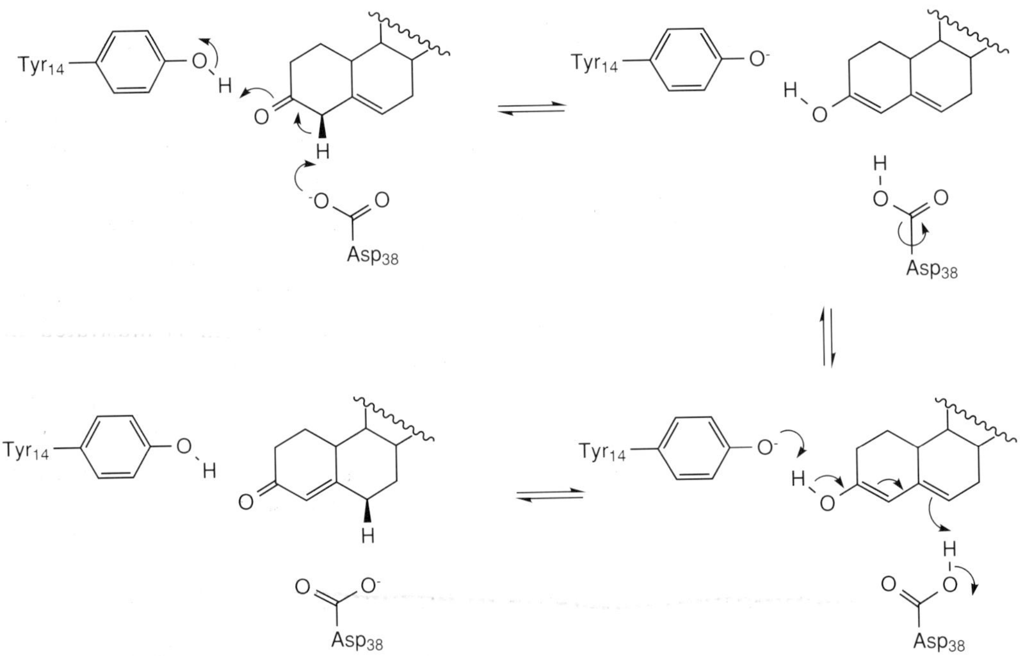

Fig. 3.14 Bifunctional catalysis in the ketosteroid isomerase reaction mechanism.

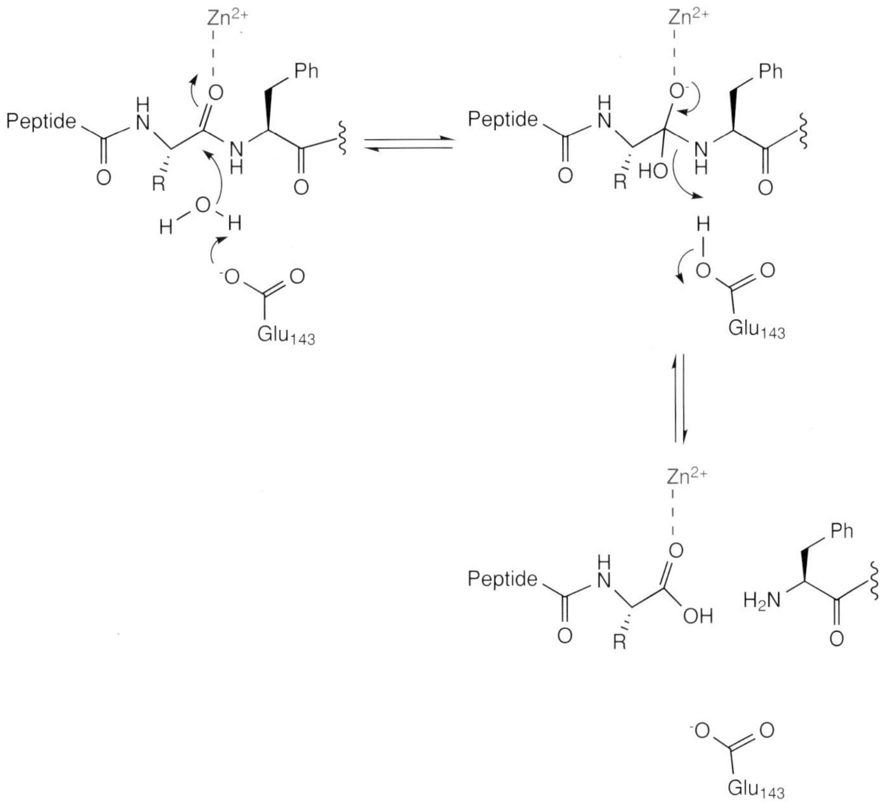

Fig. 3.15 Lewis acid catalysis in the thermolysin reaction mechanism. Glu, glutamate.

deprotonation adjacent to a ketone in solution to form an enolate species would involve removal of a proton of pK_a 18–20, which would be impractical at pH 7. However, simultaneous protonation to form an enol intermediate makes the reaction thermodynamically much more favourable.

Finally, enzymes which bind metal cofactors such as zinc and magnesium can utilize their properties as Lewis acids, i.e. electron pair acceptors. An example is the enzyme thermolysin, whose mechanism is illustrated in Fig. 3.15. In this enzyme glutamate (Glu)-143 acts as an active site base to deprotonate water for attack on the amide carbonyl, which is at the same time polarized by co-ordination by an active site zinc ion. The protonated glutamic acid probably then acts as an acidic group for the protonation of the departing amine.

3.6   Nucleophilic catalysis in enzymatic reactions

Nucleophilic (or covalent) catalysis is a type of catalysis seen relatively rarely in organic reactions, but which is used quite often by enzymes. It involves nucleophilic attack of an active site group on the substrate, forming a covalent bond between the enzyme and the substrate, and hence a covalent intermediate in the reaction mechanism. This is a particularly effective strategy for enzyme-catalysed reactions for two reasons. First, as we have seen before, an enzyme is able to position an active site nucleophile in close proximity and in correct alignment to attack its substrate, generating a very high effective concentration of nucleophile.

Second, since enzyme active sites are often largely excluded from water molecules, an enzyme active site nucleophile is likely to be 'desolvated'. Thus, a charged nucleophile in aqueous solution would be surrounded by several layers of water molecules, which greatly reduce the polarity and effectiveness of the nucleophile. However, a desolvated nucleophile at a water-excluded active site will be a much more potent nucleophile than its counterpart in solution. This effect can be illustrated in organic reactions carried out in dipolar aprotic solvents such as dimethylsulphoxide (DMSO) or dimethylformamide (DMF), in which nucleophiles are not hydrogen-bonded as they would be in aqueous solution. A consequence of this desolvation effect is that nucleophilic displacement reactions occur much more readily in these solvents.

Enzymes have a range of potential nucleophiles available to them, which are shown in Table 3.1. Probably the best nucleophile available to enzymes is the thiol side chain of cysteine, which we shall see operating in proteases and acyl transfer enzymes. The ε-amino group of lysine is used in a number of cases to form imine linkages with ketone groups in substrates, as in the example of acetoacetate decarboxylase, shown in Fig. 3.16.

This enzyme catalyses the decarboxylation of acetoacetate to acetone. Two lines of evidence were used to show that an imine linkage is formed between the ketone of acetoacetate and the ε-amino group of an active site lysine.

1   Treatment of enzyme with substrate and sodium borohydride leads to irreversible enzyme inactivation, via *in situ* reduction of the enzyme-bound

Table 3.1 Amino acid side chains used for nucleophilic catalysis in enzymatic reactions.

Amino acid	Side chain	Examples
Serine	$-CH_2-OH$	Serine proteases, esterases, lipases
Threonine	$-CH(CH_3)-OH$	Phosphotransferases
Cysteine	$-CH_2-SH$	Cysteine proteases, acyl transferases
Aspartate, glutamate	$-(CH_2)_n-CO_2H$	Epoxide hydrolase, haloalkane dehalogenase
Lysine	$-(CH_2)_4-NH_2$	Acetoacetate decarboxylase, class I aldolases
Histidine	Imidazole–NH	Phosphotransferases
Tyrosine	Ar–OH	DNA topoisomerases

Ar, aromatic group; DNA, deoxyribonucleic acid.

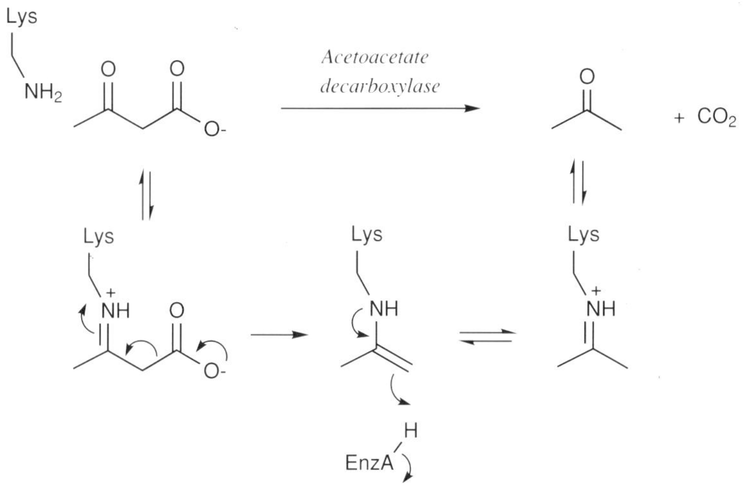

Fig. 3.16 Nucleophilic catalysis in the acetoacetate decarboxylase reaction mechanism.

imine intermediate by borohydride. Treatment of enzyme sodium with borohydride alone gives no inactivation.

2 Incubation of acetoacetate labelled with ^{18}O at the ketone position leads to the rapid exchange of ^{18}O label from this position, consistent with reversible imine formation.

As mentioned above, the pK_a of this lysine group is abnormally low at 5.9, which is sufficiently low for it to act as a nucleophile at pH 7.

The other nitrogen nucleophile available to enzymes is the versatile imidazole ring of histidine. This group is more often used for acid/base chemistry, but is occasionally used as a nucleophile in, for example, phosphotransfer reactions. Finally, enzymes have oxygen nucleophiles available in the form of the hydroxyl groups of serine, threonine and tyrosine, and the carboxylate groups of aspartate and glutamate. There are examples of each of these groups being used for nucleophilic catalysis, especially serine, which we shall see used for the serine proteases in Chapter 5. An example of the use of aspartate as a nucleophile is the enzyme haloalkane dehalogenase from *Xanthobacter autotrophicus*, which is involved in the dechlorination of organochlorine chemicals found in industrial waste. This enzyme functions by displacement of chloride by an active site aspartate residue, followed by base-catalysed hydrolysis of the covalent ester intermediate, as shown in Fig. 3.17.

This 35-kDa protein contains seven strands of β-sheet arranged centrally with intervening α-helices. This type of α, β-structure is found in many hydrolase enzymes such as the serine proteases discussed in Chapter 5. The catalytic residues Asp-124 and histidine (His)-289 are situated on loops at the ends of β-sheets. The active site cavity of volume 3.7×10^{-2} nm^3 is lined

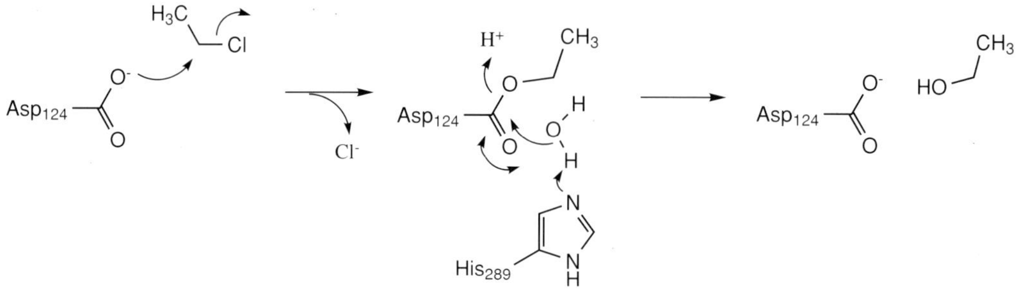

Fig. 3.17 Mechanism of haloalkane dehalogenase. His, histidine.

with hydrophobic residues, which can form favourable hydrophobic interactions with its non-polar substrates. Plate 3.1 (facing p. 152) shows a view of the protein structure, highlighting the active site catalytic residues and the surface of the active site.

3.7   The use of strain energy in enzyme catalysis

The concept of 'strain' is one that is rather difficult to explain, since it occurs very rarely in organic reactions, and there are only a few examples of enzymatic reactions in which there is evidence that it operates. Remember that the over-riding factor in achieving rate acceleration in enzyme-catalysed reactions is the difference in free energy between the enzyme–substrate (ES) complex and the transition state of the enzymatic reaction. If the enzyme can somehow bind the substrate in a strained conformation which is *closer to the transition state* than the ground state conformation, then the difference in energy between the bound conformation and the transition state will be reduced, and the reaction will be accelerated (Fig. 3.18).

How can an enzyme bind its substrate in a strained conformation? Is that not energetically unfavourable? The answer to these questions is that if the substrate is of a reasonable size, the enzyme can form a number of enzyme–substrate binding interactions and the total enzyme–substrate binding energy can be quite substantial. In some cases, in order to benefit from the most favourable overall binding interactions the substrate must adopt an

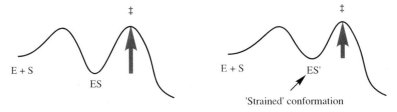

Fig. 3.18 Rate acceleration from a strained ES′ complex.

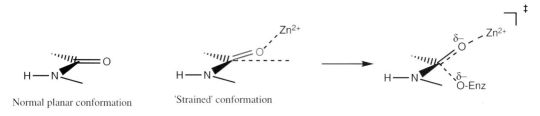

Fig. 3.19 Strained conformation observed in carboxypeptidase A.

unfavourable conformation in a *part* of the molecule. That part of the substrate may cunningly happen to be where the reaction is going to take place! In thermodynamic terms, the enzyme uses its favourable binding energy in the rest of the substrate to compensate for an unfavourable binding energy for the strained part of the molecule.

To illustrate this concept, we shall look at the example of carboxypeptidase A, a zinc-containing protease similar to thermolysin. When the X-ray crystal structure of carboxypeptidase A was solved, it was found that in order to bind the peptide substrate with the most favourable enzyme–substrate interactions, a 'twist' needed to be introduced into the scissile amide bond. In this conformation the carbonyl group of the amide being hydrolysed was bound slightly out of plane of the amide N–H bond, assisted by co-ordination to the active site zinc ion. This has the effect of reducing the overlap of the nitrogen lone pair of electrons with the carbonyl π-bond, which requires the amide bond to be planar. This makes the carbonyl much more reactive, more like a ketone group than an amide, so it is much more susceptible to nucleophilic attack, in this case by the carboxylate side chain of Glu-270. Figure 3.19 shows that in this strained conformation the carbonyl oxygen has already moved some distance towards where it will be in the transition state, so the energy difference between the bound conformation and the transition state is reduced, hence we see rate acceleration.

Thus, if an enzyme is able to bind its substrate in a less favourable but more reactive conformation, then it is able to realize additional rate acceleration in this way. Analysis of this type requires detailed insight from X-ray crystallography, so it is not surprising that there are only a few well-documented examples of this phenomenon. One other example is that of lysozyme, which we shall meet in Chapter 5.

3.8   Catalytic perfection

How fast is it possible for enzyme-catalysed reactions to proceed? Is there a limit to the rate acceleration achievable by enzymes? The answer is yes, and a small number of enzymes have achieved it. For extremely efficient

Table 3.2 Catalytic efficiencies of some diffusion-limited enzymes.

Enzyme	k_{cat}/k_m (M^{-1} s^{-1})
Triose phosphate isomerase	3.0×10^8
Acetylcholinesterase	1.4×10^8
Ketosteroid isomerase	1.3×10^8
β-Lactamase	1.0×10^8

enzymes, the rate of reaction becomes limited by the rate at which a substrate can diffuse onto its active site and diffuse away into solution—the so-called *diffusion limit*. This diffusion limit for collision of enzyme and substrate corresponds to a bimolecular rate constant of approximately 10^8 M^{-1} s^{-1}. We can compare this value with the bimolecular rate constant for reaction of free enzyme with free substrate, which is the catalytic efficiency k_{cat}/K_M (see Chapter 4, Section 4.3).

Many enzymes have catalytic efficiencies of 10^6–10^7 M^{-1} s^{-1}, but a small number have k_{cat}/k_M values which are at the diffusion limit—these are listed in Table 3.2. One of these is the enzyme acetylcholinesterase, which is involved in the propagation of nerve impulses at synaptic junctions: a process for which the utmost speed is necessary. For these enzymes the rate-determining step has become the diffusion of substrates onto the active site. As fast as a substrate diffuses onto the active site it is processed by the enzyme before the next molecule of substrate appears. These are truly wonderful catalysts.

Problems

1 Rationalize the rate accelerations observed for the phenyl esters (**8**) and (**9**) in Fig. 3.5 which contain tertiary amine groups. What type of catalysis is operating in this case?

2 Explain why the hydrolysis of the substituted benzoic acid shown below (a) at pH 4 is 1000-fold faster than the hydrolysis of the corresponding methyl ester (b) under the same conditions.

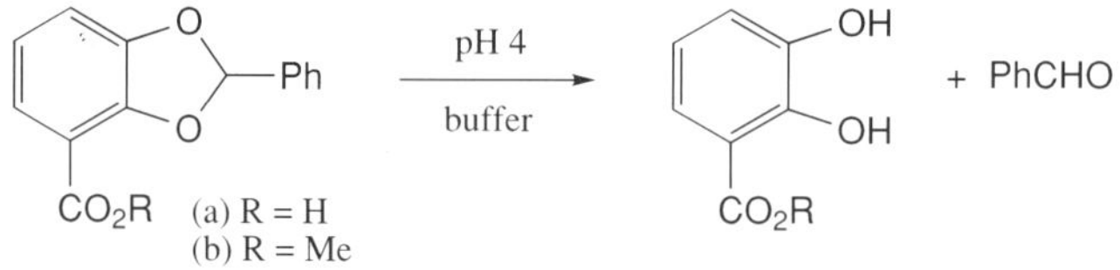

CO_2R (a) R = H
         (b) R = Me

+ PhCHO

3 (a) Incubation of compound (c), shown below, under alkaline conditions leads to a unimolecular reaction of rate constant 7.3×10^{-2} s^{-1} to give a bicyclic product—suggest a structure. Incubation of compound (d) with phenoxide (PhO⁻) leads to a bimolecular reaction of rate constant

10^{-6} M^{-1} s^{-1}. Work out the effective concentration of phenoxide in (c) and comment on the rate acceleration observed.

(c) R = OH
(d) R = H

(b) Epoxide hydrolase catalyses the hydrolysis of a wide range of epoxide substrates. Its active site contains an aspartate residue which is essential for catalytic activity. Given that the enzyme is most active at pH 8–9, propose two possible mechanisms for the enzymatic reaction, and suggest how you might distinguish them experimentally. Comparing the k_{cat} of 1.1 s^{-1} for epoxide hydrolase with the rate constant for reaction of (c) above, suggest how the enzyme might achieve its additional rate acceleration.

4 Incubation of haloalkane dehalogenase (see Fig. 3.17) with substrate under multiple turnover conditions in $H_2^{18}O$ leads to incorporation of one atom of ^{18}O into the alcohol product. However, incubation of a large quantity of enzyme with less than one equivalent of substrate in $H_2^{18}O$ revealed that the product contained no ^{18}O. Explain this observation. (Note: similar results were obtained with this type of experiment for the epoxide hydrolase enzyme in Problem 3).

Further reading

Intramolecular catalysis

Isaacs, N.S. (1987) *Physical Organic Chemistry*. Longman Scientific, Harlow.
Kirby, A.J. (1980) Effective molarities for intramolecular reactions. *Adv Phys Org Chem*, **17**, 183–278.
Menger, F.M. (1985) On the source of intramolecular and enzymatic reactivity. *Acc Chem Res*, **18**, 128–34.

Enzyme catalysis

Fersht, A. (1985) *Enzyme Structure and Mechanism*, 2nd edn. Freeman, New York.
Jencks, W.P. (1969) *Catalysis in Chemistry and Enzymology*. McGraw-Hill, New York.
Jencks, W.P. (1975) Binding energy specificity and enzyme catalysis—the Circe effect. *Adv Enzymol*, **43**, 219–410.

Haloalkane dehalogenase

Franken, S.M., Rozeboom, H.J., Kalk, K.H. & Dijkstra, B.W. (1991) *EMBO J*, **10**, 1297–1302.
Pries, F., Kingma, J., Pentenga, M. *et al.* (1994) *Biochemistry*, **33**, 1242–7.

4 Methods for Studying Enzymatic Reactions

4.1 Introduction

Having established the general principles of enzyme structure and enzyme catalysis, the remaining chapters will deal with each major class of enzymes and their associated coenzymes, and a range of enzyme mechanisms will be discussed. In this chapter we will meet the kind of experimental methods that are used to study enzymes and to elucidate the mechanisms that will be given later. There will be a brief discussion only of the biochemical techniques involved in enzyme purification and characterization, since such methods are described in much more detail in many biochemistry texts. The chapter will focus on those experimental techniques that provide insight into the enzymatic reaction and active site chemistry.

4.2 Enzyme purification

If we want to study a particular enzymatic reaction, the first thing we need to do is to find a source of the enzyme and purify it. In order to test the activity of the enzyme we must first of all have an *assay*: a quantitative method for measuring the conversion of substrate into product (Fig. 4.1). In some cases, conversion of substrate to product can be monitored directly by ultraviolet (UV) spectroscopy, if the substrate or product has a distinctive UV absorbance. Failing this, a chromatographic method can be used to separate substrate from product and hence monitor conversion. In order to quantify a chromatographic assay a radioactive label is usually required in the substrate, so that after separation from substrate the amount of product can be quantitated by scintillation counting. Such an assay is highly specific and highly sensitive, but unfortunately is rather tedious for kinetic work.

A more convenient assay for kinetic purposes is to monitor consumption of a stoichiometric cofactor or cosubstrate, for example the cofactor nicotinamide adenine dinucleotide (NADH) by UV absorption at 340 nm, or consumption of oxygen by an oxygenase enzyme using an oxygen electrode. In other cases a coupled assay is used, in which the product of the reaction is immediately consumed by a second enzyme (or set of enzymes), which can be conveniently monitored.

Once a reliable assay has been developed, it can be used to identify a rich source of the enzyme, which might be a plant, an animal tissue or a

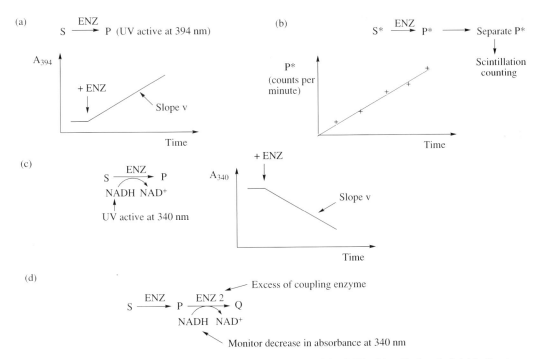

Fig. 4.1 Types of enzyme assays: (a) direct ultraviolet (UV); (b) radiochemical; (c) indirect UV; (d) coupled UV assay. A_{394}, UV absorbance at 394 nm; NADH, nicotinamide adenine dinucleotide; P, product; Q. product of coupling enzyme; S, substrate.

micro-organism. Enzymes are generally produced in the cytoplasm contained within the cells of the producing organism, so in order to isolate the enzyme we must break open the cells to release the enzymes inside (Fig. 4.2). If it is a bacterial source, the bacteria can be grown in culture media and the cells harvested by centrifugation. The bacterial cell walls are then broken by treatment in a high-pressure cell. Animal cells can be readily broken by homogenization, but plant cells sometimes require rapid freeze/thaw methods in order to break their tough cell walls.

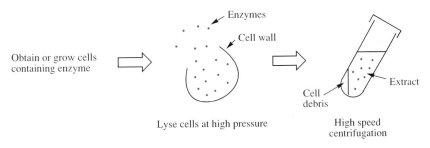

Fig. 4.2 Preparation of an enzyme extract.

Table 4.1 Enzyme purification table (see Fig. 4.3). DEAE, diethylaminoethyl.

	Volume (ml)	Enzyme activity (units ml^{-1})	Protein concentration (mg ml^{-1})	Specific activity (units mg^{-1})	Purification (-fold)
Crude extract	14	13.1	62.0	0.212	1.0
DEAE sephadex pool	19	11.6	17.0	0.684	3.2
Phenyl agarose pool	11	11.8	0.085	140	662
MonoQ anion-exchange pool	6.0	29.2	0.037	787	3710

Using the enzyme assay mentioned above, the enzyme activity can then be purified from the crude extract using precipitation methods such as ammonium sulphate precipitation, and chromatographic methods such as ion-exchange chromatography, gel filtration chromatography, hydrophobic interaction chromatography, etc. Purification can be monitored at each stage by measuring the enzyme activity, in units per millilitre, where 1 unit conventionally means the activity required to convert 1 µmol of substrate per minute. Protein concentration can also be measured using colorimetric assays, in milligrams of protein per millilitre. The ratio of enzyme activity to protein concentration (i.e. units/milligram of protein) is known as the *specific activity* of the enzyme, and is a measure of the purity of the enzyme. As the enzyme is purified the specific activity of the enzyme should increase until the protein is homogeneous and pure, which can be demonstrated using sodium dodecyl sulphate (SDS)–polyacrylamide gel electrophoresis. A purification scheme for a hydratase enzyme from the author's laboratory is shown in Table 4.1. The purification of the enzyme can be seen from the increase in specific activity at each stage of the purification. An SDS–polyacrylamide gel containing samples of protein at each stage of the purification is shown in Fig. 4.3. You can see that in the crude extract there are hundreds of protein bands, but that as the purification proceeds the 28-kDa protein becomes more and more predominant in the gel.

Why do we need pure enzyme? If we can see enzyme activity in the original extract, why not use that? The problem with using unpurified enzyme for kinetic or mechanistic studies is that there may be interference from other enzymes in the extract that use the same substrate or cofactor. There may also be enzymes that give rise to UV absorbance changes which might interfere with a UV-based assay. If the enzyme can be purified then the *turnover number* of the enzyme can be measured, which is the number of micromoles of substrate converted per micromole of enzyme per second. The turnover number can be simply calculated from the specific activity of the pure enzyme (in units per milligram of protein) and the molecular weight of the enzyme (see Problem 1, p. 69). The isolation of pure enzyme also

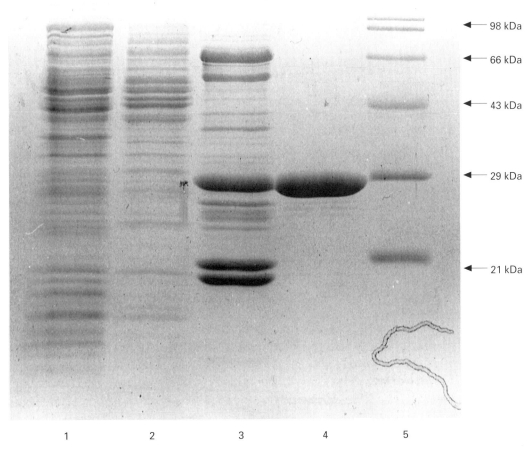

Fig. 4.3 Purification of 2-hydroxypentadienoic acid hydratase from *Escherichia coli*. The photo shows an SDS–polyacrylamide gel of samples taken from the purification of this enzyme. Lane 1, *E. coli* crude extract; lane 2, DEAE sephadex pool; lane 3, phenyl agarose pool; lane 4, monoQ anion-exchange pool; lane 5, molecular weight standards (2 × 98 kDa, 66 kDa, 43 kDa, 29 kDa, 21 kDa) (See also Table 4.1).

allows active site studies to be carried out on the homogeneous protein, and crystallization of the enzyme for X-ray crystallographic analysis.

4.3   Enzyme kinetics

It is possible to learn a great deal about how an enzyme works from a detailed kinetic study of the enzymatic reaction. For the purposes of this chapter I will give only a brief discussion of a simple model for enzyme kinetics; a full discussion of enzyme kinetics is given in texts such as Segel.

The Michaelis–Menten model for enzyme kinetics assumes that the following steps are involved in the enzymatic reaction: reversible formation of the enzyme–substrate (ES) complex, followed by conversion to product as shown in Equation 4.1.

$$E + S \xrightleftharpoons[k_{-1}]{k_1} ES \xrightarrow{k_2} E + P \tag{4.1}$$

There are several assumptions implicit in this model: (i) that the enzyme binds only a single substrate; (ii) that there is only one kinetically significant step between the ES complex and product formation; and (iii) that product formation is irreversible. Despite the fact that these assumptions are not strictly correct for most enzymes, this proves to be a useful model for a very wide range of enzymes. Derivation of a rate equation uses a kinetic criterion known as the *steady state approximation*: that the enzymatic reaction will quickly adopt a situation of steady state in which the concentration of the intermediate species ES remains constant. Under these conditions the rate of formation of ES is equal to the rate of consumption of ES. The other criterion that is used is that the total amount of enzyme in the system E_0, made up of free enzyme E and ES complex, is constant. The derivation is as follows (Equation 4.2).

Steady state approximation:
$$\text{Rate of formation of ES} = \text{Rate of breakdown of ES}$$
$$k_1 [E] [S] = k_2 [ES] + k_{-1} [ES]$$

$$\text{Total amount of enzyme } [E]_0 = [E] + [ES]$$

$$\Rightarrow \quad k_1 [E]_0 [S] - k_1 [ES] [S] = k_2 [ES] + k_{-1} [ES]$$

$$\Rightarrow \quad [ES] = \frac{k_1 [E]_0 [S]}{k_1 [S] + k_2 + k_{-1}}$$

$$\text{Rate of production of P} = k_2 [ES] = \frac{k_2 [E]_0 [S]}{\left(\dfrac{k_{-1} + k_2}{k_1}\right) + [S]} = \frac{k_{cat} [E]_0 [S]}{K_m + [S]} \tag{4.2}$$

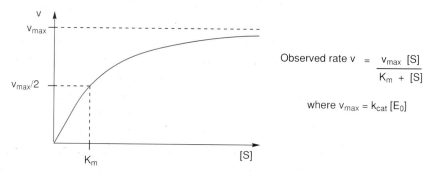

$$\text{Observed rate } v = \frac{v_{max} [S]}{K_m + [S]}$$

$$\text{where } v_{max} = k_{cat} [E_0]$$

The two kinetic constants in the Michaelis–Menten rate equation have special significance. The k_{cat} parameter is the *turnover number* mentioned above: it is a unimolecular rate constant whose units are per second (or per minute if it is a very slow enzyme!), and it represents the number of micro-moles of substrate converted per micromole of enzyme per second. Alternatively, in molecular terms it represents the number of molecules turned over by one molecule of enzyme per second, which gives a good feel for how quickly the enzyme is operating. Typically values are in the range $0.1-100 \text{ s}^{-1}$.

The K_M parameter is known as the Michaelis constant for the enzyme: its units are moles per litre (or M). In practice, the K_M is the concentration of substrate at which half-maximal rate is observed. It can be taken as a rough indication of how tightly the enzyme binds its substrate, so a substrate bound weakly by an enzyme will have a large K_M value, and a substrate bound tightly will have a small K_M. However, it must be stressed that K_M is *not* a true dissociation constant for the substrate, since it also depends on the rate constant k_2. Values of K_M are typically in the range 1 μM–1 mM.

Values of k_{cat} and K_M can be measured for a particular enzyme by measuring the rate of the enzymatic reaction at a range of different substrate concentrations. At high substrate concentrations ($[S] \gg K_M$) the rate equation reduces to $v = k_{cat} [E_0]$, so a maximum rate is observed however high the substrate concentration. Under these conditions the enzyme is fully saturated with substrate, and no free enzyme is present. So as soon as an enzyme molecule releases a molecule of product it immediately picks up another molecule of substrate. In other words the enzyme is working flat out: the observed rate of reaction is limited only by the rate of catalysis.

Under low substrate concentrations the rate equation reduces to $v = (k_{cat}/K_M)[E][S]$, so the observed rate is proportional to substrate concentration, and the reaction has effectively become a bimolecular reaction between free enzyme E and free substrate S. Under these conditions the majority of enzyme is free enzyme, and the observed rate of reaction depends on how efficiently the enzyme can bind the substrate at that concentration. The bimolecular rate constant under these conditions, k_{cat}/K_M, is known as the *catalytic efficiency* of the enzyme, since it represents how efficiently free enzyme will react with free substrate.

A schematic representation of the energetic profiles at high and low substrate concentrations is given in Fig. 4.4. At high substrate concentrations the enzyme is fully saturated with substrate, so the activation energy for the enzymatic reaction is the free energy difference between the ES complex and the transition state. At $[S] = K_M$ the enzyme is half-saturated with substrate. At low substrate concentrations the majority of enzyme is free of substrate, so the activation energy for the reaction is the free energy difference between free enzyme plus substrate and the transition state.

How do we actually determine k_{cat} and K_M? The value of k_{cat} can be roughly visualized from the plot of v versus $[S]$ by estimating the rate at high

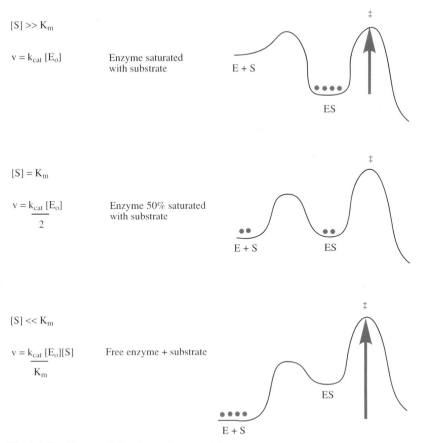

$$[S] \gg K_m$$

$$v = k_{cat} [E_o]$$

Enzyme saturated
with substrate

$$[S] = K_m$$

$$v = \frac{k_{cat} [E_o]}{2}$$

Enzyme 50% saturated
with substrate

$$[S] \ll K_m$$

$$v = \frac{k_{cat} [E_o][S]}{K_m}$$

Free enzyme + substrate

Fig. 4.4 Significance of K_m, k_{cat}. Plots are of free energy versus reaction co-ordinate. Red dots indicate the population of enzyme molecules present.

substrate concentrations. The K_M value corresponds to the substrate concentration at which half-maximal rate is observed. However, a more accurate way to determine K_M and k_{cat} from the data is to use either the Lineweaver–Burk or Eadie–Hofstee plots, shown in Fig. 4.5, both of which give straight lines from which the kinetic constants can be determined as indicated.

Enzyme inhibition takes place if a molecule other than the substrate binds at the active site and prevents the enzymatic reaction from taking place. Broadly speaking, there are two types of enzyme inhibition that are commonly observed: reversible and irreversible inhibition. Reversible inhibition occurs if an inhibitor is bound non-covalently at the active site, preventing one of the substrates from binding. The most common type of reversible inhibition observed for a single substrate enzyme reaction is competitive inhibition, where the inhibitor binds at the same site as the substrate S, and therefore competes for the same binding site. The kinetic scheme and associated kinetic behaviour observed for competitive inhibition

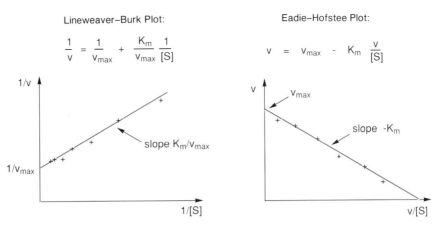

Fig. 4.5 Graphical methods for determination of K_M and k_{cat}. v, reaction velocity; [S], substrate concentration.

is shown in Fig. 4.6. If the rate of the enzyme reaction is measured at varying substrate concentrations but fixed inhibitor concentrations, apparent K_M values $((K_M)_{app})$ can be measured at varying inhibitor concentrations. If plotted on a Lineweaver–Burk plot a series of straight lines are obtained, intersecting on the y-axis. Thus, the v_{max} is unaffected by competitive inhibition, since at high substrate concentrations the substrate can competitively displace the inhibitor.

Non-competitive reversible inhibition is observed when an inhibitor, I, binds to another part of the enzyme active site leading to a non-productive EIS complex. This type of inhibition, illustrated in Fig. 4.6, typically occurs in multi-substrate reactions when the inhibitor binds to the binding site of the cosubstrate. For a full discussion of these and other types of inhibition in multi-substrate reactions, reference to texts on enzyme kinetics is recommended.

Irreversible inhibition occurs when an inhibitor first binds at the active site, then reacts with an active site group to form a covalent bond (E–I). The active site is then irreversible blocked by the inhibitor and is permanently inactivated. Irreversible inhibitors usually contain electrophilic functional groups such as halogen substituents or epoxides. The kinetic characteristic associated with irreversible inhibition is that it is time-dependent. This is because as time goes by more and more enzyme will be blocked irreversibly by conversion of the reversible EI complex to E–I. Since $[S] \gg [E]$ then, in practice, this is a unimolecular reaction, so the observed kinetic behaviour follows unimolecular reaction kinetics. Thus, if enzyme activity (i.e. v_{max}) is plotted versus time, an exponential decrease of activity is observed (Fig. 4.7).

All of the above kinetic data can be obtained by steady state kinetics, which can be observed conveniently over a 1–10-min assay period. However,

(a) *Competitive inhibition*

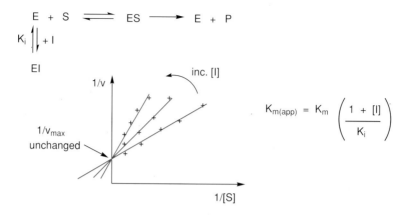

$$K_{m(app)} = K_m \left(\frac{1 + [I]}{K_i} \right)$$

(b) *Noncompetitive inhibition*

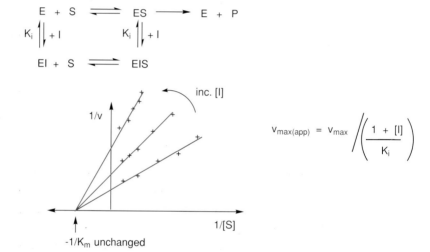

$$V_{max(app)} = V_{max} \Bigg/ \left(\frac{1 + [I]}{K_i} \right)$$

Fig. 4.6 Kinetic consequences of reversible enzyme inhibition.

if one is able to examine an enzymatic reaction before the attainment of steady state, then individual enzymatic rate constants can be measured directly using pre-steady state kinetics. If the turnover number for an enzyme is $10\,s^{-1}$, then under saturating conditions a molecule of substrate will be converted to product in 0.1 s. Therefore, in order to examine a single catalytic cycle one must examine the enzymatic reaction in the range 0–100 ms. This can be done using a stopped flow apparatus, shown in simplified form in Fig. 4.8.

This apparatus consists of two syringes containing: (i) substrate; and (ii) a stoichiometric amount (i.e. 1–100 nmol) of enzyme, connected to a rapid mixing device. The syringes are driven so as to fill the observation cuvette

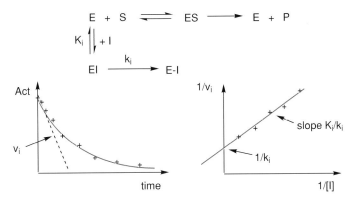

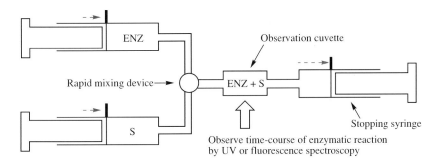

Fig. 4.7 Kinetic consequences of irreversible enzyme inhibition.

Fig. 4.8 Stopped flow apparatus.

with freshly mixed enzyme and substrate. At this point, the reaction time of
the mixture is given by the distance from the mixer and the linear flow rate.
The syringes are physically stopped, and the enzyme/substrate mixture is
allowed to react along the enzymatic reaction pathway. Changes in absorbance or fluorescence in the observation cuvette are measured on a 10–
1000-ms timescale, giving a complete time course of a single catalytic cycle
of the enzymatic reaction. Several substrate concentrations are used to
enable individual rate constants to be measured. In this way a multi-step
enzymatic reaction can, in principle, be broken down into its individual
steps, the slowest of which (the rate-determining step) should correspond to
the steady state k_{cat} value measured by steady state kinetics.

4.4   The stereochemical course of an enzymatic reaction

The vast majority of biological molecules are chiral, i.e. a molecule whose
mirror image is non-superimposable upon the original. In the case of
carbon-based compounds, if a carbon atom is surrounded by four different

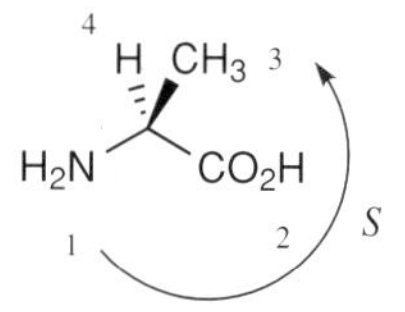

Fig. 4.9 L-alanine, a chiral
molecule.

L-alanine = 2S-alanine

groups, then it will be chiral. One simple example shown in Fig. 4.9 is that of
the amino acid L-alanine, which is a substrate for several pyridoxal-5'-
phosphate enzymes described in Chapter 9. Chiral centres can be designated
as R or S depending on the relative orientation of the four groups, prioritised
by atomic mass (see Appendix 1). Thus, L-alanine is more formally written
as 2S-alanine.

Enzyme-catalysed reactions are in general both stereoselective and
stereospecific. Stereoselectivity is the ability to select a single enantiomer of
the substrate in the presence of other isomers. Stereospecificity is the ability
to catalyse the production of a single enantiomer of the product via a specific
reaction pathway. Stereospecificity in enzyme catalysis arises from the fact
that catalysis is taking place in an enzyme active site in which the bound
substrate is held in a defined orientation relative to the active site groups
(unnatural substrates which can adopt more than one orientation in an
enzyme active site may be processed with less stereospecificity). How do we
elucidate the stereochemical course of an enzymatic reaction?

Generation of a chiral product

In many cases enzymes are able to generate a product containing one or
more chiral centres from achiral substrates—a process known as *asymmetric
induction*. For example, the aldolase enzyme fructose-1,6-diphosphate aldolase (see Chapter 7) catalyses the aldol condensation of achiral substrates
dihydroxyacetone phosphate and acetaldehyde to generate a chiral aldol
product, as shown in Fig. 4.10.

If a chiral product is formed, then its absolute configuration can be
determined in the following ways:
1 by chemical conversion to a compound of established absolute configuration, and measurement of optical activity;

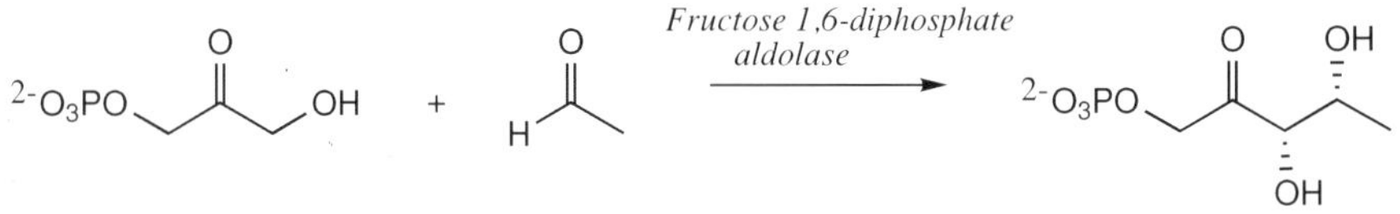

Fig. 4.10 Production of a chiral product.

2 by treatment with another known chiral reagent to make a diastereomeric derivative whose configuration can be determined by methods such as
nuclear magnetic resonance (NMR) spectroscopy or X-ray crystallography;
3 treatment with an enzyme whose reaction proceeds with known stereospecificity.

Prochiral selectivity

Carbon centres which are surrounded by XXYZ groups are known as
prochiral centres. For example, the C-1 carbon atom of ethanol is prochiral
since it is attached to two hydrogens, one methyl group and one hydroxyl
group. This is not a chiral centre, since there are two hydrogens attached, yet
these two hydrogens can be distinguished by the enzyme alcohol dehydrogenase, which oxidizes ethanol to acetaldehyde, as shown in Fig. 4.11.

How does this enzyme achieve this remarkable selectivity? If you imagine that the ethanol molecule is fixed in the plane of the page with the methyl
group pointing left and the hydroxyl group pointing right, as in Fig. 4.12,
then one of the two hydrogens is pointing up out of the page, and the other is
pointing down into the page. Since we can visualize this situation in three
dimensions, we can easily distinguish between these two hydrogens. Thus,
two hydrogen atoms attached to a prochiral centre can be distinguished *if
they are held in a fixed orientation in a chiral environment*. Enzyme active
sites satisfy both these criteria, since they are able to bind molecules in a

Hydrogens attached to prochiral centres can be designated using the
following method. The convention is that the hydrogen to be assigned is
replaced by a deuterium atom (making the centre chiral), and the chirality of
the resulting centre determined (see Appendix 1). If the resulting chiral
centre has configuration *R*, then the hydrogen atom replaced by deuterium is
labelled *proR*. Conversely, if the deuterium-containing centre is *S*, then the
hydrogen atom is labelled *proS*. In the case of alcohol dehydrogenase, the
enzyme removes stereospecifically the *proR* hydrogen. This was demonstrated by synthesizing authentic samples of $(1R-{}^2H)$- and $(1S-{}^2H)$-ethanol.
Each sample was separately incubated with the enzyme and the deuterium
content of the product analysed in each case. In the case of the $1R$ substrate
deuterium was removed by the enzyme, whereas with the $1S$ substrate
deuterium was retained in the product, as shown in Fig. 4.12.

Fig. 4.11 Alcohol dehydrogenase reaction.

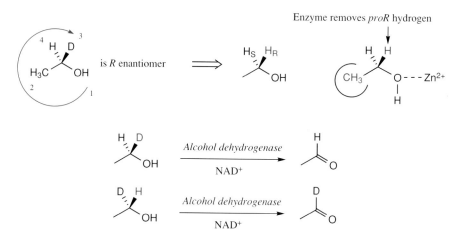

Fig. 4.12 Stereochemistry of alcohol dehydrogenase.

defined orientation using specific enzyme–substrate binding interactions, and of course the enzyme active site is chiral.

Examination of the X-ray crystal structure of alcohol dehydrogenase reveals that the C-1 oxygen substituent is bound by an active site zinc cofactor, and the methyl group is bound in such a way that the *proR* hydrogen is pointing directly at the NAD$^+$ cofactor, as shown in Plate 4.1 (facing, p. 152). The prochiral selectivity can therefore be explained easily by the orientation adopted by the substrate in the enzyme active site.

In general the stereochemical course of an enzymatic reaction is usually determined by replacement of particular atoms of the substrate by isotopes of carbon, hydrogen, oxygen and nitrogen, whose fate at the end of the reaction can then be monitored. In particular, enzymatic reactions involving prochiral centres can only be studied by replacement of one of the prochiral substituents by another isotope, thus generating a chiral centre. The isotopes available are listed in Table 4.2, together with the methods available for their analysis. In some cases such as ^{1}H, ^{13}C and ^{15}N these nuclei possess nuclear spin, which allows analysis by NMR spectroscopy. In other cases such as ^{3}H and ^{14}C the nuclei are radioactive, and their presence can be detected by scintillation counting. Isotopes such as ^{18}O which are neither radioactive nor possess nuclear spin must be detected either by mass spectrometry or by their effect on neighbouring nuclei: in the case of ^{18}O their attachment to a neighbouring ^{13}C shifts the ^{13}C NMR signal upfield by 0.01–0.05 ppm.

Enzymatic reactions which involve the substitution of one group for another group in a defined relative orientation can take place with either retention or inversion of stereochemistry. Where such reactions occur at a prochiral centre, the stereochemical course can be examined by synthesis of a stereospecifically labelled substrate. For example, the P$_{450}$-dependent

Table 4.2 Isotopes available for stereochemical elucidation.

	Isotope	Natural abundance (%)	Method of analysis
Hydrogen	^1H	99.985	NMR (I = 1/2)
	^2H	0.015	NMR (I = 1) or shift in ^{13}C NMR
	^3H	—	Scintillation counting
Carbon	^{12}C	98.9	
	^{13}C	1.1	NMR (I = 1/2)
	^{14}C	—	Scintillation counting
Nitrogen	^{14}N	99.63	
	^{15}N	0.37	NMR (I = 1/2)
Oxygen	^{16}O	99.8	
	^{17}O	0.037	NMR (I = 5/2)
	^{18}O	0.20	Mass spectrometry or shift in ^{13}C NMR

enzyme camphor hydroxylase which catalyses the hydroxylation of camphor was shown to proceed with retention of stereochemistry by use of the labelled substrate illustrated in Fig. 4.13.

Interconversions of methylene ($=CH_2$ or $-CH_2-$) groups to methyl ($-CH_3$) groups require a special type of stereochemical analysis, since the resulting methyl group contains three apparently identical hydrogen atoms (i.e. an XXXY system). However, it is possible to analyse these methylene-to-methyl interconversion using all three of the isotopes of hydrogen—^1H, ^2H and ^3H—in the form of a chiral methyl group. It is important to note that in this analysis the ^1H and ^2H substituents are present in 100% abundance, whereas only a small proportion of molecules contain ^3H (since ^3H is only available and only safe to handle in relatively low abundance). Therefore, the analysis of chiral methyl groups must focus on those molecules containing ^3H, by detecting the presence or absence of ^3H label.

Chiral methyl groups can be generated from enzymatic reactions by preparing the methylene substrate in a stereospecifically labelled form using two of the isotopes of hydrogen, and carrying out the enzymatic reaction in the presence of the third isotope. If the product can be degraded to chiral acetic acid, then the configuration of the chiral methyl group can be

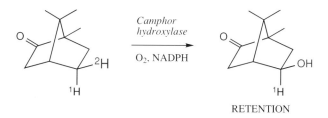

Fig. 4.13 Stereochemistry of camphor hydroxylase.

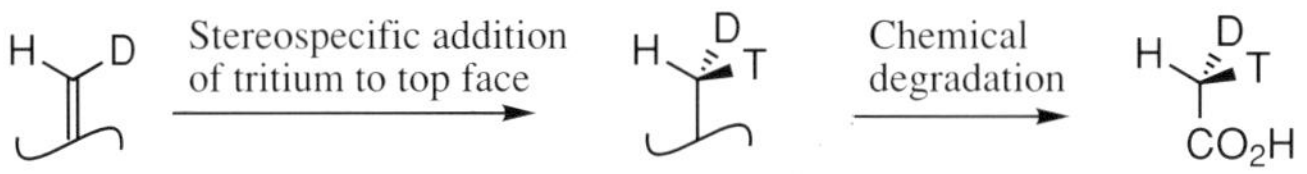

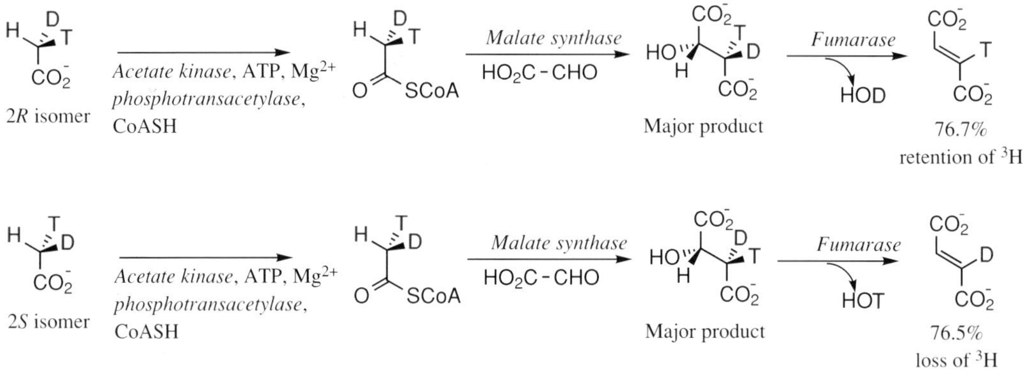

Fig. 4.14 Chiral methyl groups. ATP, adenosine triphosphate; CoASH, coenzyme A.

determined using a method developed independently by J. W. Cornforth and D. Arigoni, shown in Fig. 4.14.

The method of analysis involves conversion to chiral acetyl coenzyme A (CoA) (see Chapter 5, Section 5.4), followed by incubation with malate synthase (see Chapter 7, Section 7.2), which removes one of the hydrogens on the methyl group, and combines with glyoxalate to form malic acid. The malate synthase reaction has a preference for removal of ^1H rather than ^2H or ^3H (i.e. a kinetic isotope effect of $k_H/k_T = 2.7$), so in the majority of molecules ^1H is removed. The reaction with glyoxalate then occurs with inversion of configuration. Therefore, the 2S enantiomer of acetyl CoA is converted into the 2S,3R enantiomer of malate containing ^2H and ^3H stereospecifically at C-3, as illustrated in Fig. 4.14. Treatment with the enzyme fumarase then results in a stereospecific anti-elimination of water, the enzyme removing only the proR hydrogen at C-3. The configuration of the major product at C-3 can therefore be deduced by monitoring the fate of the ^3H, either to water or to tritiated fumaric acid. Since the stereochemistry of the fumarase and malate synthase reactions is known, the configuration of the chiral methyl group can be deduced. By this method a number of such methylene-to-methyl interconversions have been analysed.

A similar stereochemical strategy is used to analyse the stereochemistry of phosphoryl transfer reactions, since phosphates also contain three apparently identical oxygen substituents. Three isotopes of oxygen are also available: ^{16}O, ^{17}O and ^{18}O. Using skilful synthetic chemistry approaches, phosphate

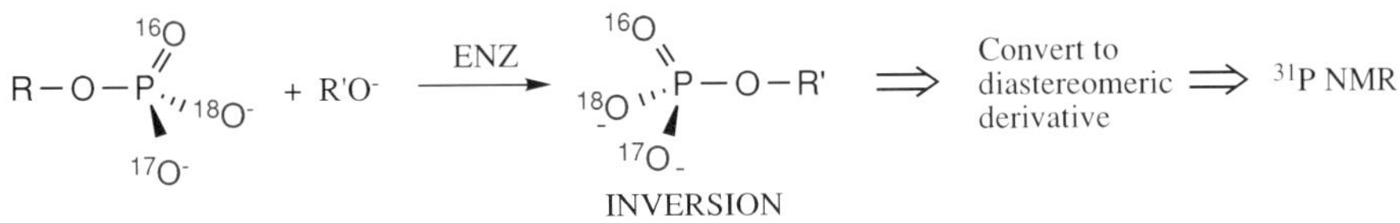

Fig. 4.15 Stereochemistry of phosphoryl transfer reaction.

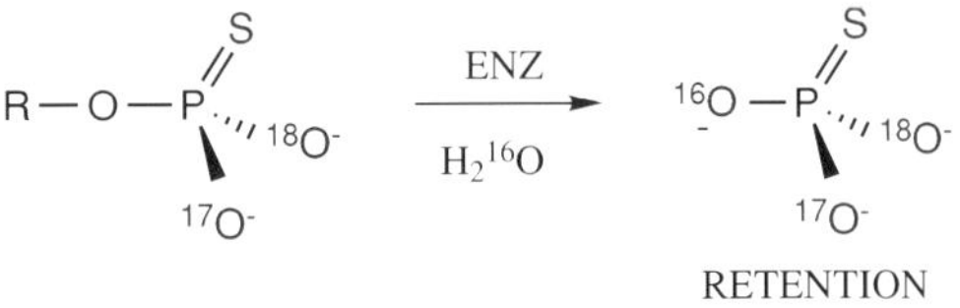

Fig. 4.16 Stereochemistry of phosphate release.

ester substrates can be prepared containing all three isotopes of oxygen. Incubation of the chiral phosphate ester substrate with the corresponding phosphotransferase enzyme generates a chiral phosphate ester product, as shown in Fig. 4.15. The configuration of the chiral product reveals whether the enzymatic reaction proceeds with retention or inversion of configuration. Analysis of the configuration of the chiral phosphate ester product is complicated, but in essence involves chemical or enzymatic conversion to a diastereomeric derivative, followed by ^{31}P NMR spectroscopic analysis.

The stereochemistry of reactions releasing inorganic phosphate presents an even more difficult problem, since there are four apparently identical oxygens to be distinguished, but only three isotopes of oxygen. This has been solved by incorporating one atom of sulphur as a substituent, since thiophosphate ester substrates are accepted by these enzymes (Fig. 4.16). Again, the configuration of the $[^{16}O, ^{17}O, ^{18}O]$–thiophosphate product can be deduced by conversion to a diastereomeric derivative followed by NMR spectroscopic analysis. For a detailed discussion of this stereochemical analysis the interested reader is referred to references at the end of the chapter.

4.5   The existence of intermediates in enzymatic reactions

Enzymatic reactions are often multi-step reactions involving a number of transient enzyme-bound intermediates (Fig. 4.17). If the enzymatic reaction is very rapid and none of the intermediates is released from the active site, how can we prove the existence of these transient intermediates?

Direct observation

Since the turnover numbers for most enzymes are greater than 1 s^{-1} then it is impractical to observe the formation of intermediates directly. However, in

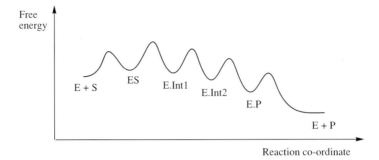

Fig. 4.17 Energy profile of a multi-step enzymatic reaction.

some cases the turnover number can be reduced by changing the temperature or pH, or by using an unnatural substrate, to such an extent that intermediates can be detected directly by NMR spectroscopy.

If no intermediate is detectable by these methods then a faster analytical method can be used in the form of stopped flow methods. Just as stopped flow methods can be used to study rapid enzyme kinetics, in the same way rapid quench methods can be used to isolate intermediates. This method involves mixing enzyme with substrate in a rapid mixing device similar to that shown in Fig. 4.8, then, after a fixed time inteval of say 100 ms, mixing with a quench reagent such as an organic solvent or a different pH solution.

This type of approach was recently used to identify a tetrahedral intermediate in the reaction of enolpyruvyl-shikimate-3-phosphate (EPSP) synthase, as shown in Fig. 4.18 (also see Chapter 8, Section 8.5). In this case the

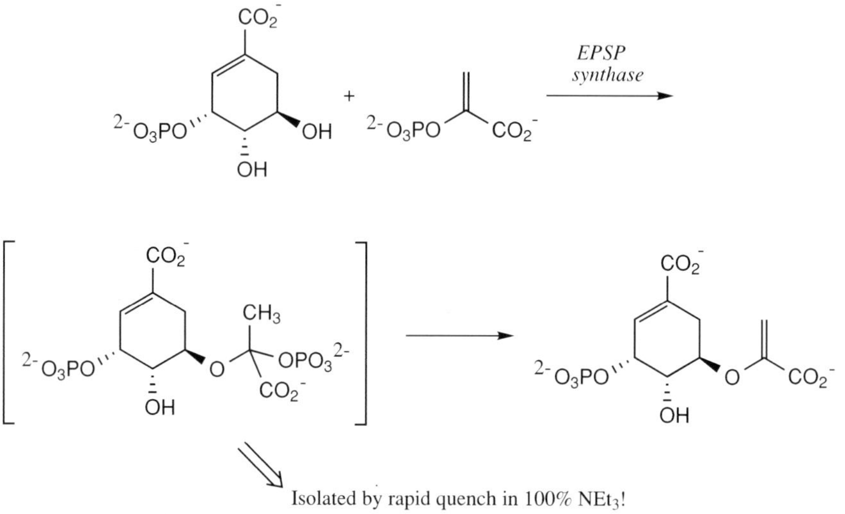

Isolated by rapid quench in 100% NEt$_3$!

Fig. 4.18 Identification of EPSP synthase reaction intermediate. NEt$_3$, triethylamine.

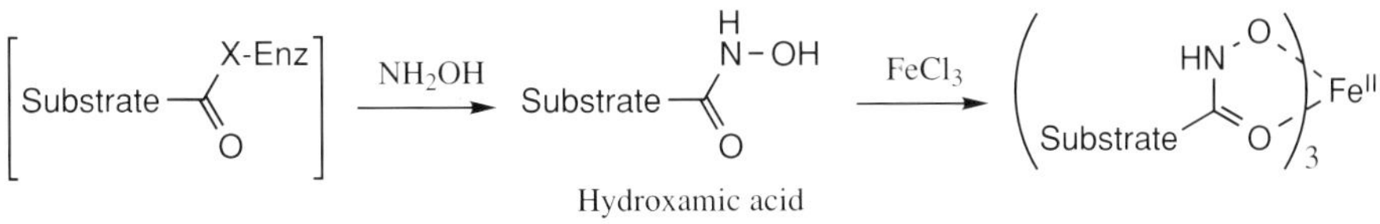

Fig. 4.19 Hydroxylamine (NH_2OH) trapping of an activated intermediate.

intermediate was isolated by quenching 50-mg quantities of enzyme and substrate with neat triethylamine, which was found to stabilize the intermediate.

Trapping

Intermediates in enzymatic reactions that possess enhanced chemical reactivity can sometimes be trapped using a selective chemical reagent. One example already mentioned is the trapping of imine intermediates formed upon reaction of active site lysine residues with carbonyl substrates with sodium borohydride. Another common example is the trapping of activated carbonyl intermediates with hydroxylamine, a potent nitrogen nucleophile. This reaction forms hydroxamic acid products which can be detected spectrophotometrically by treatment with iron (III) chloride ($FeCl_3$) solutions, as shown in Fig. 4.19.

Chemical inference

The existence of certain intermediates can be inferred by following the fate of individual atoms in the substrate. For example, the existence of an acyl phosphate intermediate in the glutamine synthetase reaction was established by incubating ^{18}O-labelled substrate with enzyme and cofactor adenosine triphosphate (ATP). The inorganic phosphate product was found to contain one atom of ^{18}O, consistent with formation of the acyl phosphate intermediate as shown in Fig. 4.20.

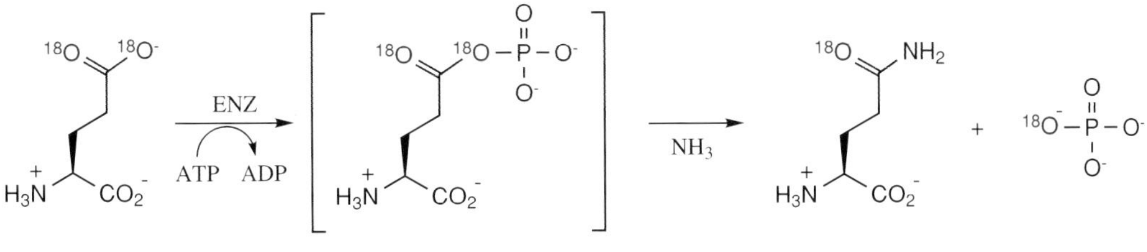

Fig. 4.20 Identification of glutamine synthetase intermediate by chemical inference. ADP, adenosine diphosphate; NH_3, ammonia.

Isotope exchange

If an enzyme reaction involves the reaction of two species to form an intermediate which is then attacked by a third species, the first step or partial reaction can be analysed in isolation using isotope exchange. Taking the glutamine synthetase reaction as an example, glutamate first reacts with ATP to form adenosine diphosphate (ADP) and an intermediate γ-glutamyl phosphate, which is attacked by ammonia to form glutamine. If the enzyme is incubated with glutamate, ^{14}C–ATP and unlabelled ADP in the absence of ammonia, then the enzyme can not complete the overall reaction, but it can convert glutamate to enzyme-bound γ-glutamyl phosphate and ^{14}C–ADP. In this case ^{14}C–ADP can be released from the enzyme without releasing the intermediate. The enzyme can then convert unlabelled ADP to ATP via the reverse reaction. Overall this process leads to the 'exchange' of ^{14}C label from ATP to ADP, as illustrated in Fig. 4.21.

This method depends on the ability of the enzyme to release ^{14}C–ADP at the intermediate stage. Since many enzymes have well-defined orders of binding of their substrates, this release may be slow or even impossible in the absence of the third substrate. A more subtle method of analysing such exchange is the method of *positional isotope exchange* developed by I.A. Rose. This method is illustrated in Fig. 4.22 for the glutamine synthetase reaction.

For this method ATP is labelled with ^{18}O at the β,γ-O bridge position. Upon formation of the reaction intermediate the γ-phosphate is transferred to glutamate, allowing the ^{18}O label to scramble amongst the β-phosphate oxygens whilst still bound to the enzyme. Upon reformation of the β,γ-O bridge in the reverse reaction the ^{18}O label will be present not only in the bridge position but also in the β-phosphate oxygens. Not only can such isotope exchanges be observed, but the rate of isotope exchange also can be measured and compared with the rate of the overall enzymatic reaction.

The final proof that a certain species is a true intermediate in an enzymatic reaction is by preparing the intermediate by independent chemical synthesis. There are two criteria that must be satisfied for a potential intermediate. The first is *chemical competence*: the intermediate must be converted by the enzyme to the product of the reaction (and also be a

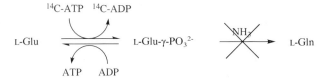

Fig. 4.21 Isotope exchange in the glutamine synthetase rection. L-Gln, glutamine; L-Glu, glutamate; L-Glu-γ-PO_3^{2-}, γ-glutamyl phosphate.

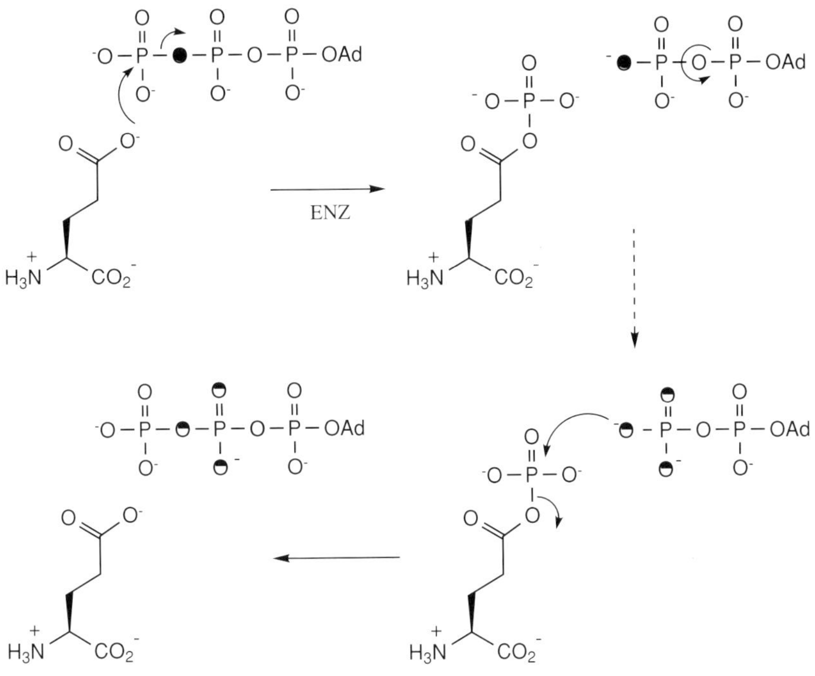

Fig. 4.22 Positional isotope exchange.

substrate for the reverse reaction if the enzymatic reaction is reversible). The second is *kinetic competence*: the rate of conversion to products must be at least as fast as the rate of the overall enzymatic reaction.

4.6   Analysis of transition states in enzymatic reactions

Enzymatic reactions are frequently multi-step reactions. In such reactions the overall rate of the enzymatic reaction is governed by the step having the highest transition state energy—the rate-determining step. If we want to determine a detailed kinetic profile for an enzymatic reaction, it is important to know which is the rate-determining step. Once again we can use isotopic substitution, this time to study the existence of kinetic isotope effects.

If a reaction involves cleavage of a C–H bond in its rate-determining step, then substitution of hydrogen for deuterium (D) leads to a reduction in rate for that step, and hence for the overall reaction. This effect arises because the C–D bond is slightly stronger than the C–H bond, because the C–D bond has a lower zero point energy, as shown in Fig. 4.23.

Substitution of a hydrogen that is removed in an enzymatic reaction with deuterium and measurement of the k_{cat} and K_M values for the deuterated

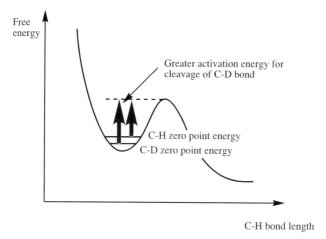

Fig. 4.23 Origin of the deuterium kinetic isotope effect.

substrate can therefore provide information about the thermodynamic profile of the reaction. Kinetic isotope effects are commonly observed also in
organic reactions, for exactly the same reasons, however the analysis of
isotope effects in enzymatic reactions can be complicated in cases where
there are a number of transition states.

Four typical scenarios are shown in Fig. 4.24.

1 If the step involving C–H cleavage has a significantly higher transition
state energy than the other steps, then substitution for a C–D bond will give
a substantial kinetic isotope effect ($k_H/k_D \sim 6-7$) (Fig. 4.24a).

2 If the step involving C–H cleavage is of similar transition state energy to
the other steps, then this step is only partially rate determining, and a kinetic
isotope effect of $k_H/k_D \sim 2-3$ is observed (Fig. 4.24b).

3 If C–H cleavage occurs after the rate-determining step, then no kinetic
isotope effect is observed (Fig. 4.24c).

4 Finally, if a partially rate-determining C–H cleavage occurs before the
rate-determining step, then a small kinetic isotope effect will be observed on
k_{cat}/K_M, but may not be observed on k_{cat} (Fig. 4.24d). Remember that
k_{cat}/K_M is the bimolecular rate constant for the reaction of free enzyme with
free substrate, whereas k_{cat} is the unimolecular rate constant for conversion
of saturated ES complex. In the latter case where the enzyme is fully
saturated with substrate an early C–H cleavage step may not be so significant
kinetically.

Kinetic isotope effects are also observed when the isotopic substitution
does not involve any of the bonds broken in the reaction, but does involve a
change of orbital hybridization in a rate-determining step. Thus, if an sp^3-
hybridized carbon bearing a deuterium atom changes to an sp^2-hybridized

(a) C-H cleavage rate determining $\dfrac{k_H}{k_D} = 6\text{-}7$

(b) C-H cleavage partially rate determining $\dfrac{k_H}{k_D} = 2\text{-}3$

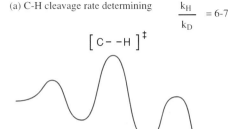

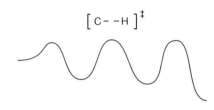

(c) C-H cleavage after rate determining step

no KIE observed

(d) C-H cleavage before rate determining step

KIE observed on k_{cat}/K_m ([S] << K_m)

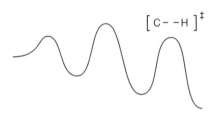

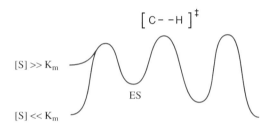

Fig. 4.24 Expression of kinetic isotope effects (KIE).

carbon bearing the deuterium atom in the rate-determining step of a reaction, a smaller *secondary deuterium isotope effect* will be observed ($k_H/k_D = 1.1$–1.4). An example is the ketosteroid isomerase reaction we met in Chapter 3 (see Fig. 3.14), illustrated in Fig. 4.25. With $4R$-[4-^{2}H]-ketosteroid as a substrate the hydrogen being abstracted is replaced by deuterium, and a primary kinetic isotope effect of 6.2 is observed on k_{cat}. However, with $4S$-[4-^{2}H]-

Fig. 4.25 Substrate kinetic isotope effects in the ketosteroid isomerase reaction.

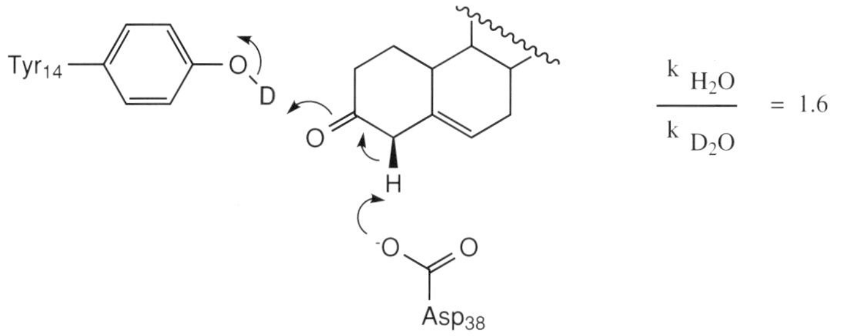

Fig. 4.26 D_2O solvent isotope effect in the ketosteroid isomerase reaction. Asp, aspartate; Tyr, tyrosine.

ketosteroid a secondary kinetic isotope effect of 1.1 is observed, due to re-hybridization of C-4 in the rate-determining step.

If an enzymatic reaction is carried out in ^2H_2O rather than ^1H_2O a solvent kinetic isotope effect is observed if there is a proton transfer from water or a water-exchangeable group in the rate-determining step. For example, in the ketosteroid isomerase reaction there is a D_2O solvent isotope effect of 1.6 on k_{cat}, due to replacement of the phenolic proton of tyrosine-14 by deuterium, as shown in Fig. 4.26.

D_2O solvent isotope effects also arise where there is base-catalysed attack of water in a rate-determining step. However, D_2O solvent isotope effects can, in some cases, arise for a number of reasons far removed from active site catalysis, so they need to be interpreted with caution. Finally, kinetic isotope effects are not restricted only to cleavage of C–H bonds: rate-determining cleavage of C–O, C–N and C–C bonds can be studied using ^{18}O-, ^{15}N- and ^{13}C-labelled substrates. In these cases the difference in zero point energy between heavy isotopes is very much smaller, so the observed effects are typically less than 1.1. For a discussion of heavy atom isotope effects the reader is referred to specialist references.

4.7   Determination of active site catalytic groups

As well as examining the molecular details of an enzymatic reaction it is equally important to study the groups present in the enzyme active site which carry out the catalysis.

One convenient method for obtaining clues regarding active site catalytic groups is to analyse the variation of enzyme activity with pH—a pH/rate profile. Thus, if there are acidic and basic groups involved in the enzyme mechanism, they must be in the correct ionization state in order for the enzyme to operate efficiently. For example, the ketosteroid isomerase reaction illustrated above has the pH/rate profile shown in Fig. 4.27, from which the pK_a value of 4.7 for the active site aspartate-38 was first deduced.

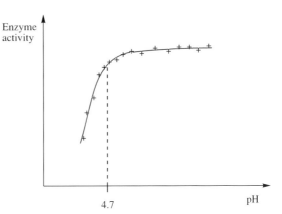

Fig. 4.27 pH/rate profile of ketosteroid isomerase.

The second method that can be used to identify active site groups is by covalent modification. There are a series of chemical reagents available which will react in a fairly specific way with different amino acid side chains, shown in Table 4.3. Thus, if an enzyme is inactivated upon treatment with diethyl pyrocarbonate, then this provides a clue that there may be an essential histidine residue at the active site of the enzyme. However, residues identified by such methods are not necessarily catalytic groups: they may simply be residues in the vicinity of the active site which when covalently modified block the entrance to the active site sufficiently to inactivate the enzyme.

A related technique involving substrate analogues is known as affinity labelling. A substrate analogue is synthesized containing a reactive functional group (e.g. halogen substituent, epoxide, etc.) in a part of the molecule. The substrate analogue is recognized by the enzyme and binds to the active site in the same way as the natural substrate, but then alkylates an

Table 4.3 Group specific reagents for active site amino acid modification.

Amino acid	Group modified	Reagent	Modification reaction
Cysteine	SH	Iodoacetate, iodoacetamide	Alkylation
		DTNB	Disulphide formation
		p-Hydroxymercuribenzoate	Metal complexation
Histidine	NH	Diethyl pyrocarbonate	Acylation
Lysine	NH_2	Succinic anhydride	Acylation
Arginine	Guanidine	Phenylglyoxal	Heterocycle formation
Aspartate/glutamate	COOH	EDC (water-soluble carbodi-imide) + amine	Amide formation
Tyrosine	Phenol	Tetranitromethane	Nitration
Tryptophan	Indole	N-bromosuccinimide	Oxidation

DTNB, 5,5'-dithiobis-(2-nitrobenzoic acid); EDC, 1-ethyl-3-(3-dimethylaminopropyl) carbodi-imide.

essential active site residue. The enzyme is then irreversibly inactivated, since the active site is blocked. The advantage of this method over the group specific reagents mentioned above is that affinity labelling is more selective in its site of action, due to the precise positioning of the substrate analogue at the enzyme active site. One example of this method is the inactivation of the serine proteases by chloromethyl ketone substrate analogues, which will be described in Chapter 5, Section 5.2.

If a successful irreversible inhibitor is found, the site of action can be determined using a radiolabelled inhibitor as shown in Fig. 4.28. The enzyme inactivated with labelled inhibitor contains a modified active site residue bearing a radioactive label. In order to identify this active site residue the inactivated enzyme is broken down into peptide fragments using a protease enzyme, and the peptide fragment containing the ^{14}C label is sequenced. At the position in the peptide sequence containing the radiolabelled inhibitor a non-standard amino acid is found and the radioactive label is released from the peptide. This method is known as peptide mapping.

The covalent attachment of an inhibitor or substrate to an enzyme can also be analysed by 'weighing' the protein. The technique of electrospray mass spectrometry can be used to determine accurately the molecular weight of pure proteins of up to 50 kDa to an accuracy of ± 1 Da! The molecular weight of a covalently modified enzyme can therefore reveal the molecular weight of the attached small molecule.

Finally, with the advent of modern molecular biology techniques it has become possible to specifically replace individual amino acids in an enzyme by site-directed mutagenesis. This method involves specifically altering the sequence of the gene encoding the enzyme in such a way that the triplet

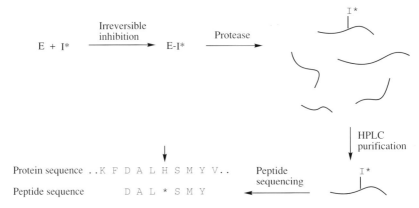

Fig. 4.28 Identification of an active site residue by peptide mapping. HPLC, high-performance liquid chromatography.

codon encoding the amino acid of interest is changed to that of a non-functional amino acid such as alanine. The mutant enzyme can then be purified and tested for enzymatic activity. In this way the precise role of individual amino acids implicated by modification studies or sequence alignments can be explored.

The ultimate solution to identifying active site amino acid groups is to solve the three-dimensional structure of the whole enzyme. This is most commonly done by X-ray crystallography, which depends on obtaining a high-quality crystal of pure enzyme that is suitable for X-ray diffraction. Recent developments in multi-dimensional NMR spectroscopy have allowed the structure determination of small proteins up to 15 kDa in size. Some of the detailed examples that we shall meet in later chapters have been analysed both by X-ray crystallography and by several of the above methods. However, although X-ray crystallography is becoming a more routine exercise, there are still only several hundred proteins whose structures have been solved, compared with tens of thousands of amino acid sequences.

Problems

1 Using the data given in Table 4.1, calculate the turnover number (k_{cat}) for the enzyme being purified. Assume that the final purified enzyme is 100% pure, and that enzyme contains one active site per monomer. The subunit molecular weight is 28 kDa.

2 From the data below, obtained from the rate of an enzyme-catalysed reaction at a range of substrate concentrations, calculate the K_m and v_{max} of the enzyme for this substrate. If 1.65 µg of enzyme was used for each assay, and if the molecular weight of the enzyme is 36 kDa, work out the turnover number for the enzyme for this substrate. Hence, calculate the catalytic efficiency (k_{cat}/K_M) for this substrate.

[S] (mM)	Rate of product formation (nmol min^{-1}; duplicate assays)
1.8	4.75, 4.44
0.9	3.55, 3.20
0.4	2.15, 2.19
0.2	1.28, 1.32
0.1	0.74, 0.76

3 Some enzymes are inhibited in the presence of high concentrations of substrate. This behaviour can be rationalized by the model below, involving the formation of a non-productive ES_2 complex. Starting from this model, use the steady state approximation for ES and ES_2 to construct a

rate equation of the form given below. By considering the behaviour at low and high substrate concentrations, sketch the expected dependence of v versus [S].

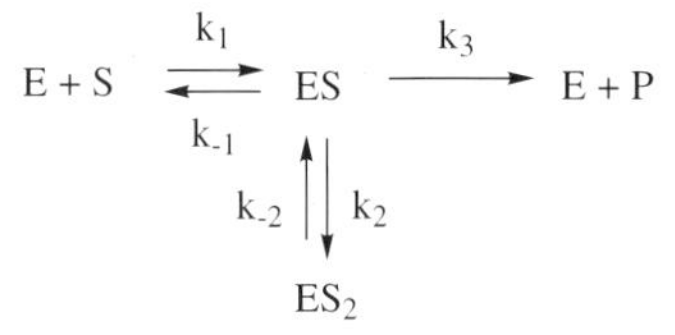

$$E + S \underset{k_{-1}}{\overset{k_1}{\rightleftharpoons}} ES \xrightarrow{k_3} E + P$$

$$ES \quad k_{-2} \updownarrow k_2$$

$$ES_2$$

$$\text{Rate} = \frac{k_3[E]_0[S]}{K_m + [S] + K_2[S]^2} \qquad \text{where } K_2 = k_2/k_{-2}$$

4 A carbon–phosphorus lyase activity has been found which catalyses the reductive cleavage of ethyl phosphonate to ethane, as shown below. The stereochemistry of this reaction was elucidated using a chiral methyl group approach. Incubation of $[1R\text{-}^2H, {}^3H]$-ethyl phosphonate with enzyme gave a ${}^3H$-labelled ethane product, which was converted via halogenation and oxidation into ${}^3H$-labelled acetic acid. Analysis using the method described in the text revealed that the acetate derivative predominantly had the S configuration. Incubation of $[1S\text{-}^2H, {}^3H]$-ethyl phosphonate with enzyme and analysis by the same method gave $2R$-acetate. Deduce whether the reaction proceeds with retention or inversion of configuration, and comment on this result.

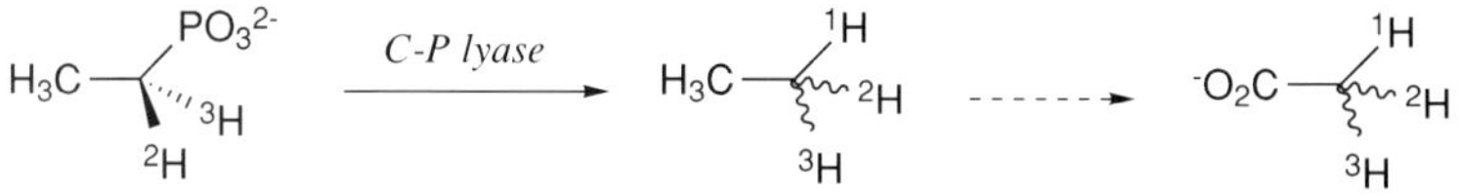

1R-ethyl phosphonate   2S acetate

5 The enzyme phosphonoacetaldehyde hydrolase catalyses the conversion of phosphonacetaldehyde to phosphate and acetaldehyde, as shown below. The enzyme requires no cofactors, but is inactivated by treatment with phosphonoacetaldehyde and sodium borohydride. Deduce which amino acid side chain is involved in the catalysis and suggest a possible mechanism.

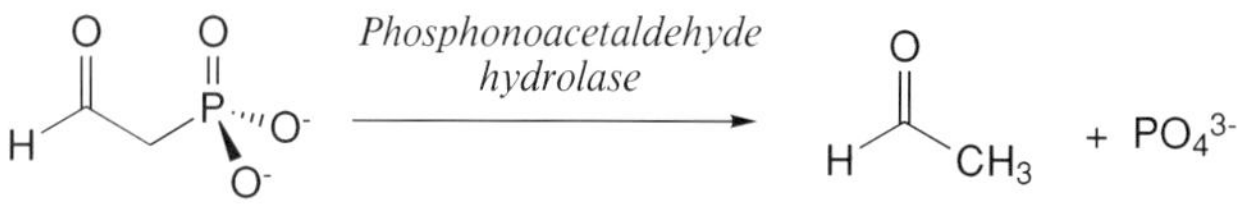

[^{17}O,^{18}O]-Thiophosphonoacetaldehyde was prepared with the stereo-chemistry shown below, and incubated with the enzyme in H$_2$^{16}O. The resulting thiophosphate was analysed and found to have the S configuration. Deduce whether the reaction proceeds with retention of inversion of configuration at the phosphorus centre. Comment on the implications for the enzyme mechanism.

The same reaction is catalysed by aniline (PhNH$_2$), but at a much slower rate. Using the labelled substrate for the aniline-catalysed process, the thiophosphate product was found to have the R configuration. Explain these observations.

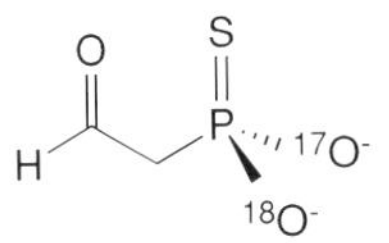

6 How would you attempt to obtain further evidence for the intermediate implied in Problem 5?

Further reading

Enzyme purification

Scopes, R.K. (1978) *Protein Purification: Principles and Practice.* Springer-Verlag, New York.

Enzyme kinetics

Fersht, A. (1985) *Enzyme Structure and Mechanism*, 2nd edn. Freeman, New York.
Segel, I.H. (1993) *Enzyme Kinetics*. Wiley-Interscience, New York.

Stereochemistry of enzymatic reactions

Overton, K.H. (1979) *Chem Soc Rev*, **8**, 447–73.
Walsh, C.T. (1979) *Enzymatic Reaction Mechanisms.* Freeman, San Francisco.

Chiral methyl groups

Floss, H.G. & Lee, S. (1993) *Acc Chem Res*, **26**, 116–22.
Floss, H.G. & Tsai, M.D. (1979) *Adv Enzymol*, **50**, 243–302.
Cornforth, J.W., Redmond, J.W., Eggerer, H., Buckel, W. & Gutschow, C. (1969) *Nature*, **221**, 1212–13.
Lüthy, J., Retey, J. & Arigoni, D. (1969) *Nature*, **221**, 1213–15.

Chiral phosphates

Gerlt, J.A., Coderre, J.A. & Mehdi, S. (1984) *Adv Enzymol*, **55**, 291–380.
Knowles, J.R. (1980) *Annu Rev Biochem*, **49**, 877–920.

Intermediates in enzymatic reactions

Anderson, K.S. & Johnson, K.A. (1990) Kinetic and structural analysis of enzyme intermediates: lessons from EPSP synthase. *Chem Rev*, **90**, 1131–49.
Boyer, P.D. (1978) Isotope exchange probes and enzyme mechanisms. *Acc Chem Res*, **11**, 218–24.
Rose, I.A. (1979) Positional isotope exchange studies on enzyme mechanisms. *Adv Enzymol*, **50**, 361–96.

Isotope effects in enzymatic reactions

Botting, N.P. (1994) *Nat Prod Reports,* **11**, 337–53.
Cleland, W.W. (1987) *Bioorg Chem*, **15**, 283–302.
Northrop, D.B. (1981) *Annu Rev Biochem*, **50**, 103–32.

Ketosteroid isomerase

Xue, L., Talalay, P. & Mildvan, A.S. (1990) *Biochemistry*, **29**, 7491–500.

Covalent modification of enzymes

Creighton, T.E. (ed.) (1989) *Protein Function—a Practical Approach*. IRL Press, Oxford.

5   Hydrolytic and Group Transfer Reactions

5.1   Introduction

Hydrolysis reactions are fundamental to cellular metabolism. In order to break down biological foodstuffs into manageable pieces that animals can utilize for energy, they must have enzymes capable of hydrolysing biological macromolecules. Thus, many of the hydrolytic enzymes that we shall meet in this chapter are involved in digestive processes. However, this is by no means the only area in which we shall encounter hydrolase enzymes.

We shall look in turn at enzymes which hydrolyse each of the three major classes of biological macromolecules: polypeptides, polysaccharides and nucleic acids. More detailed discussions of the structures and chemistry of these macromolecules can be found in most advanced chemistry or biochemistry texts. The classes of enzymes that hydrolyse these molecules are shown in Table 5.1: the polypeptides are hydrolysed by peptidase (or protease) enzymes; polysaccharides by glycosidases; and nucleic acids by nucleases.

The group transferases are related in function to the hydrolases, but carry out quite distinct reactions. In each of the above classes of hydrolases a

Table 5.1 Classes of hydrolases and group transfer enzymes.

Group transferred			Hydrolase	Transferase
Acyl	$-\!\overset{\text{O}}{\overset{\|}{\text{C}}}\!-\text{NHR}$	$\longrightarrow$ $-\!\overset{\text{O}}{\overset{\|}{\text{C}}}\!-\text{OH} + \text{RNH}_2$	Protease peptidase amidase	Transpeptidase
	$-\!\overset{\text{O}}{\overset{\|}{\text{C}}}\!-\text{OR}$	$\longrightarrow$ $-\!\overset{\text{O}}{\overset{\|}{\text{C}}}\!-\text{OH} + \text{ROH}$	Esterase lipase	Acyltransferase
Glycosyl	(glycosyl structure, R'O, HOH$_2$C, HO, OH, OR)	$\longrightarrow$ (glycosyl structure) $\text{OH} + \text{ROH}$	Glycosidase	Glycosyl transferase
Phosphoryl	$-\!\overset{\text{O}}{\underset{\text{O}^-}{\overset{\|}{\text{P}}}}\!-\text{OR}$	$\longrightarrow$ $-\!\overset{\text{O}}{\underset{\text{O}^-}{\overset{\|}{\text{P}}}}\!-\text{OH} + \text{ROH}$	Phosphatase nuclease phosphodiesterase	Kinase phosphotransferase

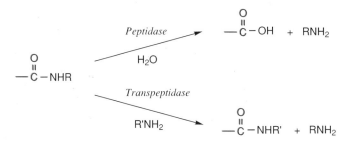

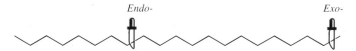

Fig. 5.1 Peptidase versus transpeptidase.

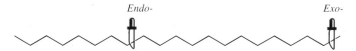

Fig. 5.2 *Exo-* versus *endo*-cleavage.

group is being cleaved and transferred to the hydroxyl group of water: for polypeptides an acyl group; for polysaccharides a glycosyl group; and for nucleic acids a phosphoryl group. Transferases simply transfer this group to an acceptor other than water. For example, there are a few transpeptidases which cleave an amide bond and transfer the acyl group to another amino group, forming a new amide bond (Fig. 5.1). In the same way glycosyl transferases and phosphoryl transferases transfer glycosyl and phosphoryl groups to acceptor substrates.

One final piece of terminology regards the position of cleavage of a very long polymeric biological macromolecule. Enzymes which cleave such biological polymers either cleave progressively from the end of the chain, which is known as *exo*-cleavage, or they cleave at specific points in the middle of the chain, which is known as *endo*-cleavage (Fig. 5.2). In the case of *exo*-cleavage the end which is cleaved is specified, for example $5' \rightarrow 3'$-exonuclease.

This chapter will deal with each of the major classes of peptidases, glycosidases and nucleases, and we will focus in more detail on the immunodeficiency virus 1 (HIV-1) protease as a recent topical example. We will also examine other examples of acyl group transfer and methyl group transfer which are of considerable biological significance.

5.2   The peptidases

Peptidases are responsible for hydrolysing the amide bonds found in the polypeptide structures of proteins, hence they are often known as proteases or proteinases. They have a very important role in the digestive systems of all animals for the breakdown of the protein content of food and are produced in large quantities in the stomach and pancreas. Different types of peptidases are produced elsewhere in the body for a large assortment of hydrolytic purposes, notably their key involvement in the blood coagulation cascade. Other proteases are produced in a wide range of species, including bacteria, yeasts and plants, of which a selection is listed in Table 5.2. Note that there is an active non-specific protease called bromelain present in fresh pineapple, which is why fresh pineapple will attack your gums if you do not brush your teeth!

There are four classes of peptidase enzymes, classified according to the groups found at their active site which carry out catalysis. They are as follows: (i) the serine (Ser) proteases; (ii) the cysteine (Cys) proteases; (iii) the metalloproteases; and (iv) the aspartyl (Asp) (or acid) proteases. Table 5.2 shows the more common commercially available proteases. The table shows to which mechanistic group they belong, whether they are *exo-* or *endo-*proteases, and what is their preferred site of cleavage. We shall now consider each class of protease in turn.

Table 5.2 Some commercially available proteases.

Name	Class	Exo/endo	Specificity	Source	pH_{opt}
Bromelain	Cys	Endo	X–X	Pineapple	6.0
Carboxypeptidase A	Metallo	Exo (C)	X–C (not Arg, Lys)	Bovine pancreas	7–8
Carboxypeptidase B	Metallo	Exo (C)	X–C (Arg/Lys)	Pig pancreas	7–9
Carboxypeptidase P	Ser	Exo (C)	X–C	*Penicillium*	4–5
Carboxypeptidase Y	Ser	Exo (C)	X–C	Yeast	5.5–6.5
Cathepsin C (dipeptidase)	Cys	Exo (N)	N–Gly/Pro–X	Bovine spleen	4–6
Cathepsin D	Asp	Endo	Phe/Leu–X	Bovine spleen	3–5
Chymotrypsin A ✳	Ser ✳	Endo	Aro–X	Bovine pancreas	7.5–8.5
Clostripain	Cys	Endo	Arg–X	*Clostridium*	7.1–7.6
Elastase	Ser	Endo	Neutral aa–X	Porcine pancreas	7.8–8.5
Factor X	Ser	Endo	Arg–X	Bovine plasma	8.3
Leucine aminopeptidase	Metallo	Exo (N)	N(not Arg/Lys)–X	Porcine kidney	7.5–9.0
Papain	Cys	Endo	Arg/Lys–X	Papaya plant	6–7
Pepsin	Asp	Endo	Hyd–X	Porcine stomach	2–4
Proteinase K	Ser	Endo	X–Aro/Hyd	*T. album*	7.5–12.0
Renin	Asp	Endo	His–Leu–X	Porcine kidney	6.0
Subtilisin Carlsberg	Ser	Endo	Neutral/acidic aa–X	*Bacillus subtilis*	7–8
Thermolysin	Metallo	Endo	X–Hyd	*B. thermoproteolyticus*	7–9
Thrombin	Ser	Endo	Arg–Gly	Bovine plasma	7–8
Trypsin	Ser	Endo	Lys/Arg–X	Bovine pancreas	8.5–8.8
V8 protease	Ser	Endo	Asp/Glu–X	*Staphylococcus aureus*	7.8

aa, amino acid; Arg, arginine; Aro, aromatic amino acid (phenylalanine (Phe)/tyrosine (Tyr)/tryptophan (Trp)); Asp, aspartic acid; C, C-terminal amino acid; Cys, cysteine; Gly, glycine; His, histidine; Hyd, hydrophobic amino acid (leucine (Leu)/isoleucine (Ile)/valine (Val)/methionine (Met)); Lys, lysine; N, N-terminal amino acid; Pro, proline; Ser, serine; X, any amino acid.

The serine proteases

These enzymes are characterized by an active site serine residue which participates covalently in catalysis. The active site serine is assisted in catalysis by a histidine (His) and an aspartate residue, the three residues acting in concert as a 'catalytic triad'. The best characterized of the serine proteases is α-chymotrypsin, a 241-amino acid endoprotease which shows specificity for cleavage after aromatic amino acids (phenylalanine (Phe), tyrosine (Tyr) or tryptophan (Trp)). This selectivity arises from a favourable hydrophobic interaction between the aromatic side chain of the substrate and a hydrophobic binding pocket situated close to the catalytic site, as illustrated in Fig. 5.3.

The catalytic mechanism of chymotrypsin is shown in Fig. 5.4. It is a classic example of covalent catalysis, in which the active site Ser-195 attacks the amide carbonyl to form a tetrahedral oxyanion intermediate. Attack of Ser-195 is made possible by base catalysis from His-57, generating an imidazolium cation which is stabilized by the carboxylate of Asp-102. Selective stabilization of the high-energy oxyanion intermediate takes place via formation of two hydrogen bonds between the tetrahedral oxyanion and the backbone amide N–H bonds of Ser-195 and glycine (Gly)-193. Note that these interactions are specific for the oxyanion intermediate and are not formed with the bound substrate, leading to reduction of the activation energy for the overall step (since this is a non-isolable high-energy intermediate that is similar to the transition state for the first step, the transition state stabilization logic of Chapter 3, Section 3.4 applies in this case).

Protonation of the departing nitrogen by His-57 gives the amine product and leads to the formation of an acyl enzyme intermediate. There is

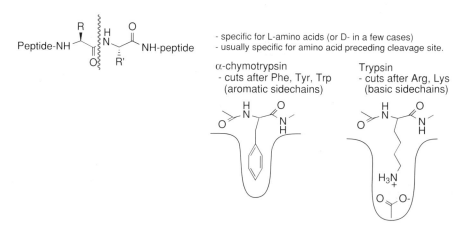

Fig. 5.3 Specificity of endopeptidases.

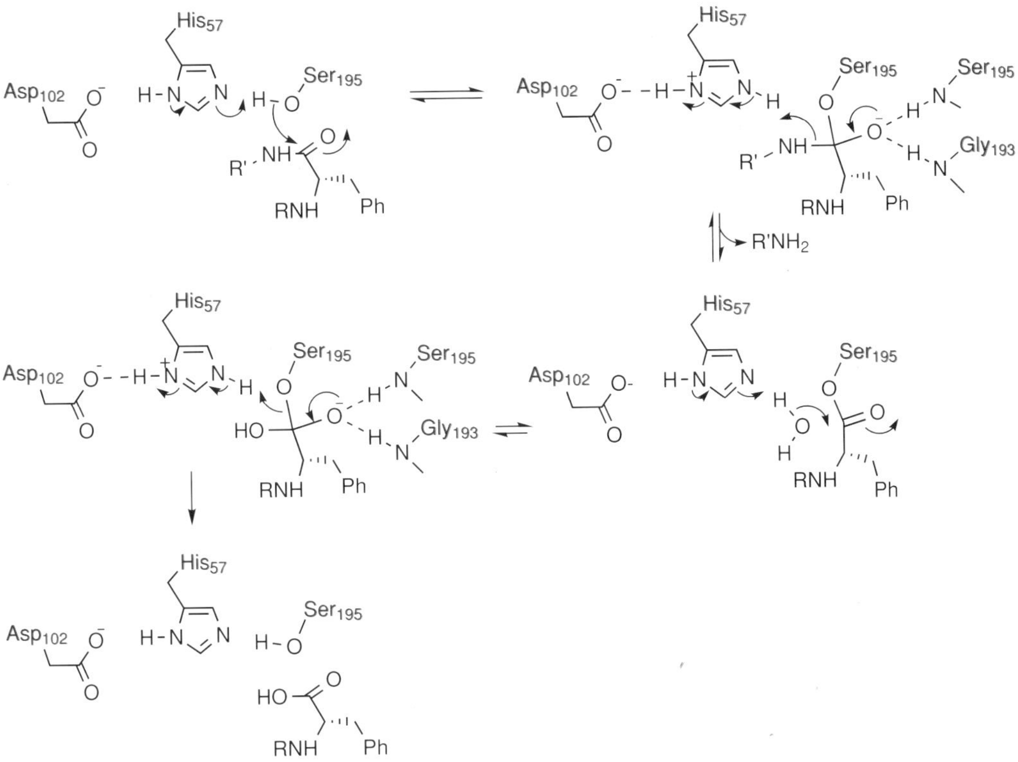

Fig. 5.4 α-Chymotrypsin mechanism.

considerable experimental evidence for the existence of this covalent inter-mediate from kinetic studies, trapping experiments and X-ray crystallo-graphy. His-57 then acts as a base to deprotonate water attacking the acyl enzyme intermediate, leading to a second tetrahedral oxyanion intermediate which is once again stabilized by hydrogen bond formation. Protonation of the departing serine oxygen by His-57 leads to release of the carboxylic acid product and completion of the catalytic cycle.

The use of the hydroxyl group as a nucleophile in this reaction is quite remarkable since an alcohol would normally have a pK_a of about 16, whereas the pK_a of histidine is normally in the range 6–8. Effectively this enzyme is using a relatively weak imidazole base to generate a potent alkoxide nucleophile. This process is made thermodynamically favourable by participation of Asp-102 in stabilizing the imidazolium cation of His-57. Mutant enzymes have been generated using site-directed mutagenesis in which Asp-102 has been replaced by asparagine (Asn)—these enzymes are 10^4-fold less active, showing the catalytic importance of this interaction.

Serine proteases are specifically inactivated by two classes of inhibitors:

organophosphorus and chloromethyl ketone substrate analogues. Mechanisms of inactivation are shown in Fig. 5.5. Organophosphorus inhibitors are attacked by the active site serine, generating stable tetrahedral phosphate esters which closely resemble the tetrahedral state of the normal enzymatic reaction, and hence are bound tightly by the enzyme. Chloromethyl ketone analogues were found to modify the active site histidine base, in this case His-57, rather than the apparently more reactive serine residue. Recently, a

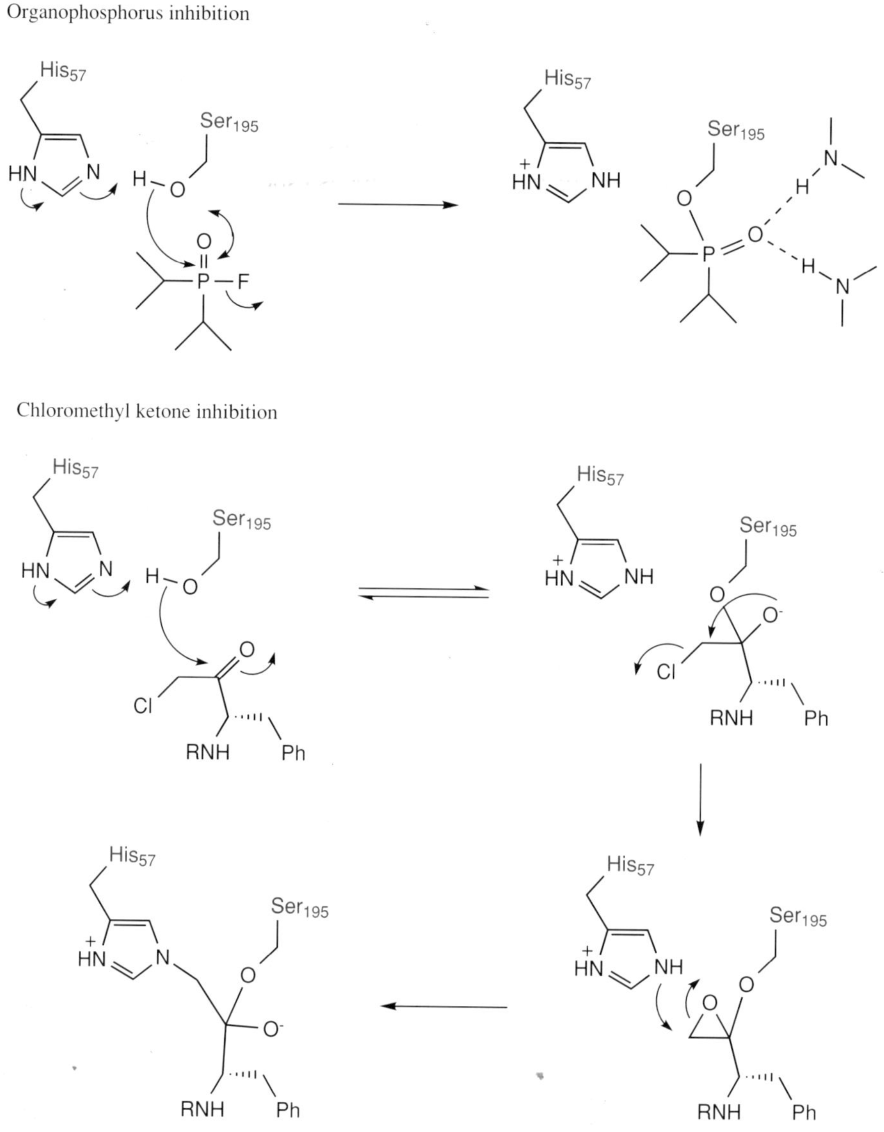

Organophosphorus inhibition

Chloromethyl ketone inhibition

Fig. 5.5 Inhibitors of serine proteases.

stereochemical analysis of this inactivation has revealed that it proceeds with retention of configuration at the chlorine-bearing carbon, suggesting that a double inversion is taking place. This can be rationalized by attack of the active site serine on the carbonyl group and displacement of chloride by the resulting oxyanion, generating an enzyme-bound epoxide. Opening of the epoxide by the neighbouring histidine leads to covalent modification of the histidine residue.

Other serine proteases such as trypsin, elastase and subtilisin have very similar active site structures and employ the same type of mechanisms, even though in the case of subtilisin the primary sequence of the enzyme bears no resemblance to that of chymotrypsin. This may be a rare example of *convergent evolution*, where the same active site structure has arisen from two evolutionary origins, suggesting that this particular three-dimensional alignment of functional groups is especially adept for this type of catalysis. The specificity of trypsin is for cleavage after basic amino acids such as lysine (Lys) and arginine (Arg). This specificity is provided by a similar specificity pocket to that of chymotrypsin in which there is the carboxylate side chain of Asp-189 at the bottom of the pocket which forms a favourable electrostatic interaction with the basic side chains of lysine and arginine-containing substrates (see Fig. 5.3).

There are several other classes of enzyme which also contain serine catalytic triads. Several members of the family of esterase and lipase enzymes discussed in Section 5.3 also contain active site serine groups, and use the same type of mechanism as chymotrypsin (which itself is capable of hydrolysing esters as well as amides).

The cysteine proteases

This family of proteins are characterized by an active site cysteine residue whose thiol side chain is, like the serine proteases, involved in covalent catalysis. The cysteine proteases are a large family of enzymes which are less commonly used for digestive purposes and are more often found as intra-cellular proteases used for post-translational processing of cellular proteins (see Table 5.2). The active site thiol is highly prone to oxidation, which means that these enzymes must be purified and handled in the presence of mild reducing agents. The active site cysteine is easily modified by cysteine-directed reagents such as *p*-chloromercuribenzoate, an organomercury compound which functions by forming a strong mercury–sulphur bond.

The best characterized member of this family is papain, a 212-amino acid endoprotease found in papaya plants. The preferred cleavage site is following basic amino acids such as arginine and lysine. The proposed mechanism for papain is shown in Fig. 5.6. There is good evidence that the active site cysteine acts as a nucleophile to attack the amide bond, generating a covalent thioester intermediate. In the case of papain, there is evidence from X-ray crystallography and modification studies that Cys-25 is deprotonated by an active site histidine base (His-159) as it attacks the amide substrate. As in the case of the serine proteases, a high-energy oxyanion intermediate is formed which is specifically stabilized by hydrogen bonding to the backbone amide N–H bonds of Cys-25 and glutamine (Gln)-17. Breakdown of the thioester intermediate by base-catalysed attack of water leads to formation of the carboxylic acid product.

Analysis of the active site histidine residue by ^{1}H nuclear magnetic resonance (NMR) spectroscopy has revealed that in the active form of the

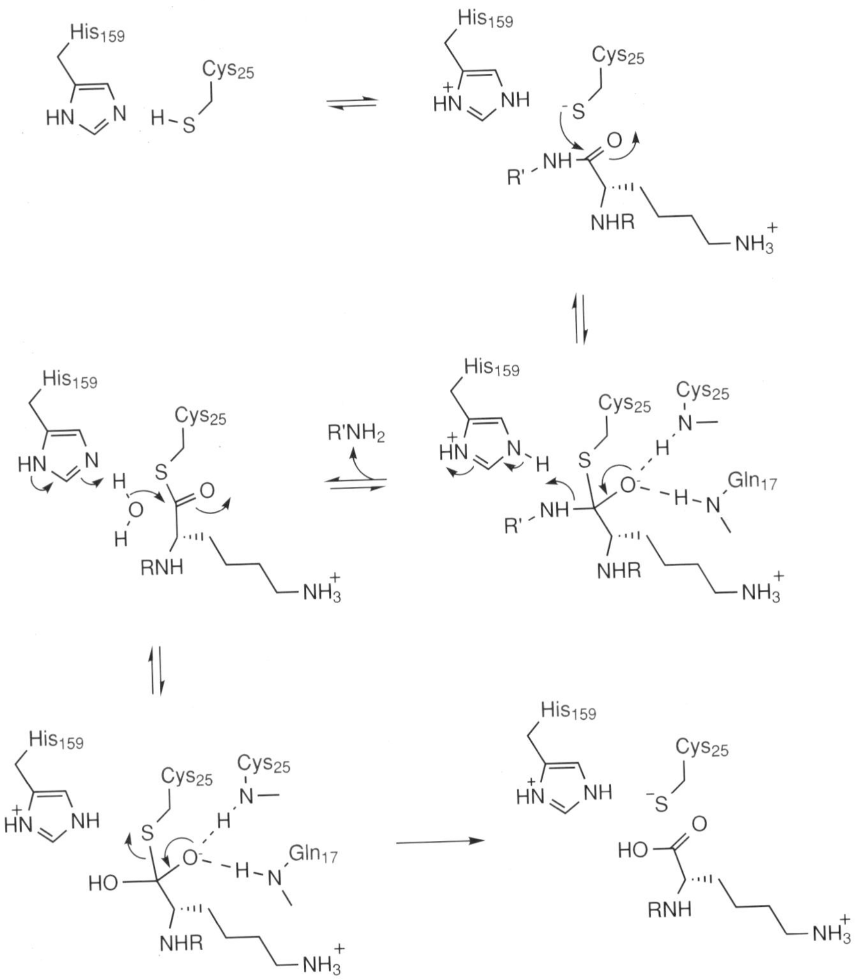

Fig. 5.6 Papain mechanism.

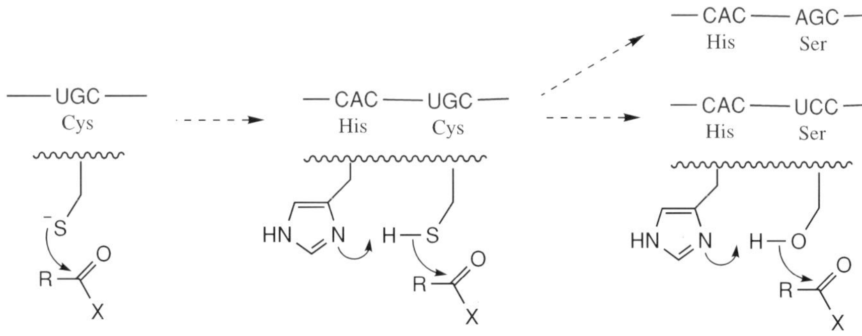

Fig. 5.7 Possible evolution of serine proteases.

enzyme the imidazole ring is in fact protonated, suggesting that in this case the resting state of the enzyme contains an imidazolium–thiolate ion pair. This is possible in the case of the cysteine proteases since the pK_a of the thiol side chain of cysteine is only 8–9, and stabilization of the ion pair by active site electrostatic interactions seems likely.

Given the similarities in mechanism between the serine proteases and the cysteine proteases, it is interesting to note that the serine proteases may have evolved from a forerunner of the cysteine proteases. This hypothesis was put forward by S. Brenner, who observed that in some cases the triplet codons encoding the active site serine have the UCX codon, whilst others have the AGC/AGU codon for serine. The key point is that these two sub-families of enzymes cannot be directly related in evolutionary terms, since two simultaneous nucleotide base changes would be required to change from one to the other, which is statistically improbable. However, they could *both* have evolved from the UGC/UGU cysteine codon via one-base changes as shown in Fig. 5.7. So just on the basis of this one observation it seems probable that a cysteine-containing enzyme was the forerunner of the serine-containing enzymes. This makes a lot of sense in terms of the catalytic chemistry, since the thiol side chain of cysteine has a lower pK_a, is more nucleophilic and is a better leaving group than the hydroxyl group of serine. So, why do the serine proteases now predominate? Perhaps, as mentioned above, because they are more stable in an oxygen atmosphere.

The metalloproteases

The metalloproteases are characterized by a requirement for an active site metal ion, usually zinc (Zn^{2+}), which is involved in the catalytic cycle. These enzymes can be readily distinguished from the other classes by treatment with metal chelating agents such as ethylenediamine tetra-acetic acid

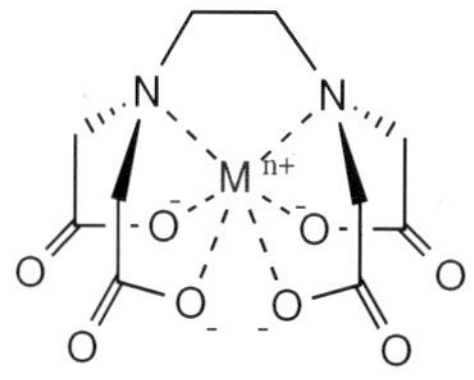

Ethylenetetra-acetic acid (EDTA) 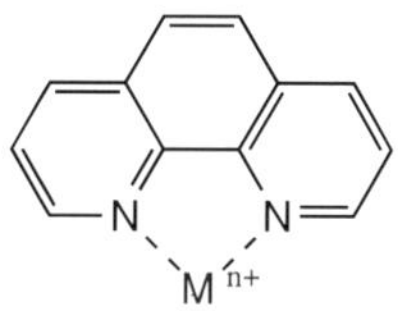   1,10-Phenanthroline

Fig. 5.8 Metal chelating agents.

(EDTA) or 1,10-phenanthroline (Fig. 5.8), leading to removal of the metal ion cofactor and inactivation.

The best characterized members of this family are carboxypeptidase A, a 307-amino acid exopeptidase from bovine pancreas which cleaves the C-terminal residue of a peptide chain (not arginine, lysine, or proline (Pro)); and thermolysin, a 35-kDa endopeptidase from *Bacillus thermoproteolyticus* which cleaves before hydrophobic amino acids such as leucine (Leu), isoleucine (Ile), valine (Val) or phenylalanine (Phe). Both enzymes contain a single Zn^{2+} ion at their active sites, which in the resting state of the enzyme is co-ordinated by three proten ligands and one solvent water molecule. There is evidence to suggest that in both enzymes when the substrate is bound the water molecule is displaced by the carbonyl oxygen of the amide bond to be hydrolysed, which is thus activated towards nucleophilic attack by Lewis acid catalysis.

The mechanism of amide bond hydrolysis has been well studied in both enzymes, with slightly different results emerging. Both enzymes contain an active site glutamate, which in theory could either act as a nucleophile to attack the amide carbonyl, or act as a base to deprotonate an attacking water molecule. X-ray crystallographic studies have revealed that glutamate (Glu)-270 in carboxypeptidase A is positioned 2.5 Å away from the carbonyl, well positioned to act as a nucleophile. Attack of Glu-270 on the amide substrate would generate an anhydride intermediate. Treatment of carboxypeptidase A with tritiated sodium cyanoborohydride $(NaB(CN)^3H_3)$ in the presence of substrate leads to the incorporation of ^3H label into the protein and the isolation of ^3H-hydroxynorvaline in the modified enzyme, consistent with the anhydride intermediate (Fig. 5.9). Direct evidence for the anhydride intermediate in the carboxypeptidase A reaction has also emerged from resonance Raman spectroscopic studies of a covalent intermediate, showing bands, at 1700–1800 cm^{-1}, characteristic of anhydrides.

In thermolysin the active site glutamate (Glu-143) is positioned 3.9 Å away from the amide carbonyl, too far away to act as a nucleophile, but far enough to accommodate an intervening water molecule. There is evidence from X-ray crystallography to suggest that Glu-143 acts as a base to

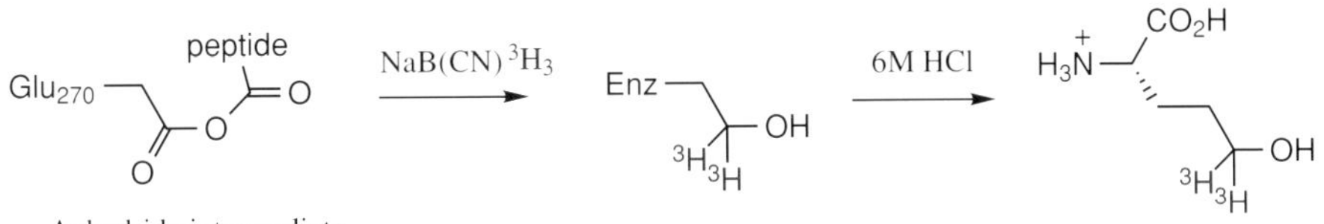

Fig. 5.9 Sodium cyanoborohydride trapping of anhydride intermediate in carboxypeptidase A.

deprotonate an attacking water molecule, forming an oxyanion intermediate which is again stabilized by specific hydrogen bonds. Breakdown of this intermediate using Glu-143 to transfer a proton to the departing nitrogen gives the hydrolysis products. The mechanisms proposed for carboxypeptidase A and thermolysin are shown in Fig. 5.10.

Thermolysin is strongly inhibited (K_i 28 nM) by a natural product phosphoramidon (Fig. 5.11), a monosaccharide derivative containing a phosphonamidate functional group. The phosphonamidate group binds to the active site Zn^{2+}, acting as an analogue of the oxyanion tetrahedral intermediate, and is therefore bound extremely tightly by the enzyme. Further phosphonamidate inhibitors have now been devised and synthesized which bind even more tightly.

The aspartyl proteases

The aspartyl (or acid) proteases are characterized by the presence of two active site aspartate residues whose carboxylate side chains are involved in catalysis. They are probably the smallest class of protease enzymes, but have recently come to prominence through the discovery that the HIV-1 virus contains an essential aspartyl protease enzyme, which we shall examine below. The aspartyl proteases are characteristic in their ability to operate at low pHs: the enzyme pepsin operates in the range 2–4, which makes it well suited for operating in the acidic environment of the stomach. This pH requirement will become apparent when we examine the mechanism of these enzymes, since one of the two aspartate residues must be protonated for activity.

This family of enzymes are also characterized by their inhibition by low levels (1 µg ml^{-1}) of pepstatin (Fig. 5.12), a modification of a naturally occurring peptide containing the unusual amino acid statine (see Fig. 5.15 for an illustration of the mechanism of inhibition by this class of inhibitor).

Case study: HIV protease

The discovery in 1983 that HIV is the causative agent of acquired immune

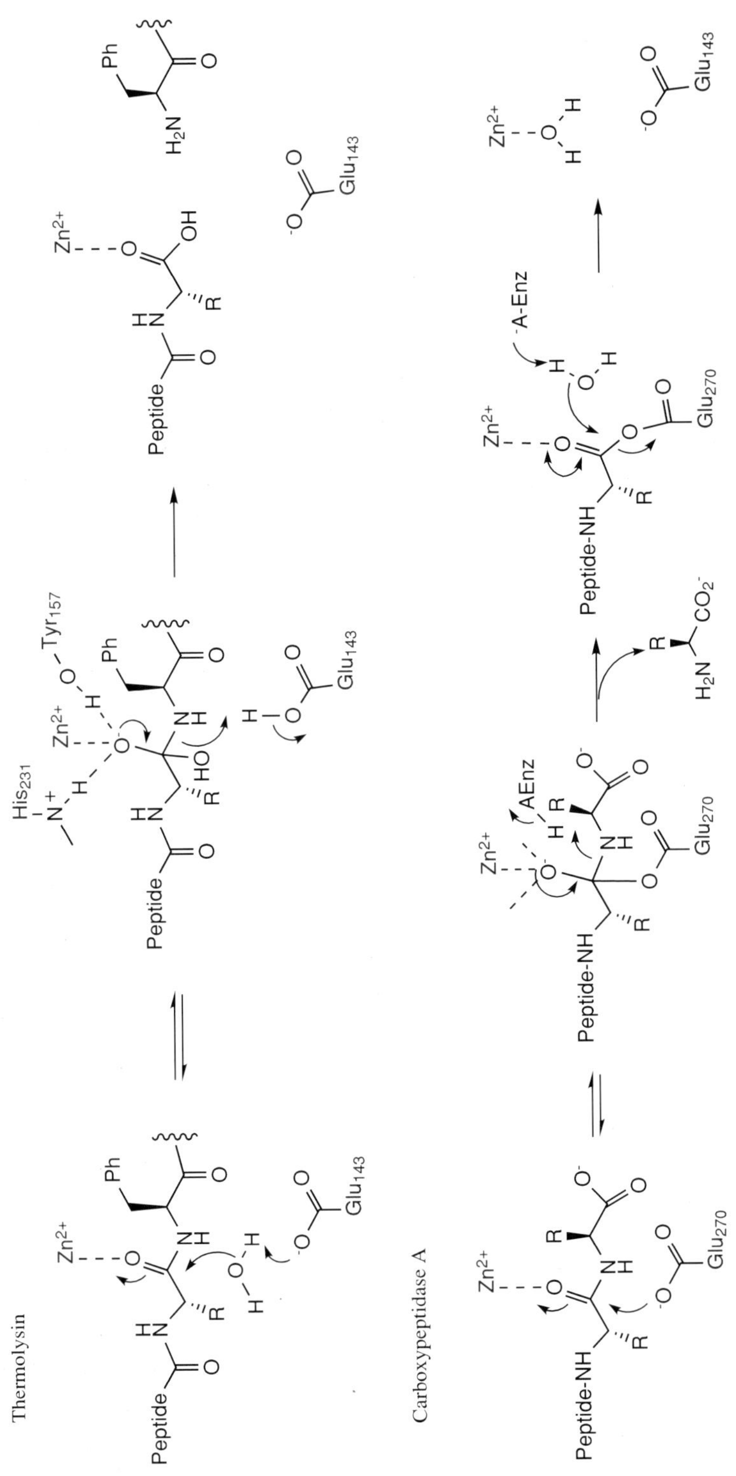

Fig. 5.10 Mechanisms of thermolysin and carboxypeptidase A.

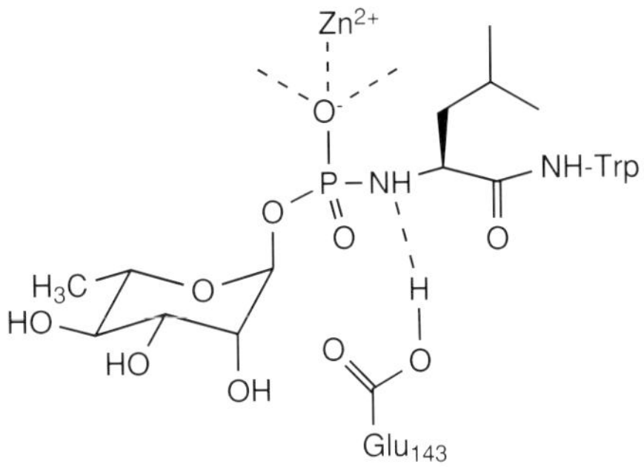

Fig. 5.11 Inhibition of thermolysin by phosphoramidin. Trp, tryptophan.

deficiency syndrome (AIDS) prompted a huge research effort into this virus. The virus was found to contain an essential aspartyl protease (known as 'HIV protease') which is required for the cleavage of two 55-kDa and 160-kDa precursor polypeptides produced from the *gag* and *pol* genes of HIV-1. The cleavage products of these precursor polypeptides are the structural proteins and retroviral replication enzymes required for the assembly of new HIV-1 virions and completion of the viral life cycle. This enzyme therefore represented an immediate target for anti-HIV therapy: if inhibitors could be devised for the HIV-1 protease, the virus would be unable to synthesize its essential proteins and the life cycle would be blocked.

The HIV-1 protease has been overexpressed, purified and crystallized, and several research groups have solved its X-ray crystal structure to high resolution. The enzyme is a 99-kDa homodimer which is similar in structure to other members of the family such as pepsin. Its active site lies at the interface of the two subunits, and the two active site carboxylate residues are Asp-25 from one subunit and the complementary Asp-25′ from the other subunit. Preferred cleavage sites for the HIV-1 protease are at Aro–proline sites, where Aro is an aromatic amino acid (tyrosine, phenylalanine, tryptophan). Using synthetic substrates containing a tyrosine–proline cleavage site, enzyme-catalysed exchange of ^{18}O from $H_2{}^{18}O$ into the amide carbonyl has been observed at up to 10% of the rate of the forward reaction. Since the reaction is in practice irreversible, this exchange is consistent with the reversible formation of a hydrated intermediate, as shown in Fig. 5.13.

D_2O solvent isotope effects of 1.5–3.2 have been measured under a range of conditions, consistent with base-catalysed attack of water being the first

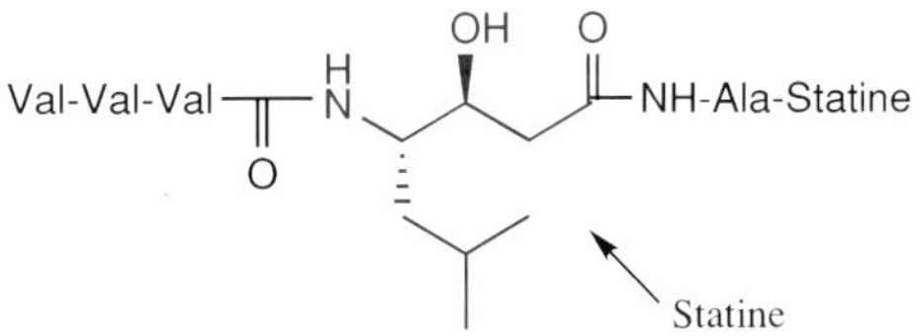

Fig. 5.12 Pepstatin.

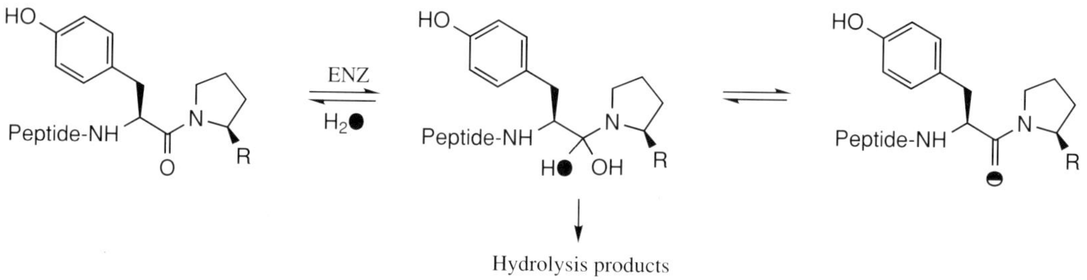

Fig. 5.13 ^{18}O exchange via a hydrate intermediate.

step of the mechanism. Finally, small inverse ^{15}N isotope effects have been measured for the departing nitrogen atom, suggesting protonation of nitrogen in the rate-determining step of the mechanism. The proposed mechanism is illustrated in Fig. 5.14.

According to this mechanism Asp-25 acts as a base to deprotonate an attacking water molecule, with Asp-25′ acting as a general acid, forming the hydrated intermediate. Breakdown of this intermediate with protonation of the departing nitrogen atom by the protonated Asp-25 completes the catalytic cycle.

Most of the research effort on the HIV-1 protease has been directed towards synthesizing potent inhibitors. Given the precedented inhibition of pepsin and other aspartyl proteases by pepstatin, a range of substrate analogues containing statine-like groups have been synthesized and found to act as potent inhibitors for the HIV-1 protease. It is thought that the statine unit mimics the tetrahedral transition state formed in the reaction, as shown in Fig. 5.15.

Administration of human T-lymphocyte cells that are infected with the HIV-1 virus with such inhibitors has demonstrated that these compunds have potent antiviral properties *in vivo*, so this stategy represents a realistic hope for development of anti-HIV therapy. Plate 5.1 (facing p. 152) shows the X-ray crystal structure of the above inhibitor bound to the active site of HIV-1 protease. This type of high-resolution data provides a good model for the development of further inhibitors of this enzyme.

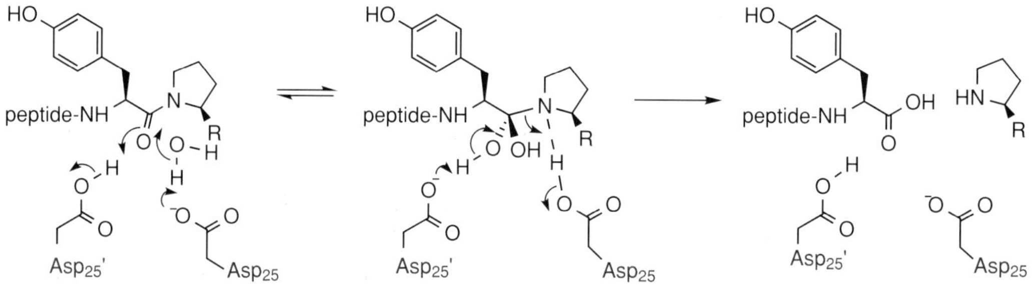

Fig. 5.14 Mechanism for HIV-1 protease.

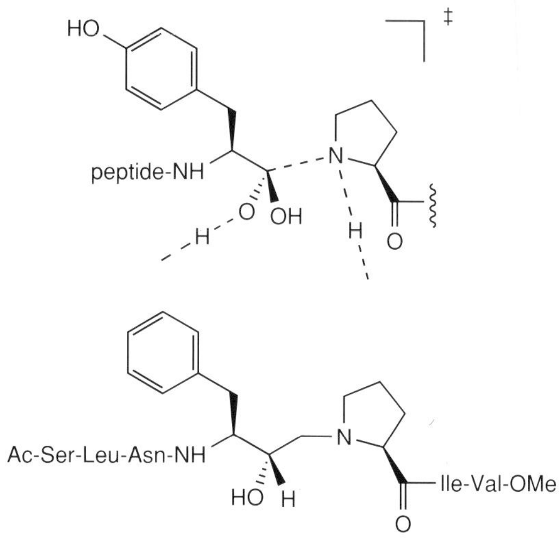

Fig. 5.15 Transition state inhibitor for HIV-1 protease. Val, valine.

5.3 Esterases and lipases

An important part of food digestion is the breakdown of fats, oils and lipid content in food. Lipids are largely made up of glycerol esters of long-chain fatty acids. The digestive system of animals contains high levels of esterase and lipase enzymes which hydrolyse the ester functional groups of fats and oils. Lipases are often fat-soluble enzymes which are able to operate at the lipid–water surface which would otherwise present a physical barrier for a soluble esterase enzyme.

Several members of the family of esterase and lipase enzymes contain active site serine groups, and proceed via the same type of mechanism as chymotrypsin (which is capable of hydrolysing esters as well as amides). Two notable examples are the enzymes pig liver esterase and porcine pancreatic lipase which are commercially available in large quantities. These enzymes have found a number of applications in organic synthesis due to their ability to hydrolyse a wide range of ester substrates with high stereospecificity. Examples of resolution reactions catalysed by these enzymes were illustrated in Chapter 3, Figs 3.2 and 3.3.

Since these enzymes proceed through a covalent acyl enzyme intermediate, they are able to catalyse transesterification reactions in which an acyl group is transferred to an alcohol acceptor. This idea has been used to develop regio- and enantiospecific acylation reactions using vinyl acetate as a solvent. Formation of the acetyl–enzyme intermediate is accompanied by formation of the enol form of acetaldehyde, which rapidly tautomerizes to give acetaldehyde, making the reaction irreversible. Attack of the alcohol functional group then gives the acetylated product. An example is illustrated

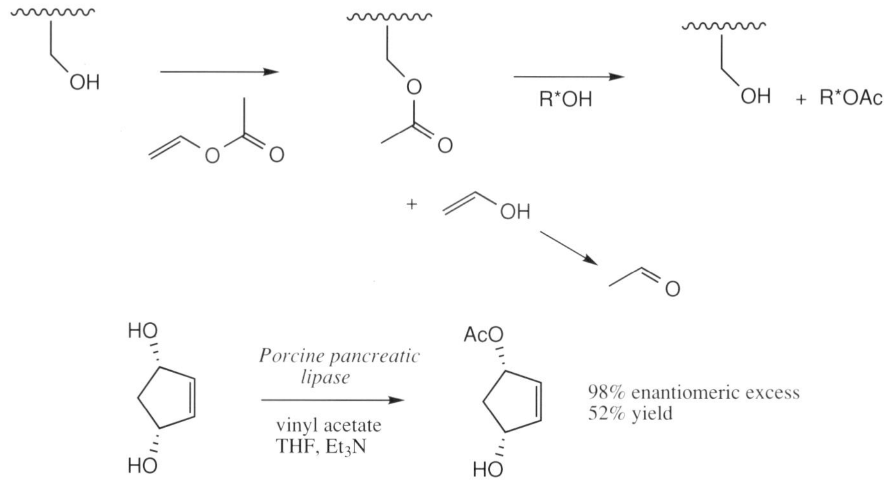

Fig. 5.16 Enantioselective acylation reaction using a serine esterase. Et₃N, triethylamine; THF, tetrahydrofuran.

in Fig. 5.16 in which an achiral *meso*-substrate is regioselectively acylated to give a single enantiomeric product.

5.4 Acyl transfer reactions in biosynthesis: use of coenzyme A (CoA)

How do living systems synthesize the amide bonds found in proteins, or the ester functional groups found in lipids, oils and other natural products? The general strategy, shown in Fig. 5.17, is to make an activated acyl derivative containing a good leaving group, and then to carry out an acyl transfer reaction.

The aminoacyl group of amino acids is activated and transferred during the assembly of the polypeptide chains of proteins by ribosomes. Amino acid activation is carried out by adenosine-5′-triphosphate (ATP)-dependent amino acyl transfer ribonucleic acid (tRNA) synthetase enzymes. Each individual amino acid is converted into an acyl adenylate mixed anhydride derivative, followed by transfer of the aminoacyl group onto a specific tRNA molecule. The aminoacyl–tRNA ester is then bound to the ribosome and the

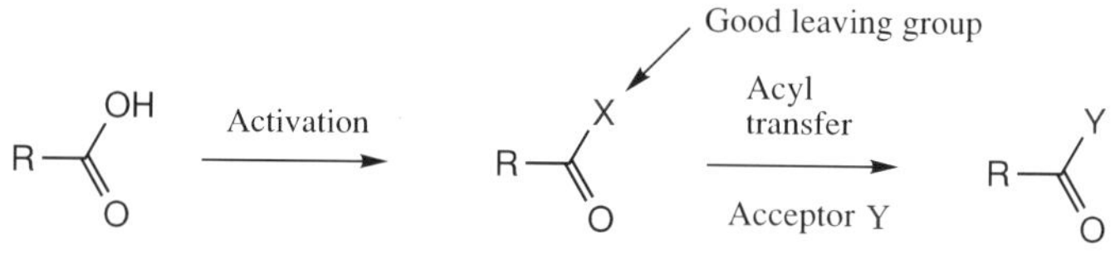

Fig. 5.17 Use of the activated acyl group for acyl transfer.

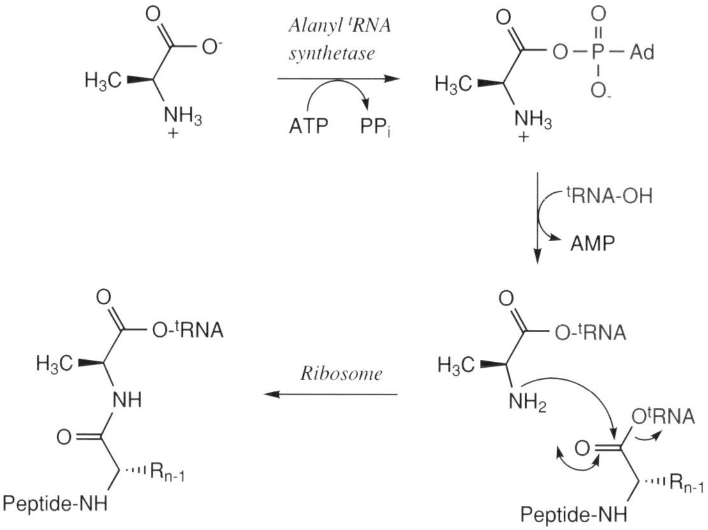

Fig. 5.18 Acyl transfer reactions in protein biosynthesis. PP_i inorganic pyrophosphate.

free amine used to form the next amide bond in the sequence of the protein (Fig. 5.18).

Activation and transfer of acyl groups is a common process found in fatty acid biosynthesis, polyketide natural product biosynthesis and the assembly of a variety of amide and ester functional groups in biological molecules. For the majority of these processes a special cofactor is used—CoA. The structure of CoA contains a primary thiol group which is the point of attachment to the acyl group being transferred, forming a thioester linkage, shown in Fig. 5.19.

CoA is well suited to carry out acyl transfer reactions, since thiols are inherently more nucleophilic than alcohols or amines. Thiols are also better leaving groups (pK_a 8–9), which explains why the hydrolysis of thioesters under basic conditions is more rapid than ester hydrolysis. Acetyl CoA is used by acyltransferase enzymes to transfer its acetyl group to a variety of acceptors, which can be alcohols, amines, carbon nucleophiles or other thiol

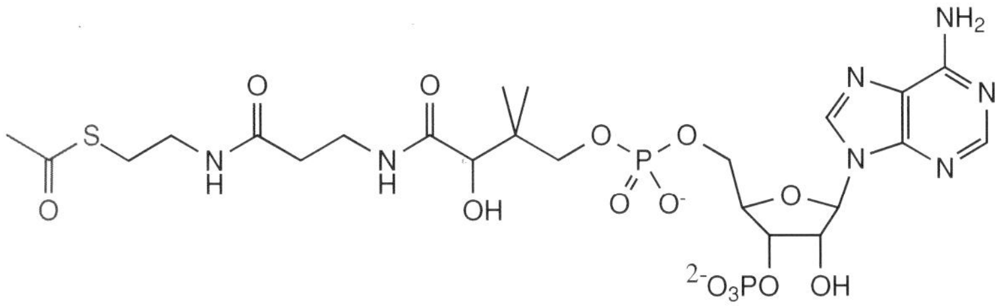

Fig. 5.19 Structure of acetyl CoA.

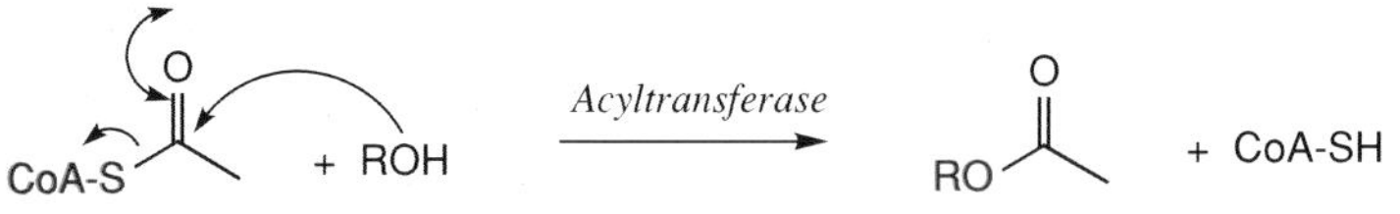

Fig. 5.20 Acyl transfer using acetyl CoA.

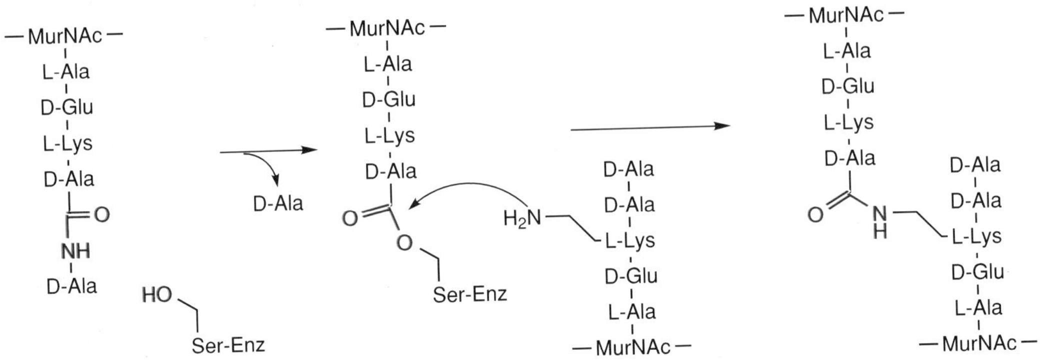

Fig. 5.21 Cross-linking of peptidoglycan catalysed by D, D-transpeptidase.

groups (Fig. 5.20). We shall encounter specific examples of the use of acetyl CoA in later chapters.

Finally, there are a small number of transpeptidase enzymes that transfer the acyl group of a peptide chain onto another amine acceptor. One important example is the transpeptidase enzyme involved in the final step of the assembly of peptidoglycan—a major structural component of bacterial cell walls. Peptidoglycan consists of a polysaccharide backbone of alternating *N*-acetylglucosamine (GlcNAc) and *N*-acetylmuramic acid (MurNAc) residues, from which extend pentapeptide chains which contain the unusual D-amino acids D-alanine and D-glutamate. The final step in peptidoglycan assembly involves the cross-linking of these pentapeptide side chains, catalysed by a transpeptidase enzyme which contains an active site serine residue analogous to the serine proteases (Fig. 5.21). A covalent acyl enzyme intermediate is formed with release of D-alanine, which can be either hydrolysed to generate a tetrapeptide side chain, or be attacked by the ε-amino side chain of a lysine residue from another chain, generating an amide cross-link. These cross-links add considerable rigidity to the peptidoglycan layer, enabling it to withstand the high osmotic stress from inside the bacterial cell.

5.5   Enzymatic phosphoryl transfer reactions

Phosphate esters are widespread in biological systems: phosphate monoesters occur as alkyl phosphates, sugar phosphates and even phosphoproteins; whilst phosphodiester linkages comprise the backbone of the nucleic acids RNA and deoxyribonucleic acid (DNA), and are found in the phospholipid components of biological membranes. Examples of phosphotriesters in biological systems are known, but they are relatively scarce in comparison. Some examples of biologically important phosphate esters are shown in Fig. 5.22.

Phosphoryl transfer is therefore of fundamental importance to biological systems for the biosynthesis and replication of nucleic acids, and in the transfer of phosphate groups between small molecules and proteins.

Mechanisms for non-enzymatic phosphate ester hydrolysis reactions have been extensively studied, revealing that the mechanistic framework for phosphoryl ester substitution is more complex than for acyl ester substitution. One might anticipate addition–elimination mechanisms for nucleophilic substitution at phosphorus, similar to nucleophilic substitution at carbonyl centres. However, in practice phosphate monoesters and diesters are resistent to hydrolysis under alkaline conditions, although phosphate monoesters are hydrolised fairly readily under acidic conditions. It has been estimated that the half-life for the hydrolysis of a phosphodiester at pH 7 and 37°C is 80 million years, which explains the extraordinary stability of DNA, even from pre-historic times. Why is this so?

Detailed mechanistic studies indicate that for substitution to occur at phosphorus there must be a good leaving group attached to phosphorus, and in the transition state the phosphoryl group should bear *two negative charges*.

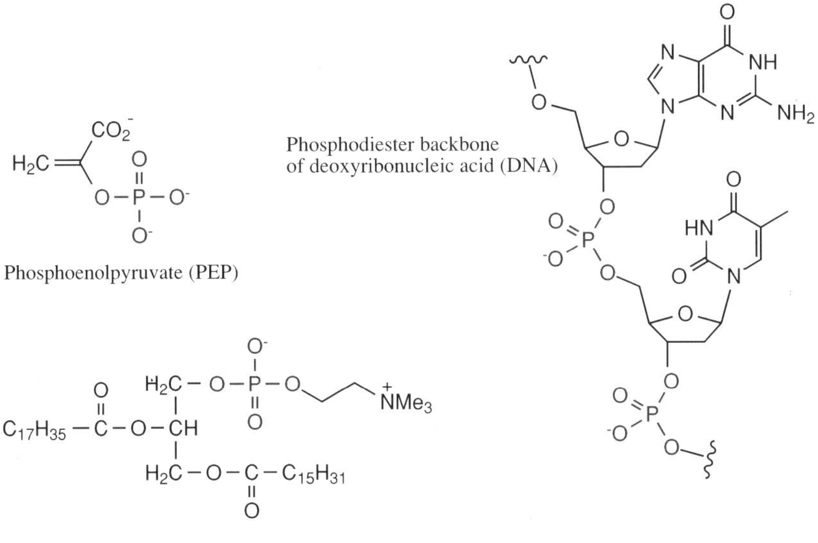

Phosphoenolpyruvate (PEP)

Phosphodiester backbone of deoxyribonucleic acid (DNA)

Phosphatidylcholine — a major component of the lipid bilayer of biological membranes

Fig. 5.22 Examples of biologically important phosphates.

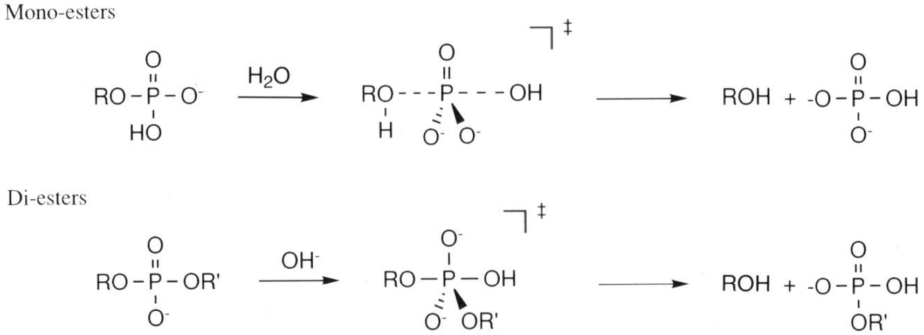

Mono-esters

Di-esters

Fig. 5.23 Transition states for alkaline hydrolysis of phosphate mono- and diesters.

This means that phosphate monoester hydrolysis occurs through a dissociative transition state in which ROH, where R is an alkyl group, departs before attack of water. Phosphodiester hydrolysis occurs through a crowded associative transition state, in which hydroxide attacks before RO^- departs (Fig. 5.23). In both cases the phosphoryl group bears two negative charges in the transition state, but the crowded nature of the latter transition state and the poor leaving group explain the sluggishness of the reaction.

Enzymatic hydrolysis of phosphate monoesters is carried out by a family of phosphatase enzymes. Of these enzymes the bovine alkaline phosphatase has been studied best, due to its availability and its broad substrate specificity: it will hydrolyse a very wide range of phosphate monoesters. Analysis of the stereochemistry of the alkaline phosphatase reaction using the chiral phosphate methodology described in Chapter 4, Section 4.4 revealed that this reaction proceeds with retention of stereochemistry at phosphorus. Rapid kinetic analysis has revealed that a stoichiometric amount of the alcohol product ROH is released prior to phosphate release, and that the k_{cat} for a range of substrates is independent of the nature of ROH. These data suggest the existence of a phospho–enzyme intermediate, whose breakdown is rate-determining. Incubation of enzyme with $RO^{32}PO_3^{2-}$ gave enzyme labelled with ^{32}P, which upon tryptic digestion and sequencing was found to be localized on a unique serine residue. This enzyme is in fact yet another member of the serine hydrolase family containing a catalytic triad. A mechanism for this enzymatic reaction is shown in Fig. 5.24.

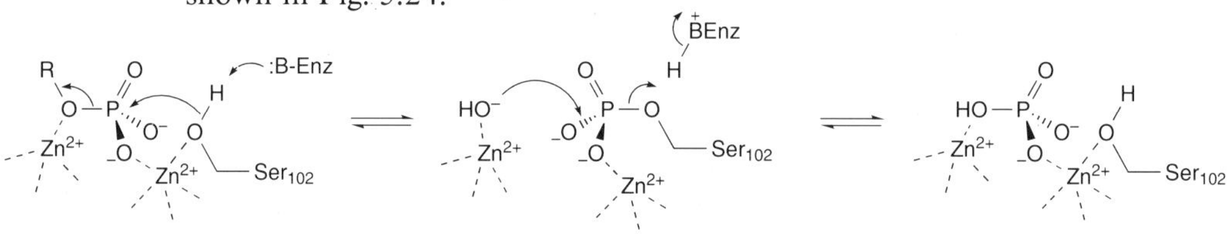

Fig. 5.24 Mechanism and stereochemistry of alkaline phosphatase.

Enzymes that cleave the phosphodiester backbone of RNA and DNA are known as nucleases. Some nucleases are relatively non-specific, for example a $3' \rightarrow 5'$ exonuclease enzyme found in snake venom which will digest single-stranded DNA from the $3'$ end successively. However, the endonuclease enzymes usually have highly specific cleavage sites, for example the restriction endonuclease *Eco*R1 (produced by *Escherichia coli*) cleaves at a specific six-base sequence $5'$-GAATTC-$3'$, making cuts on both strands of the DNA as shown in Fig. 5.25.

The reaction mechanism of restriction endonucleases is not well understood, although it is known in this case that the cleavage of the phosphodiester bond occurs with inversion at phosphorus. It is known that these enzymes are homodimers, each subunit binding to one side of the DNA strand, which explains why these enzymes make parallel cuts on palindromic six-base recognition sites.

The best characterized nuclease enzyme is ribonuclease A, a 14-kDa protein which was the first protein to be reversibly unfolded and re-folded in the absence of other proteins. This enzyme cleaves RNA at sites immediately following pyrimidine bases (cytidine or uridine), leaving a pyrimidine $3'$-phosphate product. Covalent modification studies using iodoacetate and iodoacetamide implicated two histidine residues, His-12 and His-119, as being involved in catalysis. When the X-ray crystal structure of the enzyme was solved, these residues were found on opposite sides of the active site, with the side chain of Lys-41 also occupying a prominent position. The currently accepted mechanism for ribonuclease A is shown in Fig. 5.26. The mechanism involves participation of the $2'$-hydroxyl of the pyrimidine

(a) Ribonuclease A - cleavage of RNA after pyrimidine (cytidine or uridine)

$$5'\text{-....X-p-Y-p-C-p-Z-p....-}3' \xrightarrow{\text{RNase A}} 5'\text{....X-p-Y-p-C-}3'\text{-OPO}_3^{2-} \quad + \quad 5'\text{-HO-Z-p....}3'$$

(b) Restriction endonucleases - cleavage of dsDNA at specific 6-base recognition site

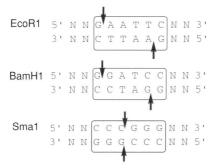

Fig. 5.25 Specificity of phosphodiesterases.

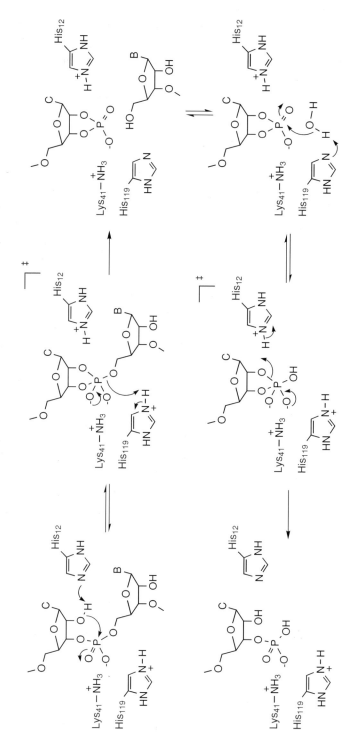

Fig. 5.26 Mechanism for ribonuclease A.

residue, which is deprotonated by His-112 and attacks the phosphodiester to form a divalent transition state stabilized by Lys-41. Breakdown of this transition state using His-119 as an acid leads to a cyclic phosphodiester intermediate which, due to internal strain, is much more reactive than the substrate. Acid–base catalysis by the two histidine groups completes the mechanism via a second divalent transition state.

5.6 Adenosine 5′-triphosphate (ATP)

Enzymatic phosphoryl transfer reactions usually involve the transfer of phosphoryl groups from a 'high-energy' phosphoric anhydride species to an acceptor, which can be an alcohol, a carboxylic acid or another phosphate. The most common source of phosphoryl groups for such transfer reactions is the coenzyme ATP. This is the nucleoside triphosphate derivative of adenosine, which is one of the components of RNA. However, in addition to its role in RNA it is used as a coenzyme by a wide range of enzymes.

ATP is a thermodynamically unstable molecule, since the hydrolysis of its phosphoric anhydride linkages is thermodynamically highly favourable, hence its designation as a 'high-energy' source of phosphate. Yet, ATP is reasonably stable in aqueous solution—why? The explanation is that although the hydrolysis of ATP is thermodynamically favourable it is kinetically unfavourable, particularly the hydrolysis of the phosphodiester groups.

The triphosphate group of ATP can be cleaved at a number of different points, shown in Fig. 5.27, leading to the transfer of either a phosphoryl group, a pyrophosphoryl group or an adenosine phosphoryl group. The most common transfer is of a single phosphoryl unit, leaving behind adenosine-5′-diphosphate (ADP). This process is used by kinase enzymes in the phosphorylation of alcohols, and is used by a number of ligase enzymes to activate carboxyl groups as acyl phosphate intermediates (Fig. 5.28).

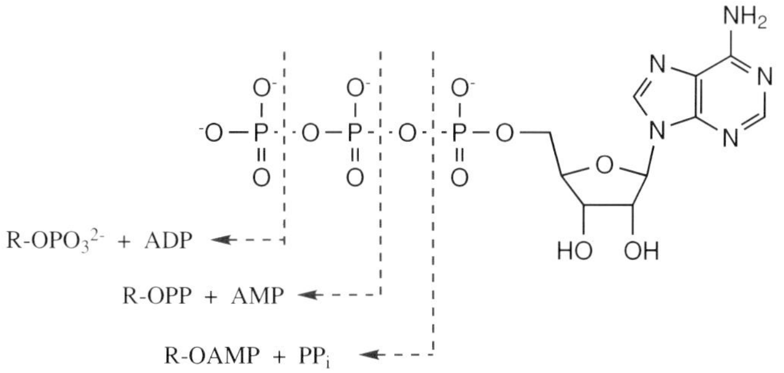

Fig. 5.27 Structure of ATP and three common modes of phosphoryl transfer. PP$_i$, inorganic pyrophosphate.

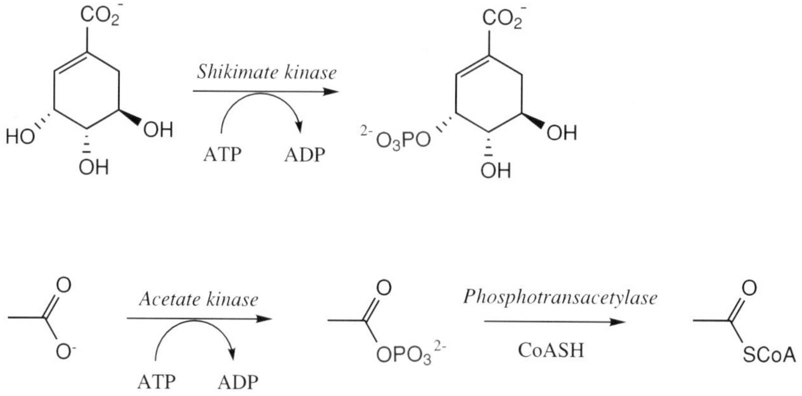

Fig. 5.28 Phosphorylation of alcohols and carboxylates by ATP. CoASH, coenzyme A.

One example of the use of ATP to activate a carboxylic acid derivative is involved in the biosynthesis of acetyl CoA. This is carried out in bacteria by the action of two enzymes: (i) acetate kinase, which activates acetate as acetyl phosphate using ATP as a cofactor; and (ii) phosphotransacetylase, which transfers the acetyl group onto CoA (Fig. 5.28). We shall meet a number of other examples of the use of ATP in later chapters.

5.7   Enzymatic glycosyl transfer reactions

Carbohydrates fulfil many important roles in biological systems: polysaccharides are important structural components of plant and bacterial cell walls, and constitute an important part of the human diet; mammalian polysaccharides such as glycogen are used as short-term cellular energy stores; whilst the attachment of carbohydrates to mammalian glycoproteins has an important role to play in cell–cell recognition processes.

Hydrolysis of polysaccharides can be achieved in the laboratory using acid hydrolysis, as shown in Fig. 5.29, since the glycosidic linkage is an acetal functional group. Glycosyl transfer enzymes also employ acid catalysis for just the same reason. It will not surprise you to learn that glycosidase enzymes are highly specific: they are specific for cleavage at the glycosidic bond of a particular monosaccharide, and they are also specific for cleavage of either an α- or a β-glycosidic linkage (Fig. 5.29).

The best studied glycosidase enzyme is lysozyme, a mammalian protein found in such diverse sources as egg whites and human tears. It functions as a mild antibacterial agent by hydrolysing the glycosidic linkage between MurNAc and GlcNAc residues in the peptidoglycan layer of bacterial cell walls. The active site contains two carboxylic acid groups which are involved in catalysis: Glu-35 and Asp-52. The mechanism is illustrated in Fig. 5.30.

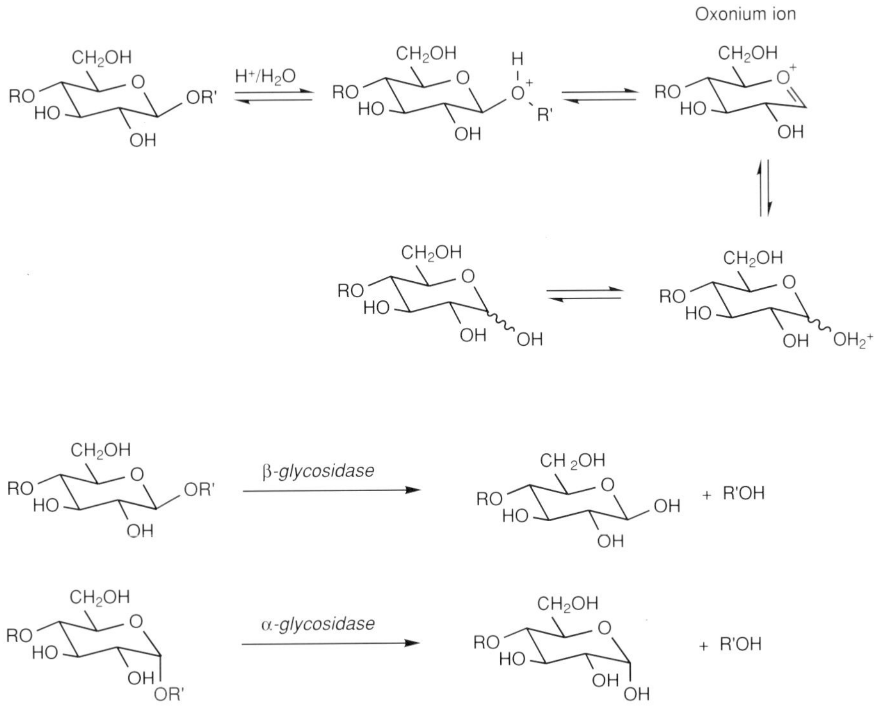

Fig. 5.29 Non-enzymatic and enzymatic glycoside hydrolysis.

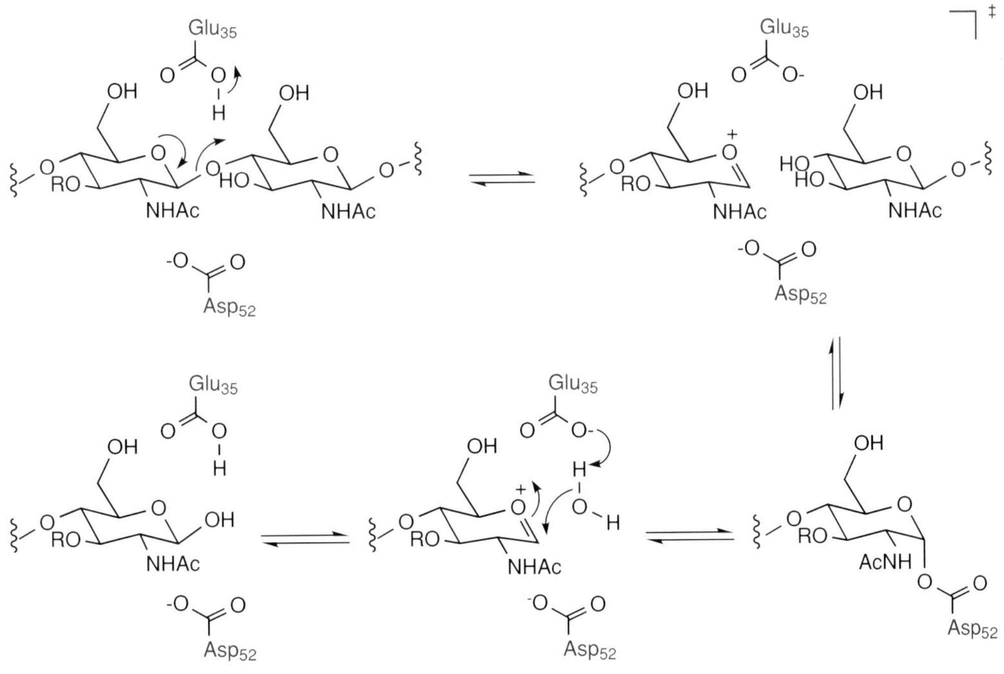

Fig. 5.30 Mechanism for lysozyme.

Upon binding of substrate, Glu-35 donates a proton to the departing GlcNAc C-4 oxygen, promoting cleavage of the glycosidic bond and formation of an oxonium ion intermediate. There has been much debate in the literature over the precise role of Asp-52: whether it stabilizes the oxonium ion via an ion pair electrostatic interaction, or whether it becomes covalently attached to the oxonium ion. There is evidence to suggest that covalent attachment does take place reversibly, and upon attack of water on the subsequent oxonium ion intermediate the product is formed with retention of stereochemistry at the glycosidic centre.

When the X-ray crystal structure of lysozyme was solved, it was found that the MurNAc residue whose glycosidic linkage was attacked could not be modelled into the active site of the enzyme—the lowest energy conformation of the substrate simply did not fit! However, if the pyranose ring of this residue adopts a somewhat flattened structure, it is able to bind to the active site, suggesting that the enzyme binds this residue in a strained but more reactive conformation. This is thought to be another example of the use of strain in enzyme catalysis, mentioned in Chapter 3, Section 3.7. Thus, the enzyme uses the binding energy of other residues in the polysaccharide chain to offset the binding of the strained ring, which is then closer in structure and free energy to the flattened oxonium ion intermediate.

There are also glycosidase enzymes which proceed via inversion of stereochemistry at the glycosidic centre. It is thought in these cases that there is no covalent intermediate, and that there is attack directly on the oxonium intermediate on the opposite face of the molecule. This is not strictly an S_N2 displacement, since acetals are not susceptible to nucleophilic attack. However, it may be viewed as a displacement occurring via a dissociative transition state, as shown in Fig. 5.31.

Glycosyl transferases are involved in the assembly of polysaccharides and oligosaccharides. Glycosidic bonds are assembled through use of an activated glycosyl unit, namely a glycosyl phosphate or glycosyl diphosphouridine derivative. For example, the disaccharide sucrose that we know as

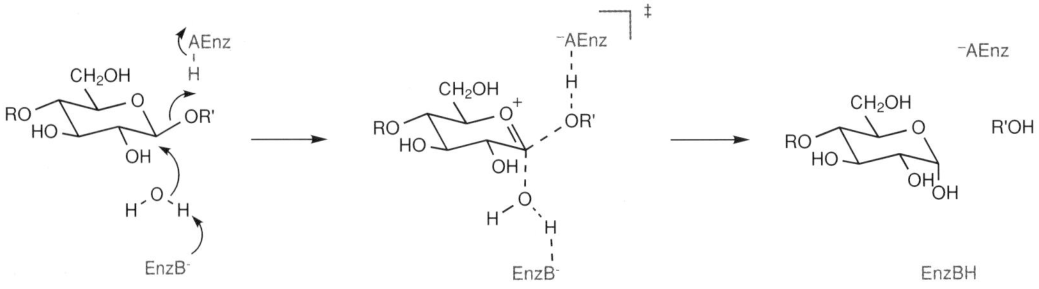

Fig. 5.31 Mechanism for an inverting glycosidase.

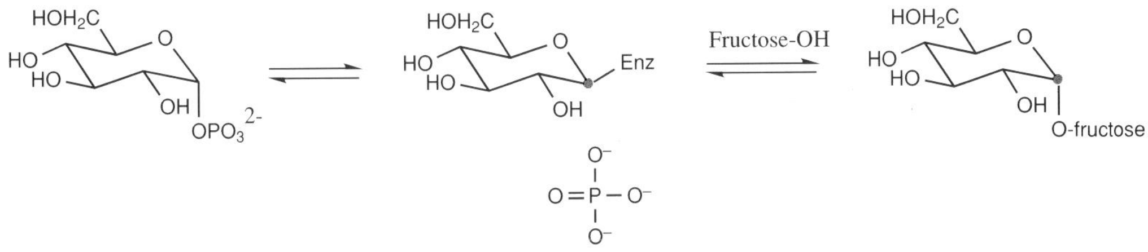

Fig. 5.32 Synthesis of sucrose by sucrose phosphorylase. Red dot indicates the position of ^{14}C label (see below).

table sugar is synthesized by the enzyme sucrose phosphorylase from glucose-1-phosphate and fructose, as shown in Fig. 5.32.

The overall reaction involves displacement of phosphate by fructose with retention of stereochemistry at the glycosyl centre, which by analogy with the above glycosidase enzymes would suggest that this is a double displacement reaction involving a covalent intermediate. This has been confirmed by incubation of the enzyme with sucrose labelled with ^{14}C at C-1 of the glucose residue, whereupon in the absence of inorganic phosphate the ^{14}C label becomes covalently attached to the enzyme. Addition of phosphate converts the ^{14}C–glucosyl–enzyme intermediate into ^{14}C-glucose-1-phosphate. Other glycosyl transferases utilize glycosyl diphosphouridine activated derivatives, which upon glycosylation release uridine diphosphate (UDP).

5.8   Methyl (CH$_3$–) group transfer: use of S-adenosylmethionine (SAM) and tetrahydrofolate coenzymes for one-carbon transfers

The final example of group transfer reaction that we shall meet is that of methyl group transfer, and more generally the transfer of one-carbon units. Many biologically important molecules contain methyl groups attached to oxygen, nitrogen and carbon substituents which have arisen by transfer of a methyl group from nature's methyl group donor—SAM. The methyl group to be transferred is attached to a positively charged sulphur atom which is a very good leaving group for such a methylation reaction. Analysis of such methyltransferase reactions using the chiral methyl group approach detailed in Chapter 4, Section 4.4 has revealed that they proceed with inversion of stereochemistry, implying that the reaction is a straightforward S$_N$2 displacement reaction (Fig. 5.33).

As well as oxygen and nitrogen nucleophiles, SAM-dependent methyltransferases also operate on stabilized carbon nucleophiles, providing many of the methyl groups found in polyketide natural products, and in the

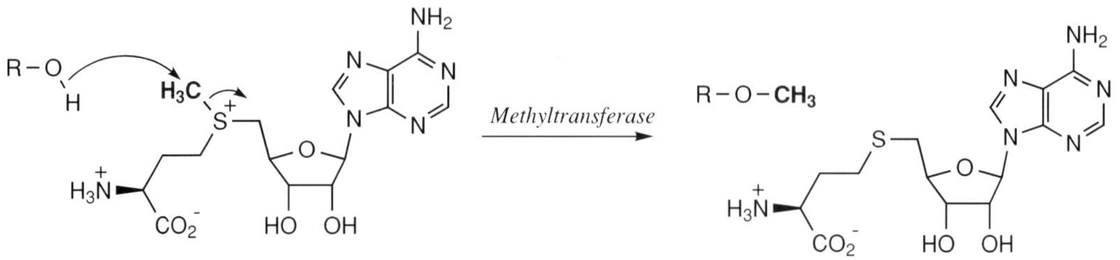

S-adenosylmethionine S-adenosylhomocysteine

Fig. 5.33 Methyl group transfer from SAM.

structure of vitamin B_{12} (see Chapter 10, Section 10.6). The byproduct of methyltransferase enzymes is S-adenosylhomocysteine, which is re-cycled to SAM via hydrolysis to adenosine and homocysteine. How is the structure of SAM assembled? It is synthesized from methionine and ATP by a very unusual displacement of triphosphate, which is subsequently hydrolysed to phosphate and pyrophosphate, as shown in Fig. 5.34.

As well as transferring methyl groups, Nature is able to transfer methylene ($-CH_2-$) groups and even methyne ($-CH=$) groups using another cofactor—tetrahydrofolate. Tetrahydrofolate is biosythesized from folic acid, which is an essential element of the human diet. The active parts of the molecule as far as one-carbon transfer is concerned are the two nitrogen atoms N-5 and N-10. It is to these atoms that the one-carbon unit is attached, and there are several different forms of the cofactor illustrated in Fig. 5.35.

The mechanism for transfer of a methylene or methyne involves nucleophilic attack on the imine formed between the one-carbon unit and either N-5 or N-10, followed by breakdown of the substrate–coenzyme intermediate. An example is the formation of 5-hydroxymethylcytidine from cytidine, which is illustrated in Fig. 5.36.

How is methylene-tetrahydrofolate synthesized in the cell? Not from free formaldehyde, which is toxic to biological systems. Conversion of tetra-

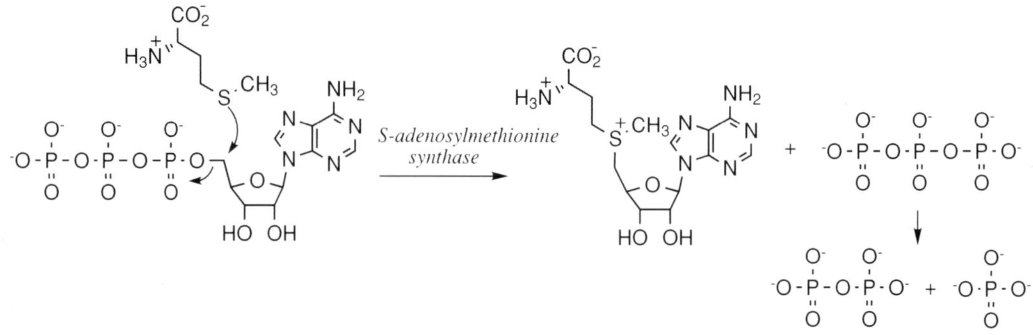

Fig. 5.34 Biosynthesis of SAM.

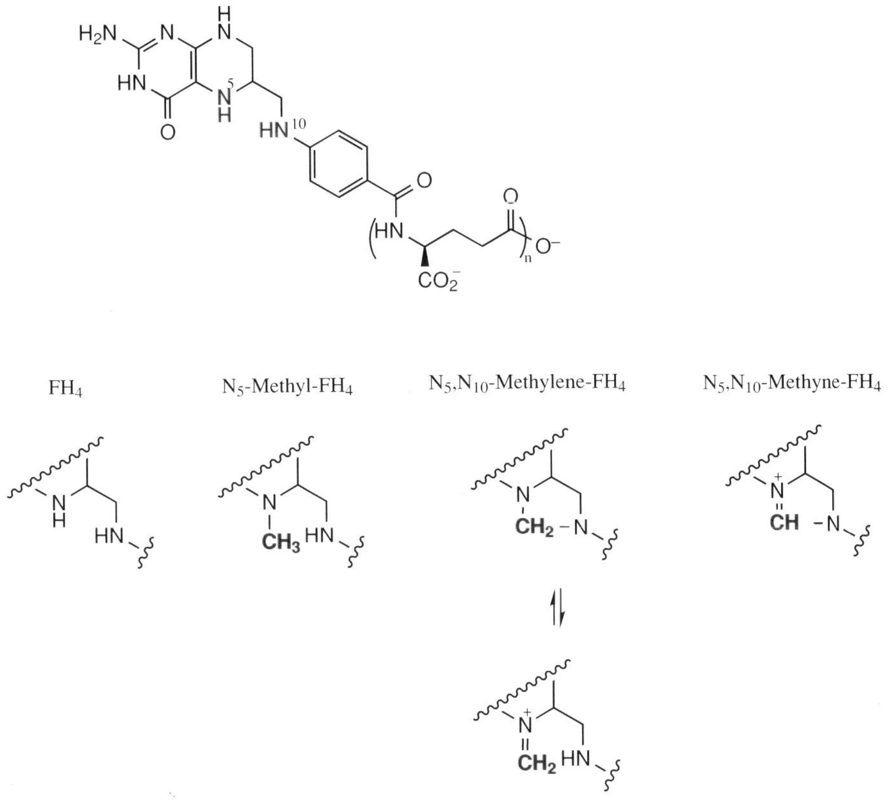

Fig. 5.35 Tetrahydrofolate (FH$_4$) and its one-carbon adducts.

hydrofolate to methylene-tetrahydrofolate is carried out by serine hydroxymethyltransferase, a pyridoxal phosphate-dependent enzyme which we shall meet in Chapter 9, Section 9.5. Suffice it to say that it is the hydroxymethyl group of serine which is transferred, generating glycine as a byproduct.

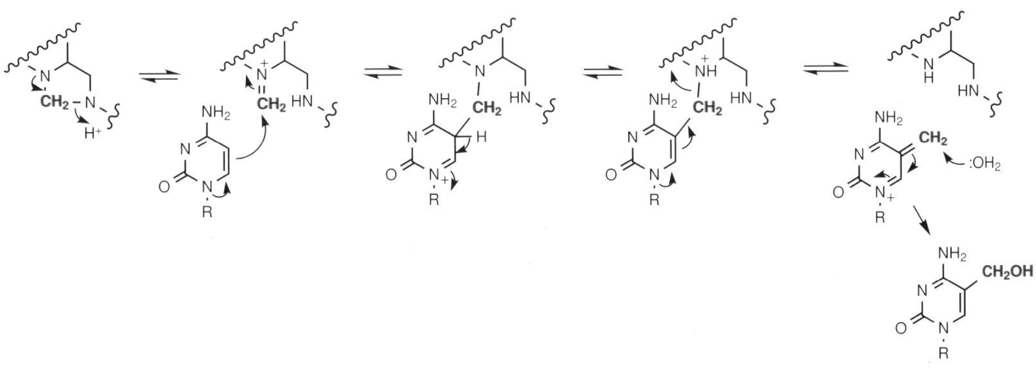

Fig. 5.36 Methylene transfer to cytidine.

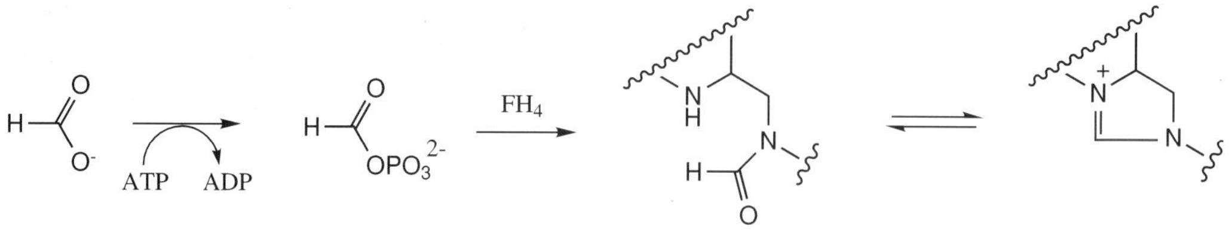

Fig. 5.37 Biosynthesis of methyne-tetrahydrofolate (FH$_4$).

Methyne-tetrahydrofolate can be synthesized either by the nicotinamide adenine dinucleotide phosphate-dependent oxidation of methylene-tetrahydrofolate, or it can be synthesized by a synthetase enzyme which uses formyl phosphate as an activated one-carbon equivalent, generated from formate by ATP (Fig. 5.37).

Thus, we have seen how enzymes are able to transfer a wide variety of carbon- and phosphorus-base groups in biological systems. These reactions are equally important for the breakdown and assembly of biological materials, and form the cornerstone of the family of enzymatic reactions found in living systems.

Problems

1 Using *p*-nitrophenyl acetate as a substrate for chymotrypsin, a rapid 'burst' of *p*-nitrophenol is observed, followed by a slower steady state release of *p*-nitrophenol. The amount of product released in the initial burst is 1 µmol of *p*-nitrophenol per micromole of enzyme. Explain.

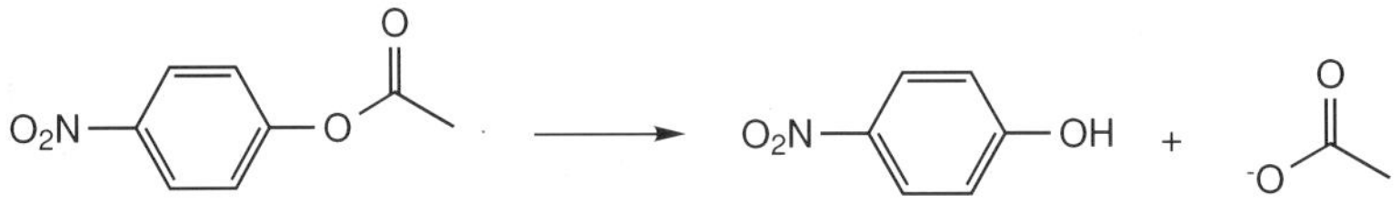

2 Acetylcholinesterase is an esterase enzyme which catalyses the hydrolysis of acetylcholine (a neurotransmitter) to choline and acetate at nerve synaptic junctions. Inhibition of acetylcholinesterase is catastrophic and usually fatal. Given that the enzyme contains a serine catalytic triad, suggest a mechanism for the enzyme.

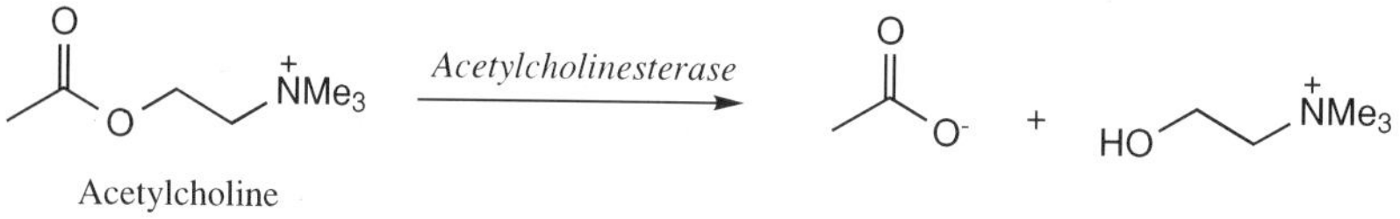

Acetylcholine

Nerve gas sarin and insecticide parathion both act on this enzyme — suggest a common mechanism of action. Suggest reasons why parathion is not so toxic to humans (median lethal dose (LD_{50}) 6800 mg kg^{-1}) as sarin.

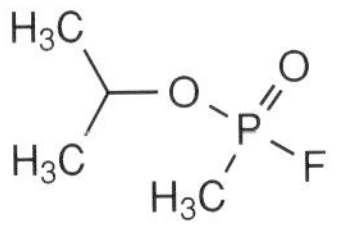

Sarin 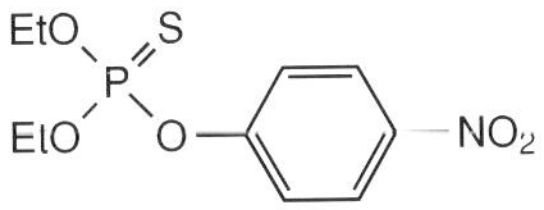   Parathion

Pyridine aldoximine methiodide is an effective antidote for organophosphorus poisoning (used at the time of the Tokyo subway incident in March 1995). Suggest a mechanism for how it works.

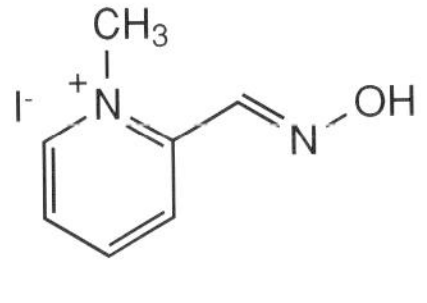

Pyridine aldoximine
methiodide (PAM)

3 Cysteine proteases are effectively inhibited by substrate analogues containing an aldehyde functional group in place of the amide targeted by the enzyme. Suggest a possible mechanism of inactivation.

4 Acetyl CoA is biosynthesized from acetate by different pathways in bacteria versus higher organisms, as shown below. In the case of the bacterial pathway, incubation of ^{18}O-labelled acetate yields one atom of ^{18}O in the phosphate product, whereas in higher organism the same experiment yields one atom of ^{18}O in the adenosine monophosphate (AMP) product. Suggest intermediates and mechanisms for the two pathways. (CoASH, coenzyme A; P_i, inorganic phosphate; PP_i, inorganic pyrophosphate.)

Bacteria:
(1) *Acetate kinase*
(2) *Phosphotransacetylase*
$$CH_3CO_2^- + ATP + CoASH \longrightarrow CH_3COSCoA + ADP + P_i$$

Higher organisms:
Acetate thiokinase
$$CH_3CO_2^- + ATP + CoASH \longrightarrow CH_3COSCoA + AMP + PP_i$$

5 Glycogen is a mammalian polysaccharide consisting of repeating α-1, 4-linked D-glucose units. It is stored in liver and muscle and is used as a carbohydrate energy source by the body, being converted to D-glucose when required by the pathway shown below. Suggest mechanisms for each of the enzymes on the pathway. What would be the medical consequences of a genetic defect in muscle glycogen phosphorylase?

Further reading

General

Abeles, R.H., Frey, P.A. & Jencks, W.P. (1992) *Biochemistry*. Jones & Bartlett, Boston.
Walsh, C.T. (1979) *Enzymatic Reaction Mechanisms*. Freeman, San Francisco.
Wong, C.H. & Whitesides, G.M. (1994) *Enzymes in Synthetic Organic Chemistry*. Pergamon, Oxford.

Serine proteases

Kraut, J. (1977) *Annu Rev Biochem*, **46**, 331–58.
Kreutter, K., Steinmetz, A.C.U., Liang, T.C., Prorok, M., Abeles, R.H. & Ringe, D. (1994) *Biochemistry*, **33**, 13 792–800.
Prorok, M., Albeck, A., Foxman, B.M. & Abeles, R.H. (1994) *Biochemistry*, **33**, 9784–90.
Craik, C.S., Roczniak, S., Largman, C. & Rutter, W.J. (1987) *Science*, **237**, 909–13.
Sprang, S., Standing, T., Fletterick, R.J. *et al.* (1987) *Science*, **237**, 905–9.
Brenner, S. (1988) *Nature*, **334**, 528–30.

Cysteine proteases

Shaw, E. (1990) *Adv Enzymol*, **63**, 271–348.
Malthouse, J.P.G., Gamesik, M.P., Boyd, A.S.F., Mackenzie, N.E. & Scott, A.I. (1982) *J Am Chem Soc*, **104**, 6811–13.
Shafer, J.A., Johnson, F.A. & Lewis, S.D. (1981) *Biochemistry*, **20**, 44–48 & 48–51.
Vernet, T., Tessier, D.C. Chatellier, J. *et al.* (1995) *J Biol Chem*, **270** 16 645–52.

Metalloproteases

Carboxypeptidase

Britt, B.M. & Peticolas, W.L. (1992) *J Am Chem Soc*, **114**, 5295–303.
Christianson, D.W. & Lipscomb, W.N. (1989) *Acc Chem Res*, **22**, 62–9.
Sander, M.E. & Witzel, H. (1985) *Biochem Biophys Res Commun*, **132**, 681–7.

Thermolysin

Bartlett, P.A. & Marlowe, C.K. (1987) *Biochemistry*, **26**, 8553–61,
Holden, H.M., Tronrud, D.E., Monzingo, A.F., Weaver, L.M. & Matthews, B.W. (1987) *Biochemistry*, **26**, 8542–53.
Matthews, B.W. (1988) *Acc Chem Res*, **21**, 333–40.

HIV protease

Erickson, J., Neidhart, D.J., VanDrie, J. *et al.* (1990) *Science*, **249**, 527–33.
Hyland, L.J., Tomaszek Jr, T.A. Roberts, G.D. *et al.* (1991) *Biochemistry* **30**, 8441–53.
Jaskolski, M., Tomasselli, A.G., Sawyer, T.K. *et al.* (1991) *Biochemistry*, **30**, 1600–9
Wlodawer, A. & Erickson, J.W. (1993) *Annu Rev Biochem*, **62**, 543–86.

Phosphoryl transfer

Knowles, J.R. (1980) *Annu Rev Biochem*, **49**, 877–920.
Frey, P.A. (1982) *Tetrahedron*, **38**, 1541–67.

Glycosyl transfer

Sinnott, M.L. (1990) *Chem Rev*, **90**, 1171–1202.

One-carbon transfer

Matthews, R.G. & Drummond, J.T. (1990) *Chem Rev*, **90**, 1275–90.

6 Enzymatic Redox Chemistry

6.1 Introduction

Oxidation and reduction involves the transfer of electrons between one chemical species and another. Electron transfer processes are widespread in biological systems and underpin the production of biochemical energy in all cells. The ultimate source of energy for all life on Earth is sunlight, which is utilized by plants for the process of photosynthesis. Photosynthesis involves a series of high-energy electron transfer processes, converting sunlight energy into high-energy reducing equivalents, which are then used to drive biochemical processes. These biochemical processes ultimately lead to the fixation of carbon dioxide and production of oxygen. Oxygen in turn serves a vital role for mammalian cellular metabolism as an electron acceptor.

Cells contains a number of intermediate electron carriers which serve as cofactors for enzymatic redox processes. In this chapter we shall examine some of these redox cofactors, and we shall also examine metallo-enzymes, which utilize redox-active metal ions to harness the oxidizing power of molecular oxygen.

First of all we need to define a scale to measure the effectiveness of these different electron carriers as oxidizing or reducing agents. How do we measure quantitatively whether something is a strong or weak oxidizing/reducing agent? We can measure the strength of an oxidizing agent electrochemically by dissolving it in water and measuring the voltage required to reduce it (i.e. add electrons) to a stable reduced form. The voltage measured under standard conditions with respect to a reference half-cell is known as the *redox potential*. The reference half-cell is the reaction $2H^+ + 2e^- \rightarrow H_2$, whose redox potential is -0.42 V at pH 7.0. A strong oxidizing agent will be reduced very readily, corresponding to a strongly positive redox potential, whilst a weak oxidizing agent will be reduced much less readily, corresponding to a less positive or a negative redox potential. A scale showing the redox potential of some important biological electron carriers is shown in Fig. 6.1. One obvious point is that the strongest available oxidizing agent is oxygen, hence the large number of enzymes which use molecular oxygen either as a substrate (oxygenases) or as an electron acceptor (oxidases).

We can use redox potentials to work out whether a particular redox reaction will be thermodynamically favourable. For example, the enzyme lactate dehydrogenase catalyses the reduction of pyruvate to lactate, using

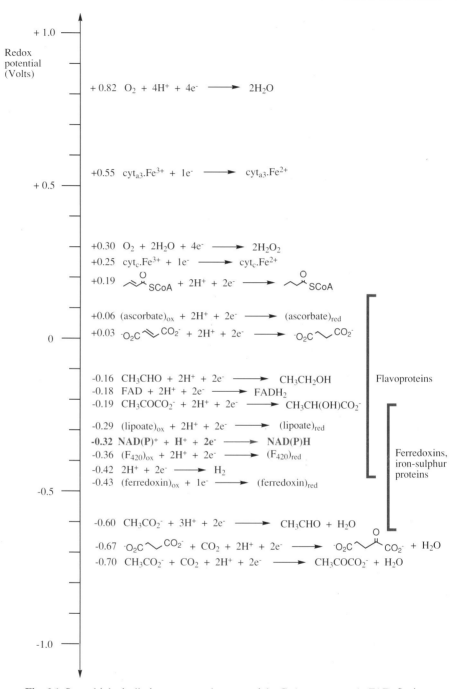

Fig. 6.1 Some biologically important redox potentials. CoA, coenzyme A; FAD, flavin adenine dinucleotide; NADP, nicotinamide adenine dinucleotide phosphate.

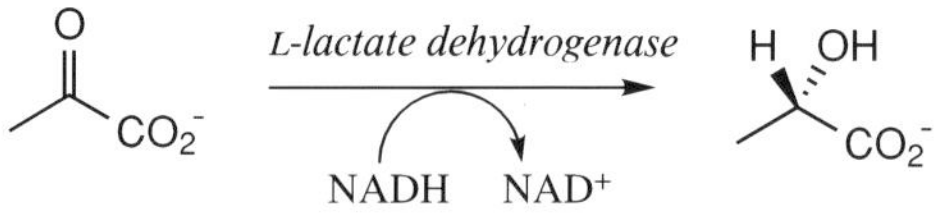

<table>
<tr><td>(1)</td><td>$NAD^+ + H^+ + 2e^-$</td><td>→</td><td>$NADH$</td><td>- 0.32 V</td></tr>
<tr><td>(2)</td><td>$CH_3COCO_2H + 2H^+ + 2e^-$</td><td>→</td><td>$CH_3CH(OH)CO_2H$</td><td>- 0.19 V</td></tr>
<tr><td>(2) - (1)</td><td>$CH_3COCO_2H + NADH$</td><td>→</td><td>$CH_3CH(OH)CO_2H + NAD^+$</td><td>+ 0.13 V</td></tr>
</table>

Fig. 6.2 Redox potentials in the lactate dehydrogenase reaction.

nicotinamide adenine dinucleotide (NADH) as a redox cofactor. NAD^+ (the oxidized form of NADH) has a redox potential of -0.32 V, whereas pyruvate has a redox potential of -0.19 V. We simply subtract the redox potentials, as shown in Fig. 6.2, giving an overall redox potential difference for the pyruvate-to-lactate reaction of $+0.13$ V. If the redox potential difference is above zero, then the reaction is thermodynamically favourable. The more positive the redox potential difference is, the more favourable the reaction is and the greater the equilibrium constant for the reaction (the equilibrium constant can be calculated from the redox potential difference using the Nernst equation — see physical chemistry texts). Remember that a highly positive redox potential means a strong oxidizing agent, whilst a highly negative redox potential means a strong reducing agent.

One final point is that an enzyme cannot change the equilibrium constant of a chemical reaction that is thermodynamically unfavourable; however, it can modify the redox potential of a cofactor *bound at its active site* by selectively stabilizing either the oxidized or the reduced form of the cofactor. For example, the redox potential for free oxidized riboflavin is -0.20 V, whereas redox potentials of between -0.45 and $+0.15$ V have been measured for flavo-enzymes in general. Just as with amino acid side chain pK_a values, enzymes are able to modify chemical reactivity at their active sites using subtle changes in micro-environment.

6.2   NAD-dependent dehydrogenases

The first class of redox enzymes that we shall meet are the dehydrogenases (see Table 6.1). These enzymes transfer two hydrogen atoms from a reduced substrate, which is usually an alcohol, to an electron acceptor. The electron acceptor in these enzymes is the coenzyme NAD, whose oxidized and reduced forms are shown in Fig. 6.3.

The redox-active part of this coenzyme is the nicotinamide heterocyclic ring. In the oxidized form NAD^+ this is a pyridinium salt, which is reduced

Table 6.1 Classes of redox enzymes.

Transformation	Enzyme class	Redox cofactors
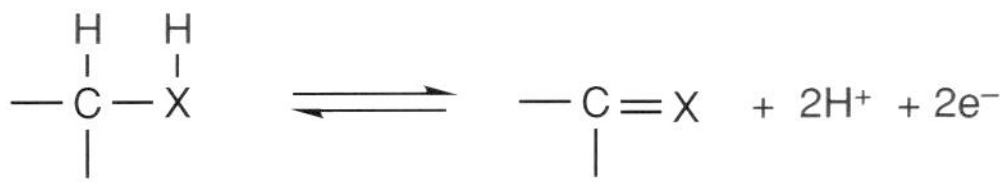 $\overset{H}{\underset{}{-}}\overset{H}{\underset{}{C}}-X \rightleftharpoons -C{=}X + 2H^+ + 2e^-$	Dehydrogenases	**1** NAD^+ (X = O, NH) **2** flavin / 1e$^-$ acceptors (X = O, NH, CHR)
$-\overset{H}{\underset{}{C}}- \longrightarrow -\overset{OH}{\underset{}{C}}-$	Oxidases	Flavin / O_2 (X = NH)
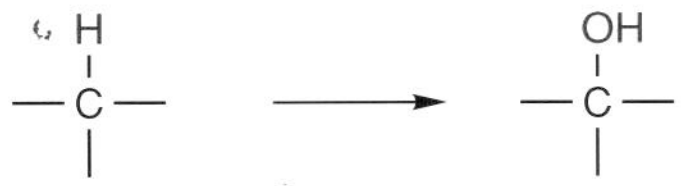 $\overset{}{C}{=}\overset{}{C} \longrightarrow \overset{HO\ \ OH}{-C{-}C-}$	Mono-oxygenases, hydroxylases	Aromatic C-H: **1** Flavin / O_2 / NADPH **2** Pterin / O_2 / NADPH Aliphatic or aromatic C-H:**1** P_{450} / O_2 / NADH **2** Non-heme metal / O_2 / e$^-$ donor **3** Fe^{2+} / O_2 / α-ketoglutarate
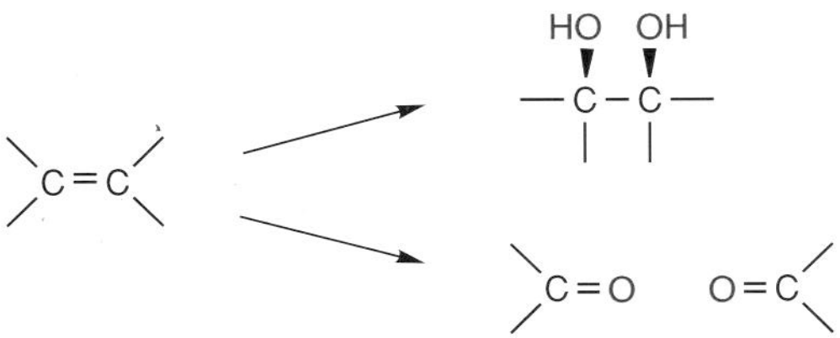	Dioxygenases (dihydroxylating)	Fe^{2+} / O_2 / flavin / NADH
$\overset{}{C}{=}\overset{}{C} \longrightarrow C{=}O \quad O{=}C$	Dioxygenases (C-C cleaving)	Fe^{2+} or Fe^{3+} / O_2
$A\overset{H}{\underset{H}{\diagdown}} \longrightarrow A + 2H^+ + 2e^-$	Peroxidases	**1** P_{450} / H_2O_2 **2** Non-heme metal / H_2O_2
$H{-}H \longrightarrow 2H^+ + 2e^-$	Hydrogenases	Ni^{2+} / factor F_{420}

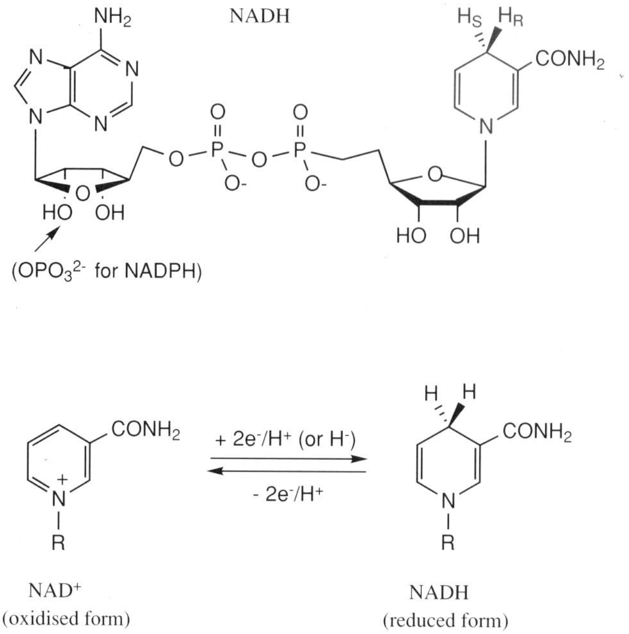

Fig. 6.3 Structures of NAD$^+$ and NADH. NADPH, nicotinamide adenine dinucleotide phosphate.

to a 1,4-dihydro-pyridine in the reduced form NADH. A phosphorylated version of the cofactor is also found, bearing a 2'-phosphate on the adenosine portion of the structure, which is written as NADP$^+$ in the oxidized form and NADPH in the reduced form. Dehydrogenase enzymes which utilize this cofactor are usually specific either for NAD$^+$/NADH or for NADP$^+$/NADPH. NAD$^+$/NADH tend to be utilized for catabolic processes (which break down cellular metabolites into smaller pieces), whereas NADP$^+$/NADPH tend to be used for biosynthetic processes.

The redox potential for NAD$^+$ is -0.32 V, consequently NADH is the most powerful of the commonly available biological reducing agents. NAD$^+$ is therefore a relatively weak oxidizing agent. NADH is most commonly used for the reduction of ketones to alcohols, although it is also sometimes used for reduction of the carbon–carbon double bond of an α,β-unsaturated carbonyl compound, as shown in Fig. 6.4.

An important point is that NADH is a stoichiometric reagent which is bound non-covalently by the enzyme at the start of the reduction and released as NAD$^+$ at the end of the reaction. The concentration of NADH in solution therefore decreases as an enzyme-catalysed reduction proceeds, which allows the reaction to be monitored by ultraviolet (UV) spectroscopy, since NADH has a strong absorption at 340 nm ($\varepsilon = 6.3 \times 10^3$ M^{-1}cm^{-1}).

We have already met the example of horse liver alcohol dehydrogenase,

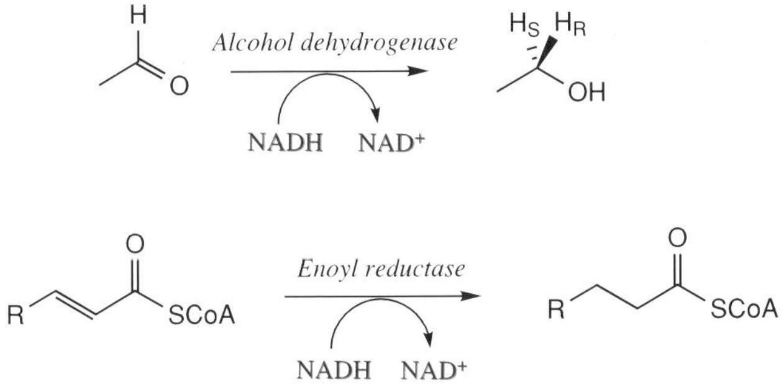

Fig. 6.4 Examples of NAD$^+$-dependent dehydrogenases. CoA, coenzyme-A.

which catalyses the NAD$^+$-dependent oxidation of ethanol to acetaldehyde (see Chapter 4, Section 4.4). In a classic series of experiments by F.H. Westheimer, this enzyme was shown to be entirely stereospecific for the removal of prochiral hydrogen atoms in both the substrate and the cofactor. These experiments are illustrated in Fig. 6.5.

Incubation of enzyme with deuterated ethanol (CH$_3$CD$_2$OH) followed by re-isolation of oxidized NAD$^+$ revealed no deuterium incorporation into the oxidized cofactor, demonstrating that the deuterium atom transferred to the cofactor was stereospecifically removed in the reverse reaction. Isolation of the deuterium-containing reduced cofactor (NADD) followed by chemical conversion to a substance of established configuration revealed that the reduced cofactor had the 4R stereochemistry. Incubation of acetaldehyde and [4R-^{2}H]-NADD with enzyme for prolonged reaction times gave no exchange of the deuterium label into acetaldehyde, indicating that the enzyme was also stereoselective for the removal of a hydrogen atom from C-1 of ethanol.

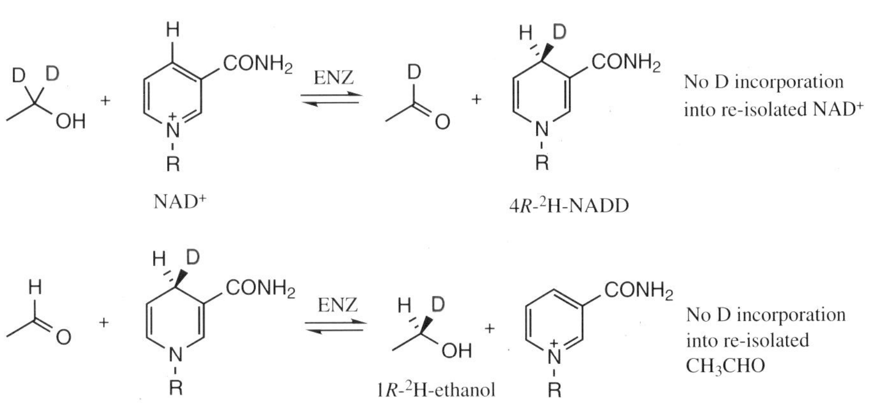

Fig. 6.5 Stereospecificity of alcohol dehydrogenase. CH$_3$CHO, acetaldehyde; NADD, deuterium-containing reduced NADH.

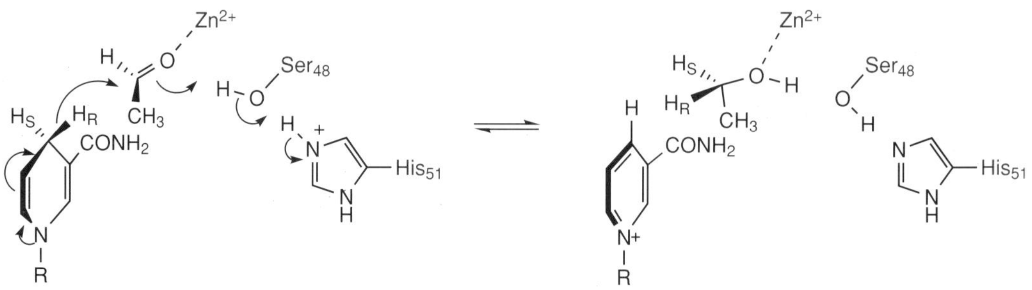

Fig. 6.6 Mechanism of alcohol dehydrogenase. His, histidine; Ser, serine; Zn²⁺, zinc.

Incubation with specifically deuterated ethanol substrates subsequently revealed that the enzyme removes specifically the *proR* hydrogen, which is transferred to the C-4 *proR* position of NAD^+. The mechanism is thought to proceed via direct hydride transfer, as shown in Fig. 6.6.

Although hydride transfer might, at first sight, seem unlikely in aqueous solution, note that this transfer is occurring at close proximity within the confines of an enzyme active site. There is also precedent for organic reactions involving hydride transfer occurring in aqueous solution, such as the Cannizzaro reaction shown in Fig. 6.7.

Deuterium-labelling studies carried out on alcohol dehydrogenase have established that the same hydrogen atom abstracted from ethanol is transferred to the nicotinamide ring. It has also been established using kinetic isotope effect studies that there is a single transition state for the enzyme-bound reaction. Although radical intermediates are in theory possible, attempts to identify such intermediates have been unsuccessful, so hydride transfer via a single transition state is the accepted mechanism.

Dehydrogenase enzymes are each specific for one of the enantiotopic C-4 protons of NADH: some are specific for the 4*R* hydrogen; others are specific for the 4*S* hydrogen. Stereospecificity of an NAD-dependent dehydrogenase can be determined readily as shown in Fig. 6.8. [1-²H₂]-Ethanol is incubated with alcohol dehydrogenase and the [4*R*-²H]-NADD product isolated. This is then incubated with the dehydrogenase of unknown stereospecificity, and the incorporation of ¹H or ²H into the reduced product is monitored. The

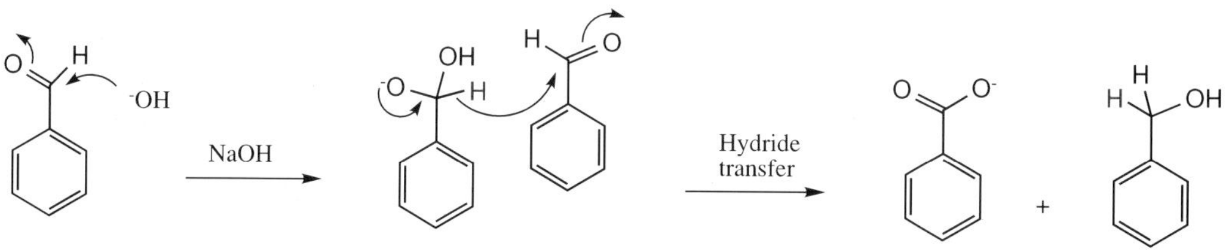

Fig. 6.7 The Cannizzaro reaction. NaOH, sodium hydroxide.

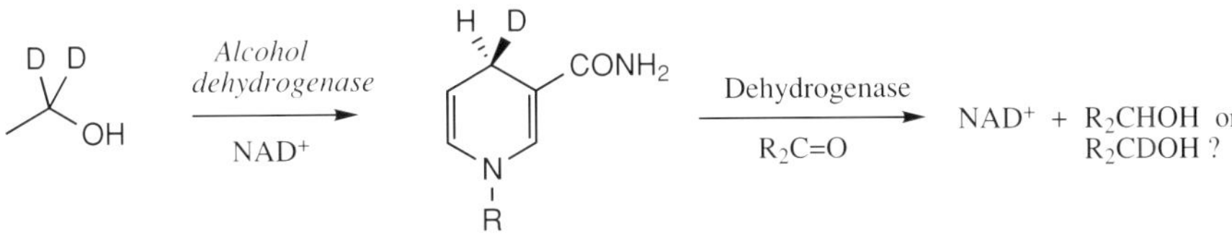

Fig. 6.8 Determination of dehydrogenase stereo-specificity.

experiment can be done more sensitively using a ^{3}H label in a similar way.

Whilst this stereospecificity is largely due to the orientation of the cofactor in the enzyme active site, there are indications that a further stereoelectronic effect may also be important. Model studies using dihydropyridine rings contained in molecules in which the two C-4 hydrogens are diastereotopic have shown that the environments of the two hydrogens are quite different (Fig. 6.9). This can be explained by a puckering of the dihydropyridine ring, causing one of the two hydrogens to adopt a pseudo-axial orientation. This C–H bond is aligned more favourably with the p-orbitals of the dihydropyridine ring and is therefore stereoelectronically better aligned for hydride transfer. It seems likely that dehydrogenase enzymes also use this stereoelectronic effect to assist the reaction.

6.3   Flavin-dependent dehydrogenases and oxidases

Riboflavin was first isolated from egg white in 1933 as a vitamin whose deficiency causes skin lesions and dermatitis. The most striking property of riboflavin is its strongly yellow–green fluorescence, a property conveyed onto those enzymes which bind this cofactor. The structure of riboflavin consists of an isoalloxidine heterocyclic ring system, which is responsible for its redox activity. Attached to N-10 is a ribitol side chain, which can be phosphorylated in the case of flavin mononucleotide (FMN) or attached

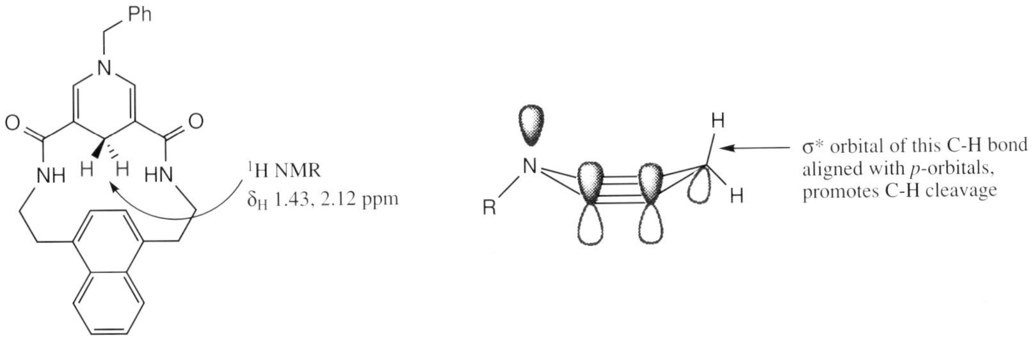

Fig. 6.9 Puckering of the dihydropyridine ring of NADH. NMR, nuclear magnetic resonance.

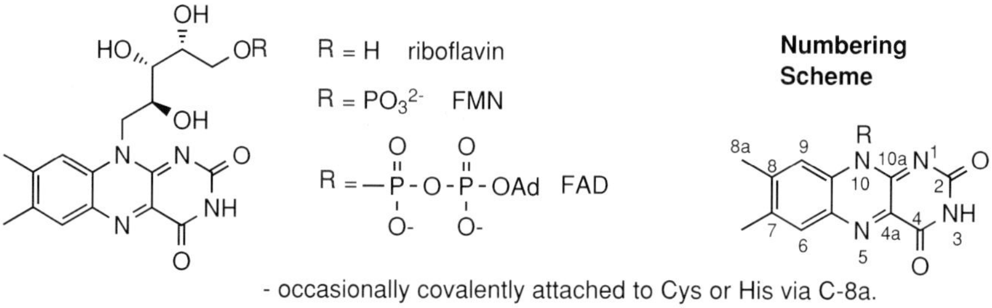

Fig. 6.10 Structures of flavin redox cofactors. Cys, cysteine; His, histidine.

through a diphosphate linkage to adenosine in flavin adenine dinucleotide (FAD), as shown in Fig. 6.10.

The first important difference between NAD and flavin is that enzymes which use flavin bind it very tightly, sometimes covalently (attached to cysteine or histidine (His) through C-8a), such that it is not released during the enzymatic reaction but remains bound to the enzyme throughout. Consequently the active form of the cofactor must be re-generated at the end of each catalytic cycle by external redox reagents. The next important difference is that flavin can exist either as oxidized FAD, or reduced $FADH_2$, or as an intermediate semiquinone redical species $FADH^{\bullet}$, as shown in Fig. 6.11. Therefore, flavin is able to carry out one-electron transfer reactions, whereas NAD is restricted to two-electron hydride transfers. This seemingly minor point has far-reaching consequences, since it allows flavin to react with the most powerful oxidizing agent in biological systems: molecular oxygen.

In the reactions of the flavin-dependent dehydrogenases and oxidases, a pair of hydrogen atoms is transferred from the substrate to the flavin nucleus, generating reduced $FADH_2$ (or $FMNH_2$). A few examples of reactions catalysed by these enzymes are shown in Fig. 6.12. Since oxidized FAD is required for the next catalytic cycle, the enzyme-bound $FADH_2$ must be oxidized *in situ*. In the flavin-dependent dehydrogenases this is done by external oxidants, which *in vivo* are electron carriers such as cytochromes (Table 6.1). *In vitro* the reduced flavin can be oxidized by chemical oxidants such as benzoquinone.

In the case of the flavin-dependent oxidases the re-generation of oxidized

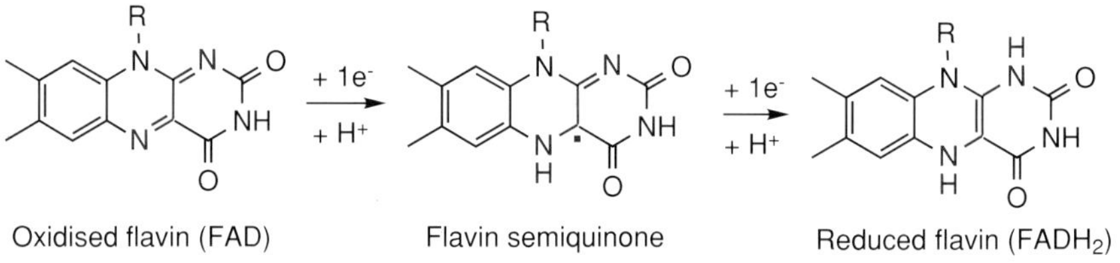

Oxidised flavin (FAD)   Flavin semiquinone   Reduced flavin ($FADH_2$)

Fig. 6.11 Redox states of riboflavin.

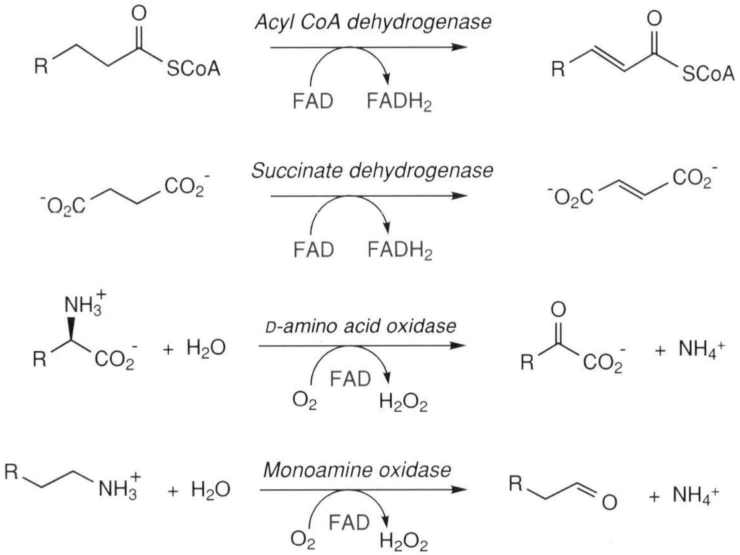

Fig. 6.12 Flavin-dependent dehydrogenases and oxidases. CoA, coenzyme A; H_2O_2, hydrogen peroxide; NH_4^+, ammonium.

flavin is carried out by molecular oxygen, which is reduced to hydrogen peroxide. Since the ground state of molecular oxygen contains two unpaired electrons (in its $\pi_{2px,y}$ molecular orbitals) it is spin-forbidden to react with species containing paired electrons. However, reduced flavin is able to transfer a single electron to di-oxygen to give superoxide and flavin semi-quinone. Recombination of superoxide with the flavin semiquinone followed by fragmentation of the peroxy adduct generates oxidized flavin and hydrogen peroxide, as shown in Fig. 6.13.

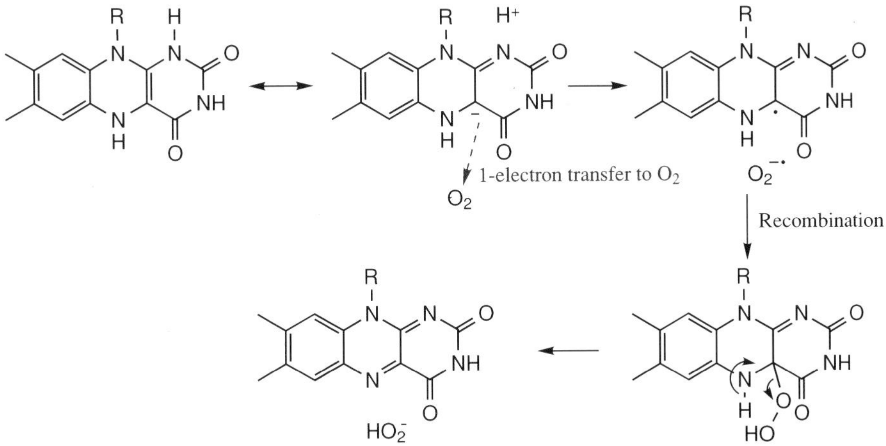

Fig. 6.13 Regeneration of oxidized flavin by molecular oxygen.

How does the flavin nucleus carry out the dehydrogenation reaction? There are many possible mechanisms that can be (and have been) written for this type of reaction with flavin, which fall into three categories:

1 hydride transfer;

2 nucleophilic attack by the substrate on the flavin nucleus;

3 radical mechanisms involving single electron transfers.

Although hydride transfer can occur from NAD(P)H onto flavin in many redox enzymes, most of the available evidence that will be discussed below suggests that single electron transfer is involved in substrate dehydrogenation. Nucleophilic attack at the C-4a position is well precedented in flavin model systems, but there is no firm evidence for the existence of covalent substrate–flavin adducts in flavin-dependent dehydrogenases. We shall examine the available evidence for the enzymes acyl coenzyme A (CoA) dehydrogenase and monoamine oxidase.

There is good evidence in the reaction of acyl CoA dehydrogenase that the first step of the mechanism involves deprotonation of the substrate adjacent to the thioester carbonyl, since the enzyme catalyses the exchange of a proton from the α-carbon with ^2H_2O. However, there is equally convincing evidence for an α-radical intermediate, since the enzyme is inactivated by a substrate analogue containing a β-cyclopropyl group. This inhibitor is a metabolite of hypoglycine A, which is the causative agent of Jamaican vomiting sickness. The proposed mechanism of inactivation is shown in Fig. 6.14. Upon formation of the α-radical a rapid opening of the strained cyclopropyl ring takes place, giving a re-arranged radical intermediate which re-combines with the flavin semiquinone irreversibly.

A possible mechanism for acyl CoA dehydrogenase consistent with both observations is shown in Fig. 6.15. Deprotonation to form an α-carbanion is followed by a one-electron transfer from the substrate to FAD, generating a

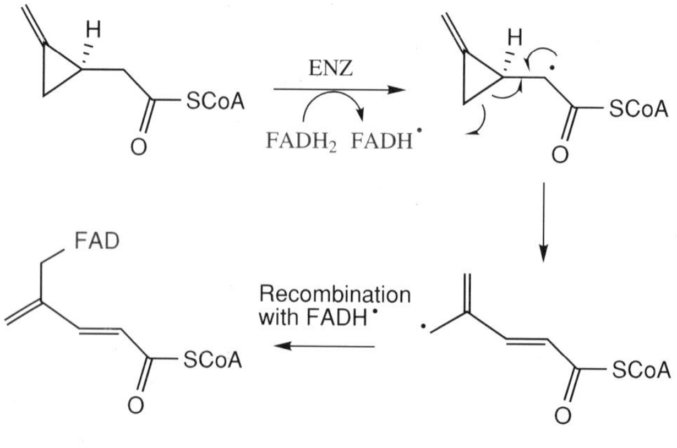

Fig. 6.14 Inactivation of acyl CoA dehydrogenase by hypoglycin A.

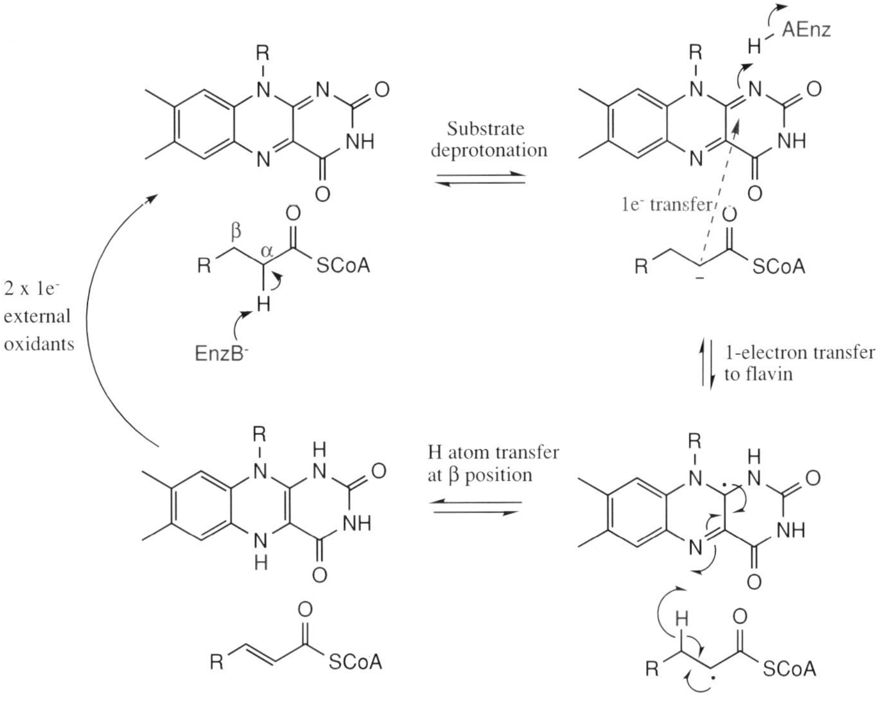

Fig. 6.15 Probable mechanism of acyl CoA dehydrogenase.

substrate radical and the flavin semiquinone. The reaction can be completed by abstraction of $H^{\bullet}$ at the β-position by the flavin radical, generating reduced $FADH_2$. In support of the final proposed step in the mechanism there are isolated examples of flavo-enzymes which catalyse hydrogen-atom transfer from $FADH_2$ to the β-position of α,β-unsaturated carbonyl substrates (see Problem 4, p. 132).

The monoamine oxidase reaction is a very important medicinal target for the mammalian nervous system, since it is responsible for the deamination of several neuro-active primary amines. Inhibitors of monoamine oxidase therefore have potential clinical applications as antidepressants. Monoamine oxidase is rapidly inactivated by *trans*-2-phenylcyclopropylamine (marketed as the antidepressant Tranylcypromine) via a radical mechanism, shown in Fig. 6.16. This behaviour indicates that an amine radical cation intermediate is formed in the mechanism via single electron transfer to FAD. This amine radical cation undergoes a further one-electron oxidation to give an iminium ion, which is then hydrolysed to give the corresponding aldehyde, as shown in Fig. 6.16. The second one-electron transfer could proceed either by hydrogen-atom transfer (i.e. loss of $H^{\bullet}$ from the α-position, route (a)) or by loss of a proton from the α-carbon followed by single electron transfer (route (b)). There is evidence from further 'radical trap' inhibitors

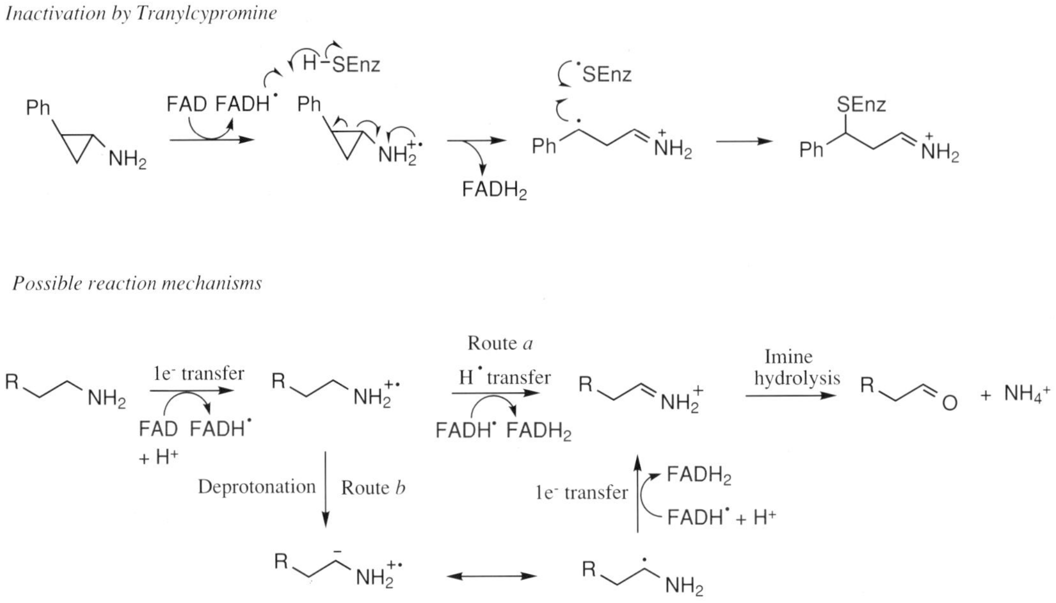

Inactivation by Tranylcypromine

Possible reaction mechanisms

Fig. 6.16 Monoamine oxidase inactivation and reaction mechanisms.

for the existence of an α-radical intermediate, consistent with route (b).

Thus, both the flavin-dependent dehydrogenases and oxidases appear to follow radical mechanisms involving single electron transfers to flavin. In the case of the amine oxidases, there is good chemical precedent for this type of mechanism in the form of electrochemical single electron oxidation of amines to the corresponding imines. Similar radical mechanisms can be written for succinate dehydrogenase and D-amino acid oxidase. Note that in acyl CoA dehydrogenase deprotonation occurs adjacent to a thioester carbonyl to form a stabilized enolate anion, whereas in succinate dehydrogenase deprotonation would occur adjacent to a carboxylate anion. As discussed in Chapter 7, Section 7.2, a proton adjacent to an ester or a carboxylic acid is much less acidic than a proton adjacent to a ketone or thioester. The mechanism by which this problem is resolved is not well understood, but may involve protonation of the carboxylate and/or electrostatic stabilization of the carbanion intermediate. This issue is also relevant to the cofactor-independent amino acid racemases which will be discussed in Chapter 10, Section 10.2.

6.4   Flavin-dependent mono-oxygenases

The ability of flavin to react with molecular oxygen, seen above with oxidases, also makes possible a number of *mono-oxygenase* reactions, in which one atom from molecular oxygen is incorporated into the product (see Table 6.1). The most common flavin-dependent mono-oxygenases are the phenolic hydroxylases, in which a hydroxyl group is inserted into the *ortho*- or *para*-position of a phenol (or aniline). As in the flavin-dependent oxidases, oxygen reacts with reduced $FADH_2$ to form a peroxy flavin adduct, which then acts as an electrophilic species for attack of the phenol. The mechanism for *p*-hydroxybenzoate hydroxylase is shown in Fig. 6.17.

In this case $FADH_2$ is generated from FAD by reduction with NADPH. Presumably the mechanism for this involves hydride transfer to either N-1 or N-5 of the flavin nucleus. Thus, flavin has the ability to accept two electrons from NADPH, but then transfer them via two one-electron transfers. It is apparent from this mechanism why hydroxylation occurs *ortho* or *para* to a phenol, because the phenolic hydroxyl group activates the *ortho*- and *para*-positions for attack.

There is also at least one example of an enzyme in which the flavin hydroperoxide intermediate can act as a nucleophile, rather than an electrophile. This is the enzyme cyclohexanone mono-oxygenase, which catalyses the oxidation of cyclohexanone to the corresponding seven-membered

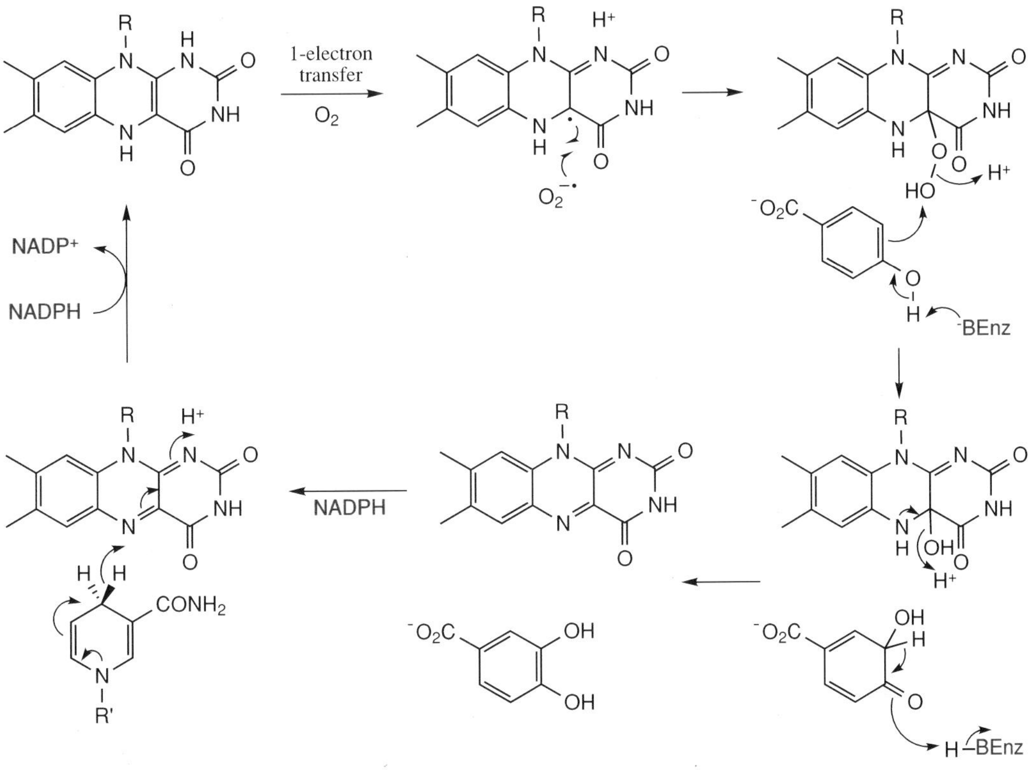

Fig. 6.17 Mechanism for *p*-hydroxybenzoate hydroxylase.

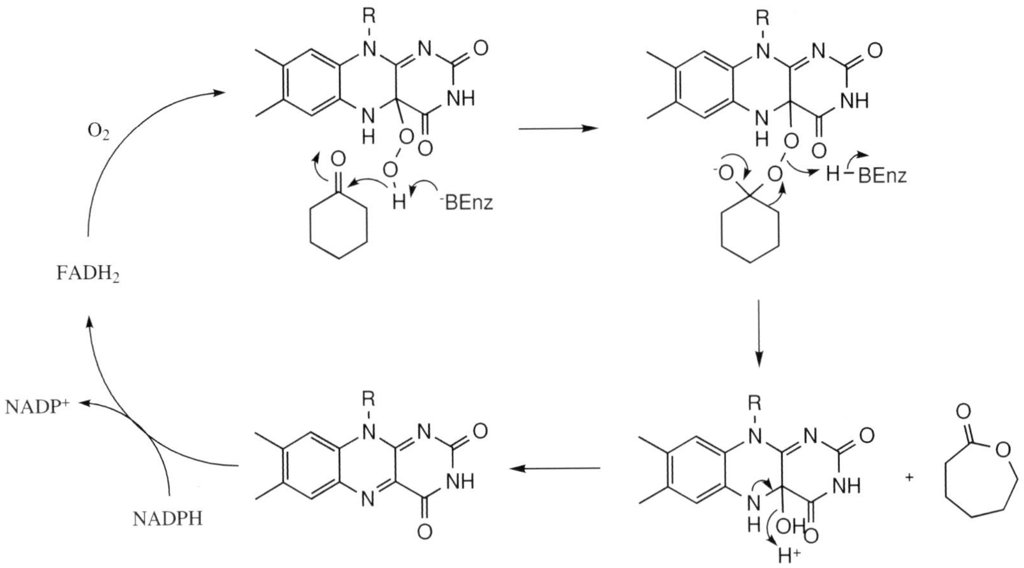

Fig. 6.18 Mechanism for cyclohexanone mono-oxygenase.

lactone. This reaction is analogous to the well-studied Baeyer–Villiger oxidation which uses a peracid to achieve the same transformation. A likely mechanism for this reaction is shown in Fig. 6.18. Again, NADPH is used to generate reduced flavin for reaction with oxygen.

This enzyme is used by *Pseudomonas* for the biodegradation of aliphatic hydrocarbons containing cyclohexane rings. The lactone product is subsequently hydrolysed and broken down into small fragments which can be utilized for growth.

6.5   Case study: glutathione and trypanothione reductases

Glutathione (GSH) is a tripeptide formed from glutamate (Glu), cysteine (Cys) and glycine (Gly) (L-Glu-γ-L-Cys-Gly) which is found at concentrations of about 1 mM in all mammalian cells. Its function is to protect the cell against 'oxidative stress', or the presence of activated oxygen species which would otherwise have harmful effects upon the cell. In the course of its reaction with these activated oxygen species, the thiol side chain of GSH is converted into oxidized disulphide (GSSG). Reduced GSH is re-generated in the cell by the enzyme glutathione reductase (Fig. 6.19), which uses NADPH as reducing equivalent and contains one equivalent of tightly bound FAD.

Glutathione reductase is a dimer of identical 50-kDa protein subunits. Each subunit contains FAD- and NADPH-binding domains which are

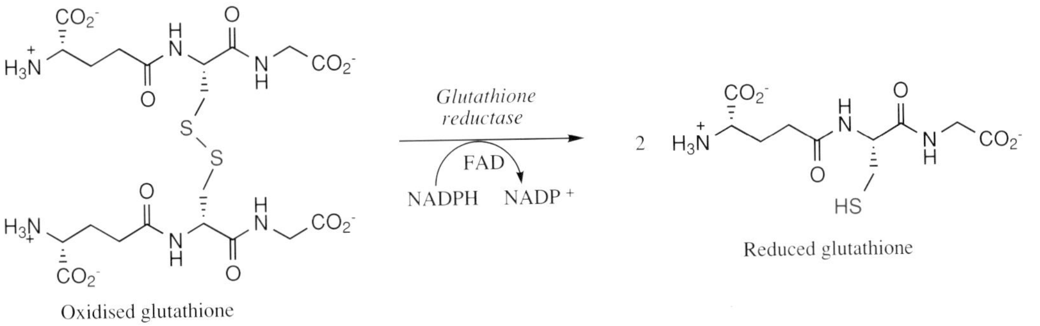

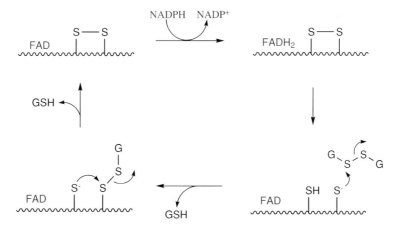

Fig. 6.19 Glutathione reductase reaction.

composed of a βαβαβ secondary structural motif common to many nucleotide-binding proteins. The active site of the enzyme lies at the interface of the two proteins subunits. On one face of the flavin cofactor is the NADPH cofactor responsible for generation of reduced $FADH_2$. On the other face of the flavin cofactor are two active site cysteine residues (Cys-46 and Cys-41) which form an active site disulphide at the start of the catalytic cycle.

The mechanism of the glutathione reductase catalytic cycle is shown in Fig. 6.20. NADPH reduces the bound flavin to $FADH_2$, which in turn reduces the active site disulphide into two reduced cysteine residues. Attack of a free cysteine thiol onto the disulphide linkage of oxidized GSH generates one equivalent of reduced GSH. Attack of the second free cysteine thiol generates a second equivalent of reduced GSH and re-generates the active site disulphide.

A high-resolution X-ray crystal structure has been determined for the human glutathione reductase, which has made possible extensive protein engineering studies. Close examination of the NADPH binding site reveals

Fig. 6.20 Mechanism of glutathione reductase.

that the 2-'phosphate of NADPH is bound by three positively charged residues: arginine (Arg)-218, histidine (His)-219 and Arg-224 (Plate 6.1b facing p. 152). Arg-218 and Arg-224 are strictly conserved in the amino acid sequences of other flavoprotein disulphide oxidoreductases which use NADPH. However, the amino acid sequence of NADH-specific lipoamide dehydrogenase contains methionine and proline, respectively, at these positions, which are incapable of forming electrostatic interactions. Might it be possible to change the specifity for NADPH by selectively mutating these residues?

Mutation of either of these two arginine residues in the *Escherichia coli* glutathione reductase enzyme (to methionine and leucine, respectively) gave mutant enzymes whose k_{cat}/K_M values for NADPH were reduced by approximately 100-fold, whilst a mutant enzyme containing both mutations had a 500-fold reduced k_{cat}/K_M. Mutation of four additional residues identified in the NADPH binding site to the corresponding residues in the NADH-specific lipoamide dehydrogenase gave a mutant enzyme with an eightfold preference for NADH over NADPH. The wild type enzyme in contrast has a 2000-fold preference for NADPH over NADH. This type of study shows that enzyme characteristics such as cofactor specificity can in principle be modified rationally using protein engineering.

A closely related enzyme trypanothione reductase (TR) has been found in *Trypanosoma* and *Leishmania* parasites which cause human diseases such as sleeping sickness and Chagas' disease. These parasites use a modified form of GSH called trypanothione in which the C-terminal glycine carboxylates are connected by a spermidine linker, as shown in Fig. 6.21. How is the selectivity for trypanothione achieved?

Examination of the amino acid sequence of the *T. congolense* enzyme revealed that three of the amino acid residues involved in substrate binding in human glutathione reductase are modified in the parasite enzyme: Arg-347 (found as alanine (Ala) in TR), Ala-34 (found as glutamine in TR) and Arg-37 (found as tryptophan in TR). Mutation of these three residues in the

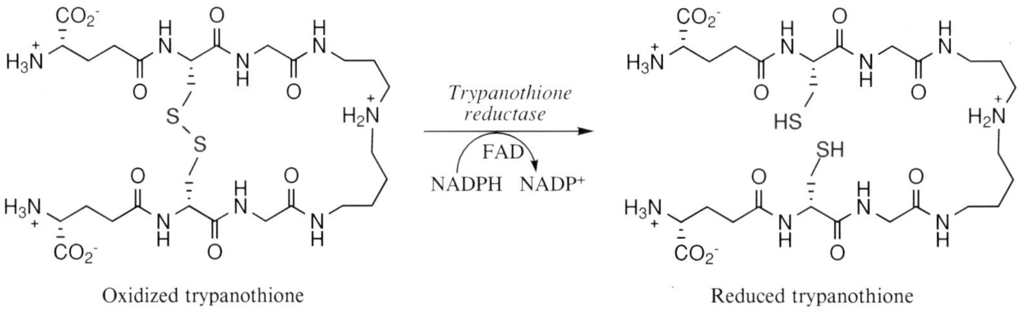

Fig. 6.21 Trypanothione reductase.

parasite TR to the corresponding residues in the human glutathione reductase gave a mutant enzyme with 10^3-fold lower TR activity and 10^4-fold higher glutathione reductase activity (Plate 6.1c, facing p. 152). The specificity of the parasite enzyme for trypanothione over GSH offers the potential for selective antiparasite activity via inhibition of TR.

6.6   Deazaflavins and pterins

Nicotinamide and riboflavin are by far the most common carbon-based redox cofactors used by enzymes, but there are several other heterocyclic redox cofactors used by particular enzymes.

Factor F_{420} was isolated in 1978 from methanogenic bacteria (strictly anaerobic bacteria which ferment acetate to methane and carbon dioxide) in yields of up to 100 mg kg^{-1} cells. It was found to have a structure similar to that of riboflavin, except that N-5 is replaced by a carbon atom. The discovery of this deazaflavin prompted an investigation into the properties of other deaza-analogues of riboflavin, which are shown in Fig. 6.22. Both

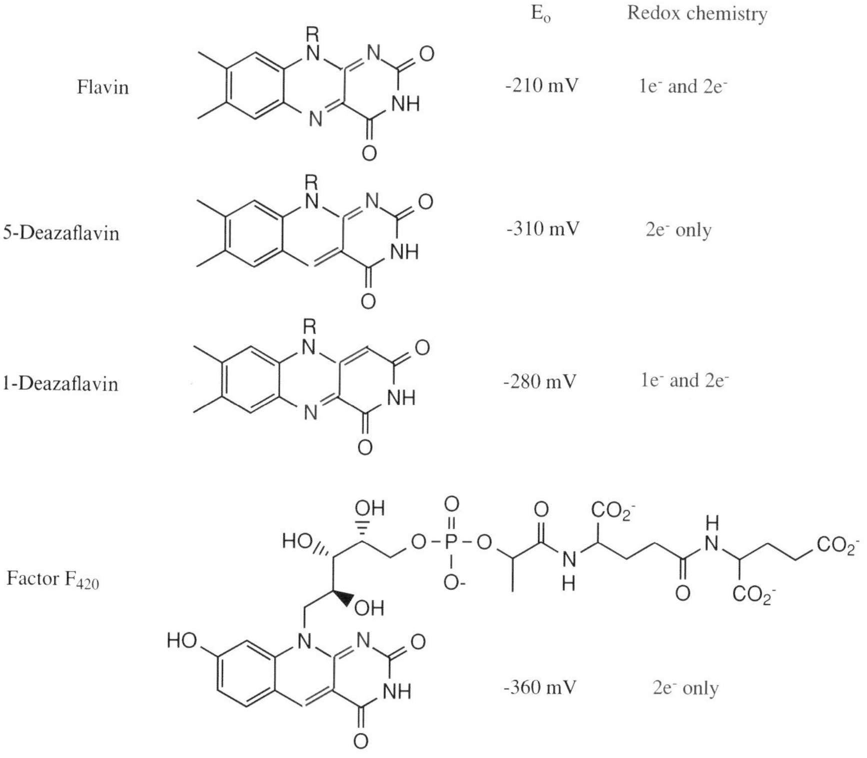

Fig. 6.22 Structures and redox potentials of natural and synthetic deazaflavins.

factor F_{420} and 5-deazaflavin have much lower redox potentials than ribo-flavin itself, but neither has a stable semiquinone form, so are capable of only two-electron transfer. In this respect their chemistry is more similar to nicotinamide than riboflavin. 1-Deazaflavin, on the other hand, behaves more like riboflavin in terms of its redox potential and its ability to carry out one-electron redox chemistry.

Factor F_{420} is used as a cofactor in a nickel (Ni^{2+})-dependent hydrog-enase enzyme found in methanogenic bacteria which uses hydrogen gas to reduce carbon dioxide to methane. Its role appears to be as one component of a complex chain of electron carriers in this multi-enzyme complex. The redox potential of F_{420} is ideally suited for its role in this enzyme, since at -0.36 V it is higher than the redox potential for hydrogen (-0.42 V), but lower than the redox potentials of other redox cofactors such as NADH and flavin. Therefore, F_{420} is able to accept electrons from hydrogen and transfer them to NAD^+ or FAD.

The pterin cofactor is used in a number of redox enzymes, in particular a small family of mono-oxygenase enzymes which hydroxylate aromatic rings. Its structure contains a two ring heterocyclic system resembling that of riboflavin. Phenylalanine hydroxylase catalyses the conversion of L-phenyl-alanine to L-tyrosine, using tetrahydropterin as a cofactor. The enzyme incorporates one atom of oxygen from dioxygen into the product, similar to the flavin-dependent mono-oxygenases. However, unlike the flavin-dependent mono-oxygenases, there is no hydroxyl group present in the ortho- or para-positions of the substrate.

Conversion of phenylalanine labelled with deuterium at C-4 of the ring by phenylalanine hydroxylase gives 3-^{2}H-tyrosine, indicating that a 1,2-shift (known historically as the 'NIH shift', since this startling result was discov-ered at the National Institute of Health research laboratories) is taking place during the reaction. It is likely that a pterin hydroperoxide is formed upon reaction with dioxygen, as found with flavin. The mammalian phenylalanine hydroxylase requires iron(II) for activity; however, the bacterial enzyme contains no iron cofactor, and to date there is no firm evidence that there is a difference in mechanism between the two varieties of enzyme. The NIH shift could result upon formation either of an epoxide intermediate or a carbonium ion intermediate, shown in Fig. 6.23. Although re-arrangement is observed with the [4-^{2}H]-substrate, there is no kinetic isotope effect ob-served with this substrate, implying that the re-arrangement occurs after the rate-determining step of the reaction.

6.7   Iron–sulphur clusters

We have seen in the case of flavin how single electron transfer is an

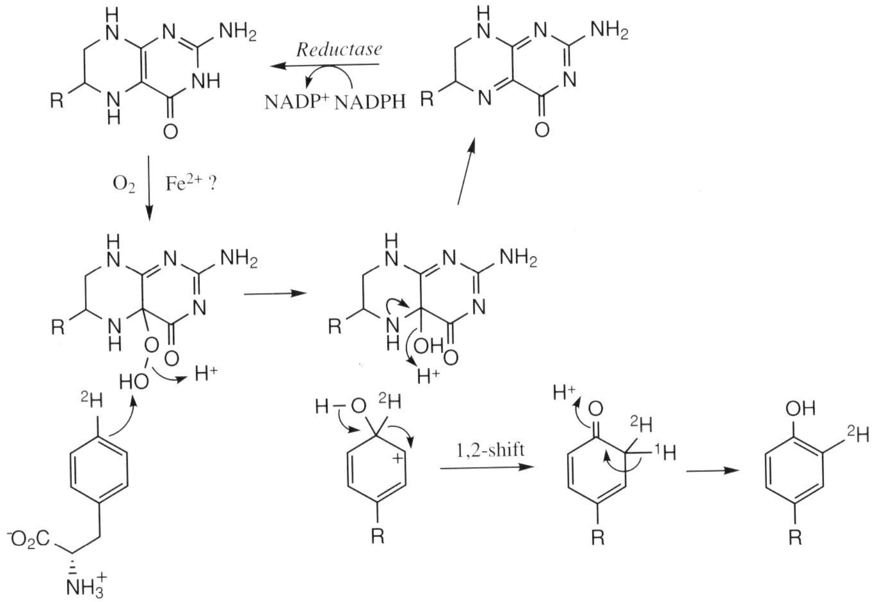

Fig. 6.23 Mechanism for phenylalanine hydroxylase reaction. R, $CH(OH)CH(OH)CH_3$.

important process in biological systems. The most common type of one-electron carrier found in biological systems is the family of iron–sulphur clusters. They are inorganic clusters of general formula $(FeS)_n$, where n is commonly 2 or 4. The Fe_2S_2 and Fe_4S_4 clusters shown in Fig. 6.24 are commonly found in biological electron carriers known as ferredoxins, and are also found in a number of redox enzymes. They have the ability to accept a single electron from a single electron donor such as flavin, and transfer the single electron to another electron carrier or to the active site of a redox enzyme. We shall meet one such example in Section 6.10. Redox potentials for ferredoxins are typically in the range -0.2 to -0.6 V (see Fig. 6.1), implying that these are strongly reducing biological redox agents.

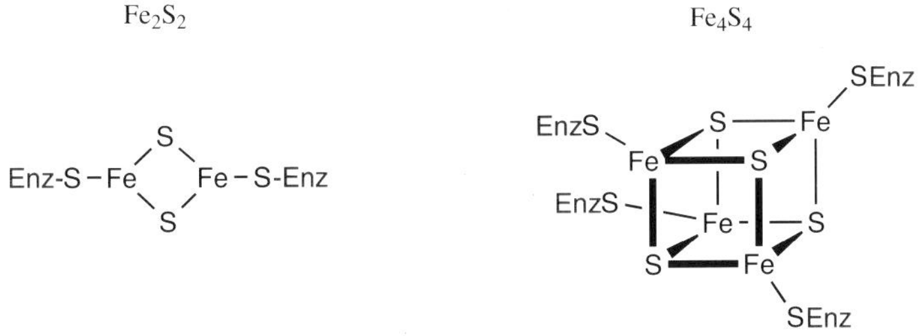

Fig. 6.24 Iron–sulphur clusters.

6.8   Metal-dependent mono-oxygenases

Enzymatic hydroxylation of organic molecules is a remarkable example of an enzymatic reactions for which there is little precedent in organic chemistry. We have seen above how flavin is able to utilize dioxygen to carry out the hydroxylation of a phenolic aromatic ring. However, more remarkably there is a class of mammalian enzymes which are able to catalyse the specific hydroxylation of unactivated alkanes. These enzymes are known as the P_{450} mono-oxygenases, due to the presence of a haem cofactor which upon treatment with carbon monoxide gives a characteristic absorbance at 450 nm. At the centre of the haem cofactor is an iron centre which in the resting enzyme is in the $+3$ oxidation state, but which is reduced to the $+2$ oxidation state upon substrate binding. This reduction is carried out by a reductase subunit of the enzyme which contains a flavin cofactor, itself reduced by NADPH.

The mechanism of hydroxylation by P_{450} enzymes shown in Fig. 6.25 is explained as follows. Just as reduced flavin is able to donate a single electron to dioxygen in the case of the flavin-dependent mono-oxygenases, so the reduced iron(II) is able to donate a single electron to dioxygen, forming iron(III)-superoxide. Transfer of a second electron from the flavin reductase subunit via the iron centre generates iron(III)-peroxide. Protonation of the peroxide generates a good leaving group for cleavage of the O–O bond, giving an iron(V)-oxo species. This species is thought to carry out substrate hydroxylation by abstraction of H˙ from the substrate, forming a substrate radical species and iron(IV)-hydroxide. Homolytic cleavage of the Fe–O bond and transfer of HO˙ to the substrate radical gives the hydroxylated

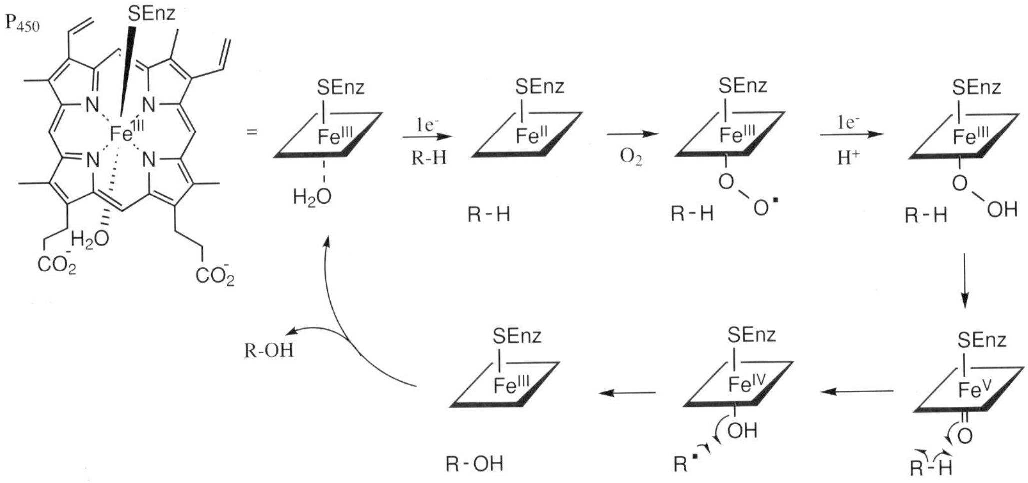

Fig. 6.25 Mechanism of P_{450}-dependent hydroxylation.

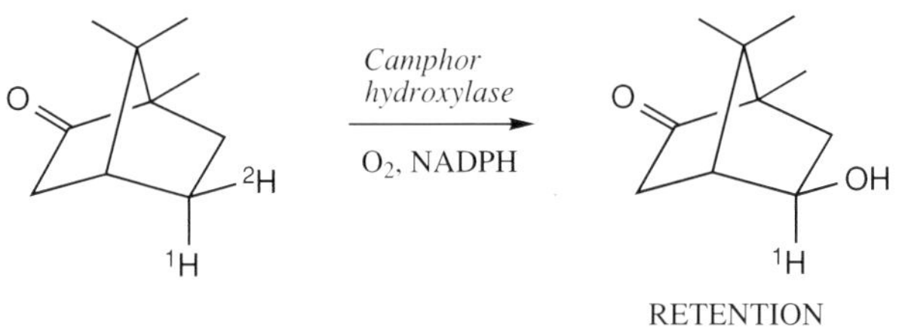

Fig. 6.26 Stereochemistry of camphor hydroxylase.

product and regenerates iron(III).

The best characterized enzyme of this class is camphor hydroxylase, which catalyses the stereospecific hydroxylation of camphor. This reaction has been shown by deuterium labelling to proceed with retention of stereo-chemistry at the position of hydroxylation, as shown in Fig. 6.26. Retention of stereochemistry is generally found in other P_{450}-dependent hydroxylases.

P_{450}-dependent enzymes also catalyse other types of oxidative reactions: one common example is demethylation. This reaction can be rationalized, as shown in Fig. 6.27, by hydroxylation of the terminal methyl group, generating a labile hemi-acetal which breaks down, liberating the free alcohol and formaldehyde.

Other P_{450}-dependent enzymes utilize hydrogen peroxide instead of dioxygen to access the iron(III)-peroxide intermediate directly—these enzymes are known as peroxidases. An important function of peroxidases in plants, where they are widely found, is to initiate lignin formation by abstraction of H˙, as will be discussed in Chapter 7, Section 7.4. Finally, the chloroperoxidases are a family of P_{450}-independent enzymes found in marine organisms which catalyse the formation of C–Cl bonds in organic molecules. These enzymes also use hydrogen peroxide to access the iron(III)-peroxide intermediate, which reacts with a chloride ion to generate an electrophilic species (i.e. a 'Cl^{+}' equivalent). Formation of a C–Cl bond with a phenol substrate is shown in Fig. 6.28.

Iron is the only metal utilized in P_{450}-dependent mono-oxygenases. However, metal-dependent mono-oxygenases are found which involve other redox-active metal ions such as copper, which can access the +1 and +2 oxidation states, and hence also has the ability to interact with molecular oxygen.

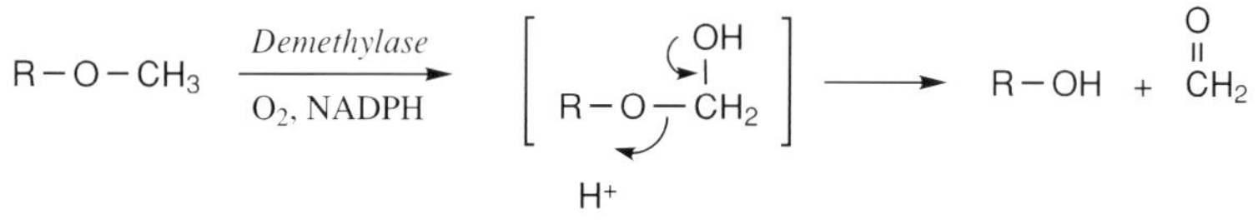

Fig. 6.27 P_{450}-dependent demethylation reactions.

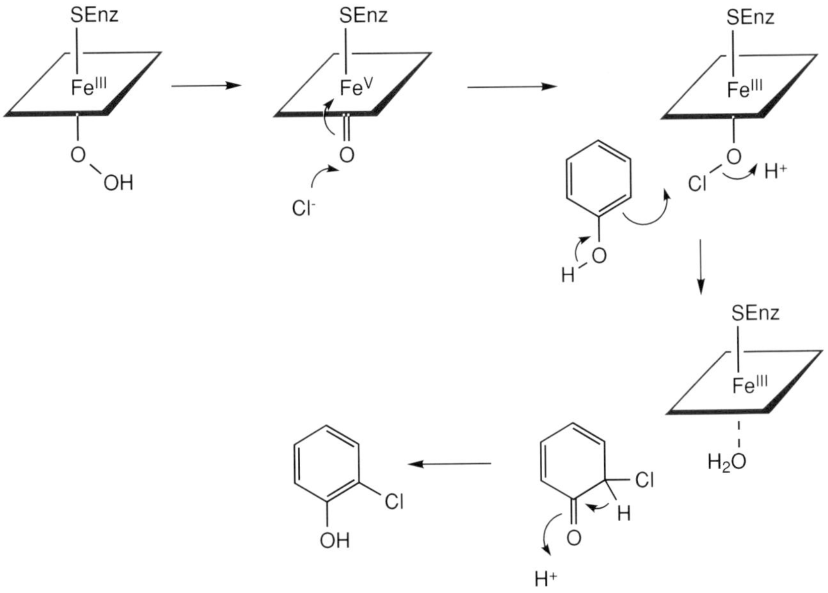

Fig. 6.28 Mechanism for a chloroperoxidase reaction, with phenol.

6.9 α-Ketoglutarate-dependent dioxygenases

Dioxygenases are enzymes which incorporate both atoms of dioxygen into the product(s) of their enzymatic reactions. The first class of dioxygenases that we shall consider catalyse hydroxylation reactions, which at first glance seem similar to the P_{450}-dependent mono-oxygenases. However, there are two differences:

1 these enzymes use α-ketoglutarate as a cosubstrate, which they convert into succinate and carbon dioxide;

2 they use non-haem iron rather than the P_{450} cofactor.

As in the haem enzymes, the non-haem iron(II) can donate one electron to oxygen, forming iron(III)-superoxide which can react with the carbonyl group of α-ketoglutarate, forming a cycle peroxy intermediate. Decarboxylation of this intermediate releases succinate and produces an iron(IV)-oxo species. This species can carry out the hydroxylation reaction via a radical mechanism, shown in Fig. 6.29, which is similar to the P_{450} hydroxylase mechanism.

An important member of this family of enzymes is prolyl hydroxylase, which catalyses the hydroxylation of proline amino acid residues in developing collagen fibres to 4-hydroxyproline. Collagen is a very important structural protein found in skin, teeth, nail and hair, and the presence of hydroxyproline is necessary for maintenance of the tertiary structure of the

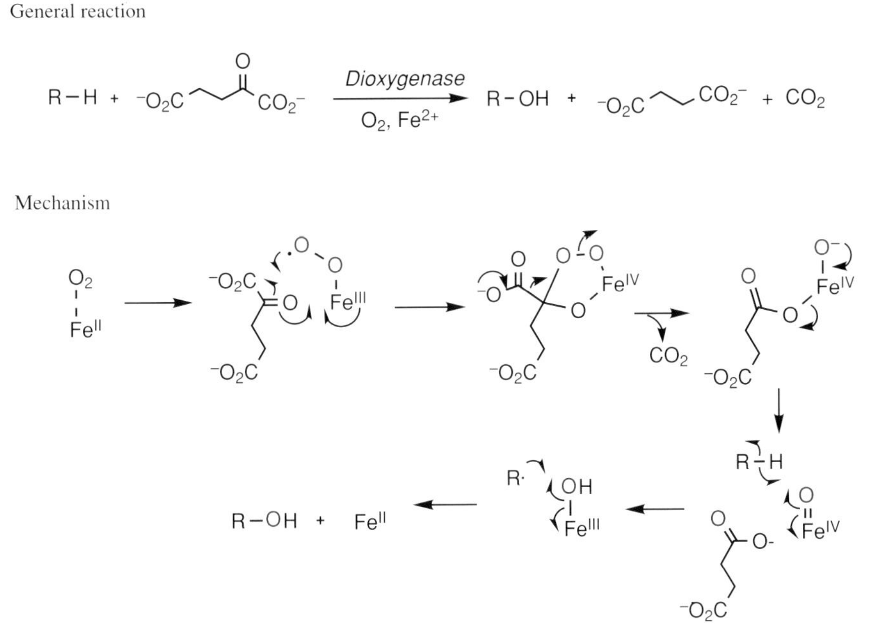

General reaction

Mechanism

Fig. 6.29 α-Keto-glutarate-dependent Fe(II) dioxygenases.

protein. This enzyme also uses ascorbic acid—vitamin C—as a cofactor, and this is one of the major functions of vitamin C in the body. The overall reaction apparently does not require any further redox equivalents, but ascorbic acid is required in non-stoichiometric amounts for full activity. The probable role of ascorbic acid is to maintain the iron cofactor in the reduced iron(II) state in cases where the catalytic cycle is not completed. For example, if superoxide is released from the enzyme active site an inactive iron(III) form of the enzyme is generated. Ascorbic acid has the ability to

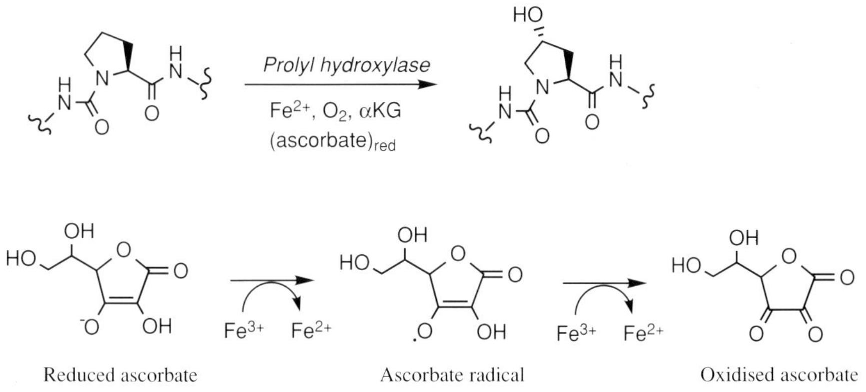

Reduced ascorbate   Ascorbate radical   Oxidised ascorbate

Fig. 6.30 Prolyl hydroxylase and the role of ascorbate as a reducing agent. αKG, α-ketoglutarate.

reduce iron(III) to iron(II), generating a stable ascorbate radical, shown in Fig. 6.30. Thus, in the absence of vitamin C the body's machinery for making collagen is impaired, leading to the symptoms of the dietary deficiency disease scurvy.

6.10   Non-haem iron-dependent dioxygenases

There are two further classes of non-haem iron-dependent dioxygenase enzymes that we shall mention: the catechol dioxygenases and the aromatic dihydroxylating dioxygenases. Both these classes of enzyme are involved in the bacterial degradation of aromatic compounds in the environment. As shown in Fig. 6.31, the dihydroxylating dioxygenases catalyse the oxidative conversion of an aromatic substrate into the corresponding *cis*-dihydrodiol. The *cis*-diol is then oxidized by a class of NAD^+-dependent dehydrogenases into the corresponding aromatic diol, or catechol. Oxidative cleavage of the aromatic ring is then carried out in one of two ways. Cleavage of the C–C bond between the two hydroxyl groups and insertion of both atoms of dioxygen is catalysed by non-haem iron(III)-dependent intradiol dioxygenases. Alternatively, cleavage of a C–C bond adjacent to the two hydroxyl groups and insertion of two oxygen atoms can be catalysed by non-haem iron(II)-dependent extradiol dioxygenases.

The dihydroxylating dioxygenases are multi-component enzymes consisting of three subunits. Two electrons are transferred by a flavoprotein reductase subunit from NADH to $FADH_2$. The two electrons are then transferred singly to iron–sulphur clusters contained in a ferredoxin subunit. These electrons are then transferred to the active site of the dioxygenase subunit, as shown in Fig. 6.32.

The mechanisms of these non-haem iron-dependent dioxygenases are not well understood, but all presumably involve activation of dioxygen by electron transfer from the iron cofactor. In case of the dihydroxylating

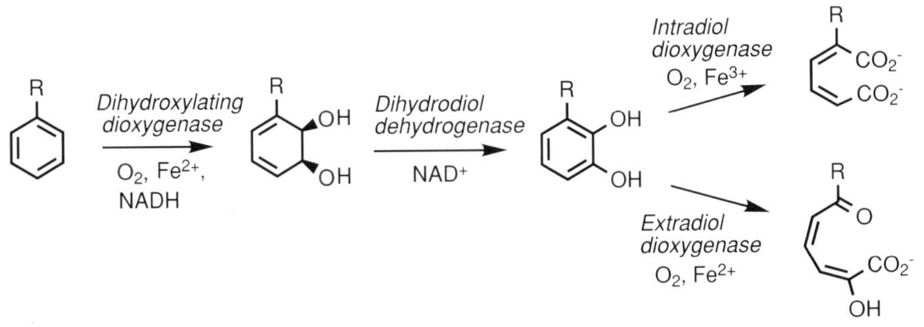

Fig. 6.31 Aromatic hydroxylation/cleavage reactions.

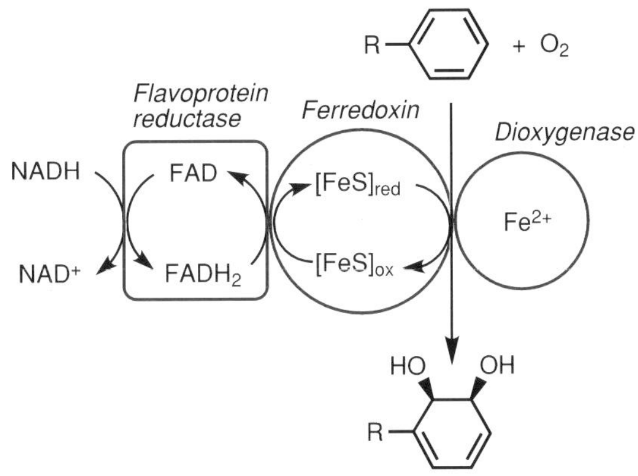

Fig. 6.32 Electron transfer in dihydroxylating dioxygenases.

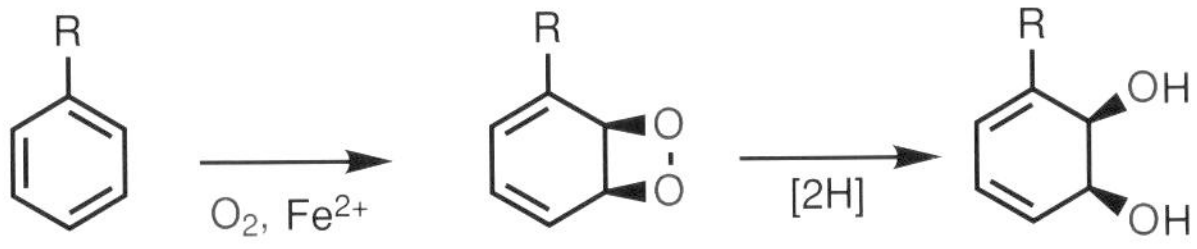

Fig. 6.33 Dioxetane mechanism for dihydroxylating dioxygenases.

dioxygenases, formation of a four-membered dioxetane intermediate has been proposed, followed by reduction to generate the *cis*-diol product, as shown in Fig. 6.33. This mechanism would explain the *cis*-stereochemistry of the product and the incorporation of ^{18}O into both hydroxyl groups from $^{18}O_2$. However, the formation of dioxetane intermediates from ground state oxygen is endothermic and, therefore, apparently unlikely. No evidence for any alternative mechanism has been found.

Dioxetane intermediates were also proposed for the intradiol and extradiol catechol dioxygenases, based on $^{18}O_2$ incorporation experiments. However, recent evidence suggests that 1,2-re-arrangements (similar to the

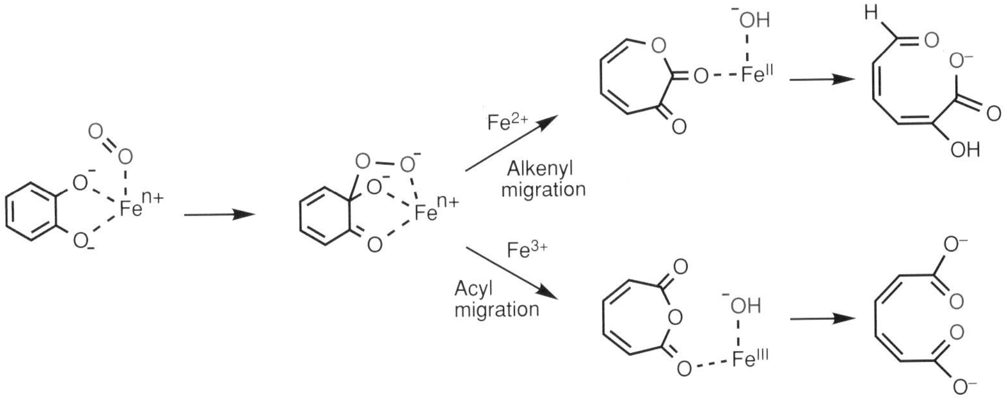

Fig. 6.34 Peroxide rearrangements proposed for catechol dioxygenases.

Baeyer–Villiger oxidation of ketones) of peroxy intermediates are taking place in these reactions, leading to anhydride and lactone intermediates, respectively, as shown in Fig. 6.34.

Many of the redox reactions in the second half of this chapter have little precedent in organic chemistry, although in some cases inorganic model complexes have been prepared which mimic the action of metallo-enzymes. Thus, utilizing only a small selection of organic redox coenzymes and metal redox cofactors as electron carriers, an extraordinary range of enzyme-catalysed oxidation/reduction chemistry is possible.

Problems

1 Work out the redox potential differences for the following enzymatic reactions, using the data in Fig. 6.1. In example (c) what can you deduce about the redox potential of the enzyme-bound flavin?

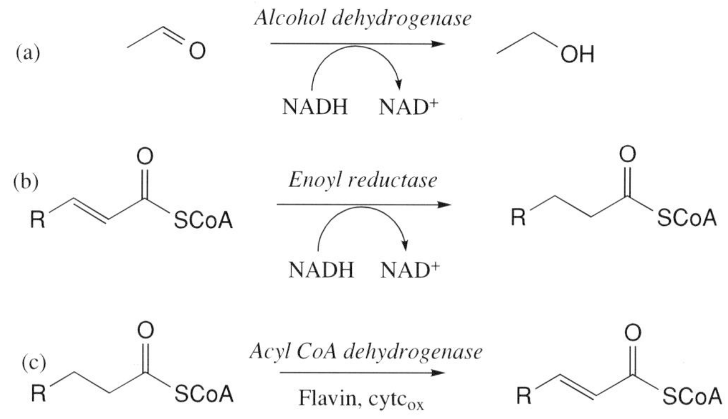

2 In the enoyl reductase reaction illustrated in Problem 1(b), incubation of $4S$-^{2}H-NADD with enzyme and crotonyl CoA gives no incorporation of deuterium in the butyryl CoA product. Incubation of $4R$-^{2}H-NADPD with enzyme and crotonyl CoA gives $3R$-^{2}H-butyryl CoA. Incubation of enzyme, NADH and crotonyl CoA in ^{2}H$_2$O gives $2S$-^{2}H-butyryl CoA. Explain these results, and write a mechanism for the enzyme reaction.

3 The enzyme which catalyses the conversion below has been purified and requires a *catalytic* amount of NAD$^+$ for activity. Given that H* is transferred intact as shown, suggest a mechanism for the enzymatic reaction.

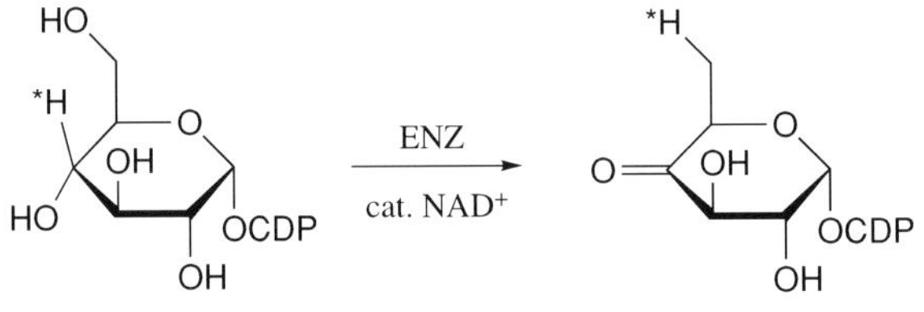

4 The reductase enzyme which catalyses the reaction shown below contains a stoichiometric amount of tightly bound FAD, which is reducible during catalytic turnover, and utilizes NADPH as a reducing equivalent. Incubation of $4S$-^{2}H-NADPD with enzyme and substrate gives product containing one atom of deuterium in the β position. Incubation with unlabelled NADPH in ^2H_2O gives product containing one atom of deuterium in the α position. Suggest a mechanism for the enzymatic reaction consistent with these data.

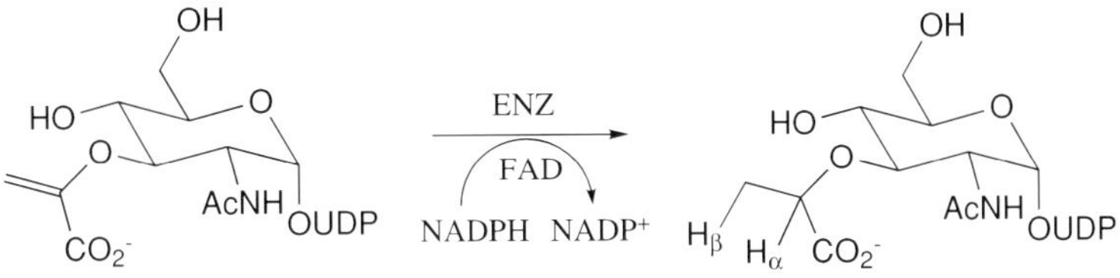

5 A flavo-enzyme has been purified which catalyses the conversion of p-nitrophenol into hydroquinone and nitrite. The enzyme contains tightly bound FAD, and each catalytic cycle consumes one equivalent of dioxygen and *two* equivalents of NADH. Suggest a mechanism.

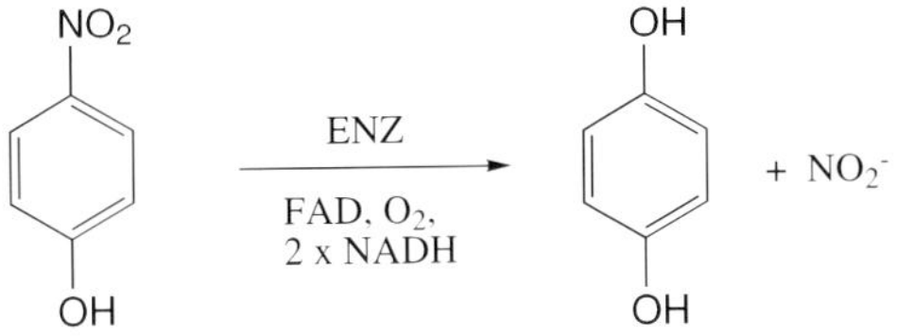

6 Thymine hydroxylase is an α-ketoglutarate (α-KG)-dependent iron(II) dioxygenase which catalyses three successive oxidations of thymine, as shown below.
(a) Write a mechanism for the first oxidation.
(b) 5-Ethynyluracil is a potent inhibitor of this enzyme. Inactivation results in covalent modification of the enzyme with a stoichiometry of one adduct/enzyme subunit, and also generates a byproduct 5-carboxyuracil. Suggest a possible mechanism of inactivation which would account for these observations.

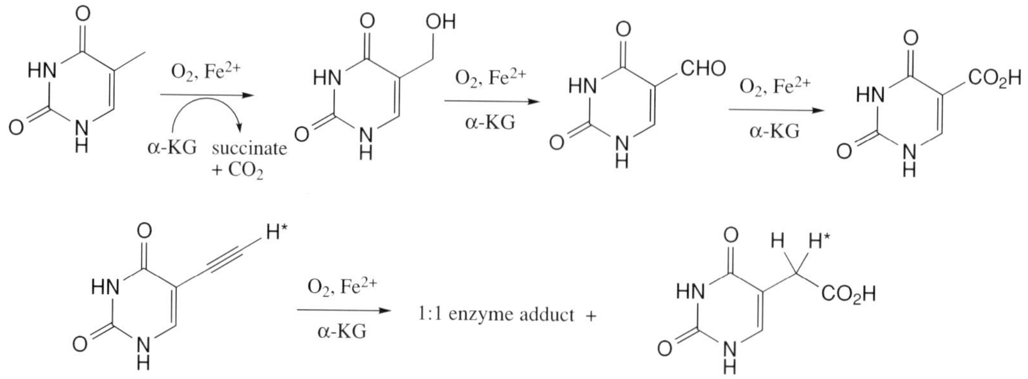

O_2, Fe^{2+} α-KG   succinate + CO_2

O_2, Fe^{2+} α-KG

O_2, Fe^{2+} α-KG

O_2, Fe^{2+} α-KG   1:1 enzyme adduct +

H*

Further reading

General

Abeles, R.H., Frey, P.A. & Jencks, W.P. (1992) *Biochemistry.* Jones & Bartlett, Boston.
Walsh, C.T. (1979) *Enzymatic Reaction Mechanisms.* Freeman, San Francisco.
Wong, G.H. & Whitesides, G.M. (1994) *Applications for Organic Synthesis: Enzymes in Synthetic Organic Synthesis.* Pergamon, Oxford.

NAD-dependent dehydrogenases

Everse, J. & Kaplan, N.O. (1973) *Adv Enzymol*, **37**, 61–134.
Westheimer, F.H., Fisher, H.F., Conn, E.E. & Vennesland, B. (1951) *J Am Chem Soc*, **73**, 2403.
Loewus, F.A., Westheimer, F.H. & Vennesland, B. (1953) *J Am Chem Soc*, **75**, 5018–23.

NADH models

Almarsson, O., Karaman, R. & Bruice, T.C. (1992) *J Am Chem Soc*, **114**, 8702–4.
Rob, F., van Ramesdonk, H.J., von Gerresheim, W., Bosma, P., Scheele, J.J. & Verhoeven, J.W. (1984) *J Am Chem Soc*, **106**, 3826–32.

Flavin-dependent enzymes

Bruice, T. (1980) *Acc Chem Res*, **13**, 256–62.
Walsh, C. (1980) *Acc Chem Res*, **13**, 148–55.

Acyl CoA dehydrogenase

Ghisla, S., Thorpe, C. & Massey, V. (1984) *Biochemistry*, **23**, 3154–61.
Lai, M., Li, D., Oh, E. & Liu, H.W. (1993) *J Am Chem Soc*, **115**, 1619–28.

Monoamine oxidase

Silverman, R.B. (1995) *Acc Chem Res*, **28**, 335–42.

Mono-oxygenases

Massey, V. (1994) *J Biol Chem*, **269**, 22 459–62.

Deazaflavins

Walsh, C.T. (1986) *Acc Chem Res*, **19**, 216–21.

Glutathione/trypanothione reductase

Schulz, G.E., Schirmer, R.H. Sachsenheimer, W. & Pai, E.F. (1978) *Nature*, **273**, 120–4.
Scrutton, N.S., Berry, A. & Perham, R.N. (1990) *Nature*, **343**, 38–43.
Sullivan, F.X., Sobolov, S.B., Bradley, M. & Walsh, C.T. (1991) *Biochemistry*, **30**, 2761–7.

Pterin-dependent mono-oxygenases

Benkovic, S.J. (1980) *Annu Rev Biochem*, **49**, 227–52.
Carr, R.T., Balasubramanian, S., Hawkins, P.C.D. & Benkovic, S.J. (1995) *Biochemistry*, **34**, 7525–32.

Iron–sulphur clusters

Holm, R.H., Ciurli, S. & Weigel, J.A. (1990) *Prog Inorg Chem*, **38**, 1–74.
Sweeney, W.V. & Rabinowitz, J.C. (1980) *Annu Rev Biochem*, **49**, 139–62.

Metal-dependent oxygenases

Akhtar, M. & Wright, J.N. (1991) P_{450} enzymes. *Nat Prod Reports*, **8**, 527–52.
Bertini, I., Gray, H.B., Lippard, S.J. & Valentine, J.S. (1994) *Bio-inorganic Chemistry*. University Science Books, Mill Valley, California.
Feig, A.L. & Lippard, S.J. (1994) Non-heme iron (II)-dependent oxygenases. *Chem Rev*, **94**, 759–805.

Prolyl hydroxylase

Cardinal, G.J. & Udenfriend, S. (1974) Prolyl hydroxylase. *Adv Enzymol*, **41**, 245–300.
Kivirikko, K.I., Myllyla, R. & Pihlajaniemi, T. (1989) *FASEB J*, **3**, 1609–17.

Catechol dioxygenases

Que, L. Jr (1985) Catechol dioxygenases. *J Chem Ed*, **62**, 938–43.
Sanvoisin, J., Langley, G.J. & Bugg, T.D.H. (1995) *J Am Chem Soc*, **117**, 7836–7.

7   Enzymatic Carbon–Carbon
Bond Formation

7.1   Introduction

The formation of carbon–carbon bonds is central to *biosynthesis*, which is the assembly of carbon-based compounds within living cells. Most of these compounds are *primary metabolites*—molecules such as amino acids, carbohydrates and nucleic acids which are necessary to support life. Many organisms also produce *secondary metabolites*—molecules whose presence is not essential for the survival of the cell, but which often have other biological properties such as defence against micro-organisms or communication with other organisms. In this chapter we shall analyse the types of enzymatic reactions used in the assembly of the carbon skeletons of primary and secondary metabolites, and also carbon–carbon cleavage reactions involved in their degradation.

A carbon–carbon bond consists of a pair of electrons contained within a filled molecular orbital. Formation of a carbon–carbon bond can be achieved either by donation of a pair of electrons from one carbon atom to an empty orbital on another carbon atom, or by the combination of two single electron species. Examples of these processes are illustrated in Fig. 7.1, involving:

1 nucleophilic attack of a carbanion onto an electron-accepting carbonyl group;
2 electrophilic attack of a carbocation onto an electron-rich alkene;
3 recombination of two phenoxy radicals.

We shall meet examples of each of these types of carbon–carbon formation reactions in this chapter.

One general point to note is that carbanions and carbocations are high-energy species which can usually only be generated under strenuous reaction conditions in organic chemistry. How then are they generated by enzymes which work at neutral pH with relatively weak acidic and basic groups? The answer is that carbanion and carbocation intermediates in enzyme-catalysed reactions must be highly stabilized, either by neighbouring groups in the substrate molecule, or by the enzyme active site, using the type of enzyme–substrate interactions mentioned in Chapter 2, Section 2.7. This stabilization will be explained where possible in the following examples, however in some cases the means by which high-energy intermediates are stabilized is not fully understood.

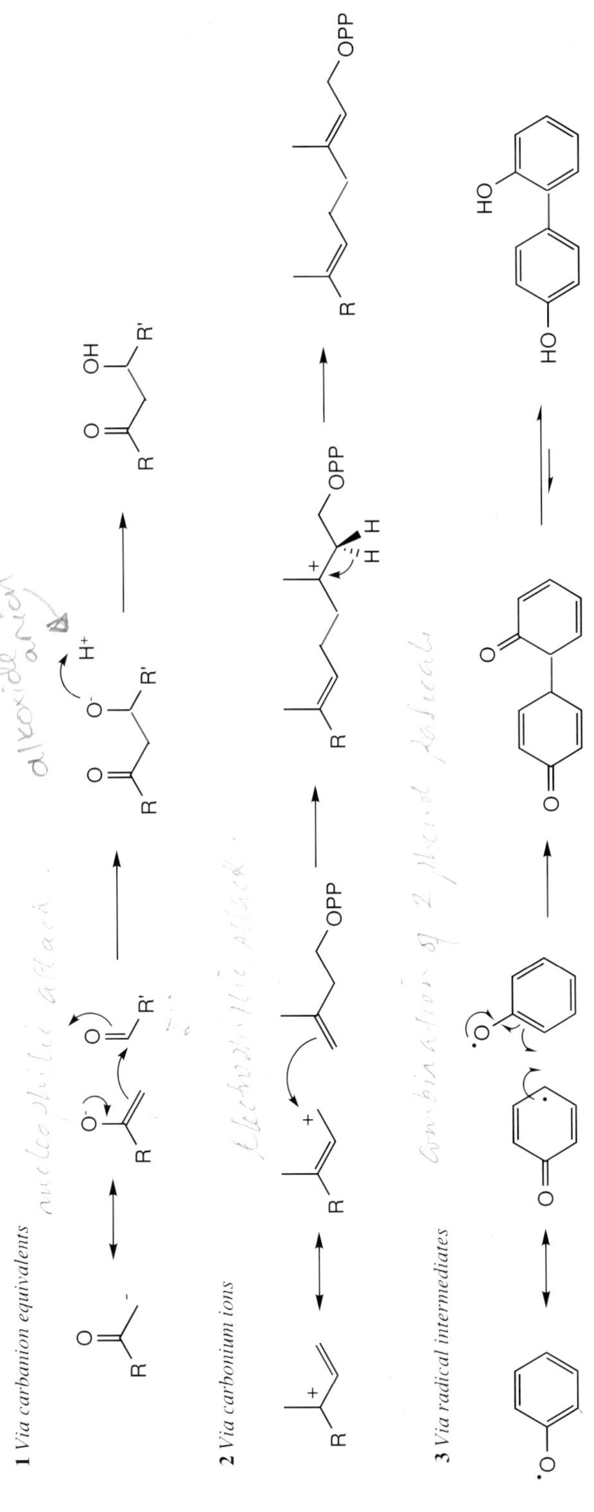

Fig. 7.1 Formation of carbon–carbon bonds.

7.2   Carbon–carbon bond formation via carbanion equivalents

Aldolases

The aldol reaction involves the condensation of two carbonyl compounds via an enolate intermediate. The reaction is illustrated in Fig. 7.2 for the case of the self-condensation of acetone in alkaline aqueous solution. A wide range·of aldol reactions occur in biological systems which are used for the formation of carbon–carbon bonds. Cleavage of carbon–carbon bonds by the reverse reaction is also found.

The enzymes which catalyse these aldol reactions are known as aldolases, and they are divided into two families based on their mechanism of action. The class I aldolases function by formation of an imine linkage between one carbonyl reagent and the ε-amino group of an active site lysine residue, followed by deprotonation of the adjacent carbon to generate an enamine intermediate. Enamines are well-known enolate equivalents in synthetic organic chemistry: they can be formed under mild conditions by condensation of the carbonyl compound with a primary or secondary amine, and they react with carbonyl compounds also under mild conditions.

The class II aldolases do not proceed through enamine intermediates but instead use a metal ion to assist catalysis. The metal ion is usually a divalent metal ion such as magnesium (Mg^{2+}), manganese (Mn^{2+}) or zinc (Zn^{2+}). The two classes are exemplified by the enzyme fructose-1,6-bisphosphate aldolase, which is found in mammals as a class I enzyme, and in bacteria as a class II enzyme.

Case study: fructose-1,6-bisphosphate aldolase

Fructose-1,6-bisphosphate aldolase catalyses the reversible reaction of di-hydroxyacetone phosphate (DHAP) with glyceraldehyde-3-phosphate (G3P) to give fructose-1,6-bisphosphate, shown in Fig. 7.3.

The reaction catalysed by the class I enzyme from rabbit muscle has been shown to proceed via formation of an imine linkage between the ε-amine group of an active site lysine residue and the C-2 carbonyl of DHAP (Fig. 7.4). The imine linkage can be reduced by sodium borohydride to give

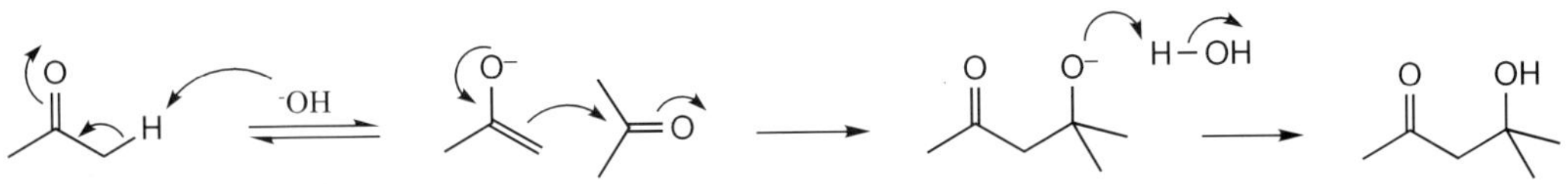

Fig. 7.2 Aldol reaction.

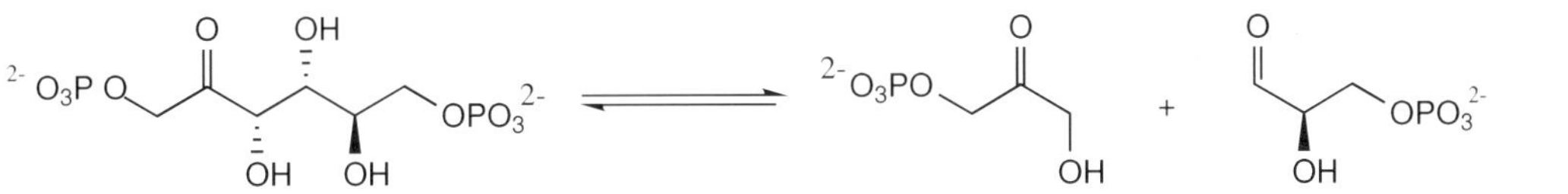

Fig. 7.3 Fructose-1,6-bisphosphate aldolase-catalysed reaction.

Fructose 1,6-bisphosphate (FBP)   Dihydroxyacetone-1-P (DHAP)   Glyceraldehyde-3-P (G3P)

an irreversibly inactivated secondary amine. Incubation of enzyme with radiolabelled G3P and sodium borohydride followed by proteolytic digestion of the inactivated enzyme has identified the imine-forming active site residue as lysine (Lys)-229.

Incubation of enzyme and DHAP in ^2H_2O (in the absence of G3P) leads to stereospecific ^2H exchange at C-3 to give $3S$-[3-^2H]-DHAP. This indicates that the *proS*-proton at C-3 is specifically removed following imine formation, giving an enamine intermediate. In order to generate the observed stereochemistry at C-3 and C-4 of the product the enamine intermediate must react from its *si*-face (see Appendix 1) with the *si*-face of the aldehyde group of G3P. The resulting imine intermediate is hydrolysed to yield the product fructose-1,6-bisphosphate.

An X-ray crystal structure was determined for rabbit muscle fructose-1,6-bisphosphate aldolase in 1987. The tertiary structure of the protein subunit is an αβ-barrel consisting of nine parallel β-sheets interconnected by α-helices. The active site lies in the centre of the barrel formed by the β-sheets, as shown in Plate 7.1 (facing p. 152). Close to the imine-forming Lys-229 in the centre of the active site cavity lies the amino group of another lysine residue, Lys-146. Mutation of Lys-146 to alanine or glutamine gives mutant enzymes which are more than 10^6-fold less active, whilst a histidine mutant enzyme is only 2000-fold less active than the wild type enzyme, suggesting that Lys-146 fulfils an acid–base function. A likely possibility is that the close proximity of Lys-146 lowers the pK_a of the ε-amino group of Lys-229, favouring the deprotonated form of Lys-229 which is required for nucleophilic attack on the substrate. Lys-146 could then protonate the carbinolamine intermediate involved in imine formation, and perhaps also act as the base for enamine formation. The active site cavity and the two catalytic lysine residues are highlighted in Plate 7.1(a).

The fructose-1,6-bisphosphate aldolase enzyme from yeast is a class II enzyme dependent upon Zn^{2+} for activity. The Zn^{2+} cofactor can be removed and replaced with other divalent metal ions such as Mn^{2+}, retaining catalytic activity. Nuclear magnetic resonance (NMR) spectroscopic studies on the Mn^{2+}-containing enzyme have revealed that the Mn^{2+} ion is situated 7.6 Å away from the carbonyl group of DHAP, too far away

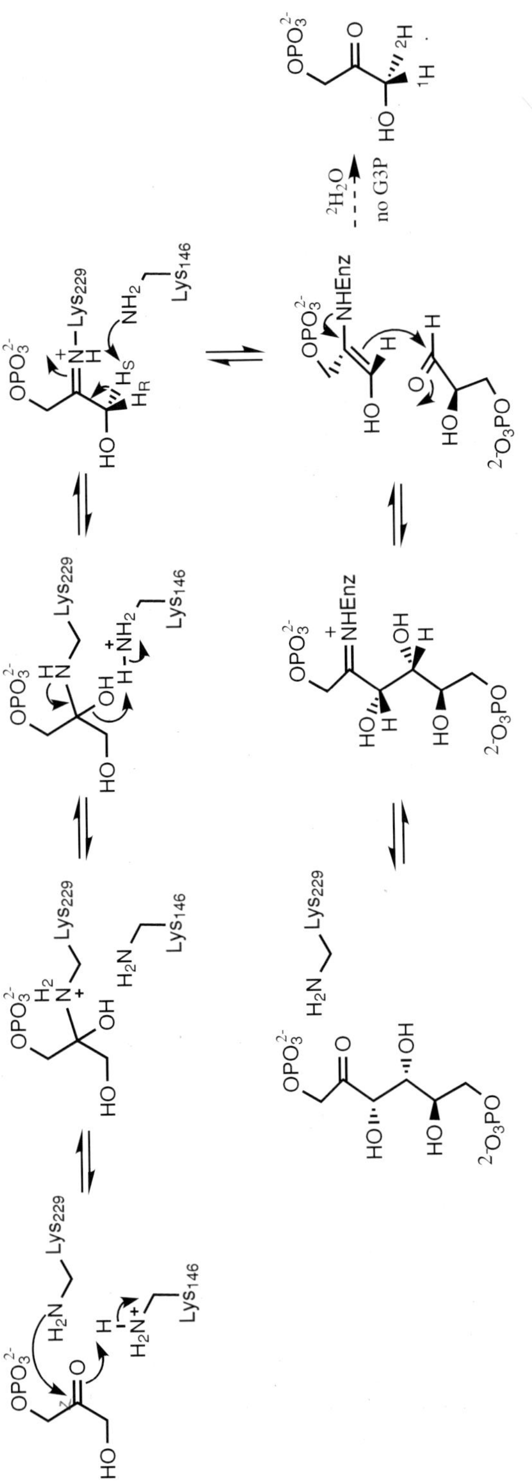

Fig. 7.4 Mechanism of rabbit muscle class I fructose-1,6-bisphosphate aldolase. Lys, lysine.

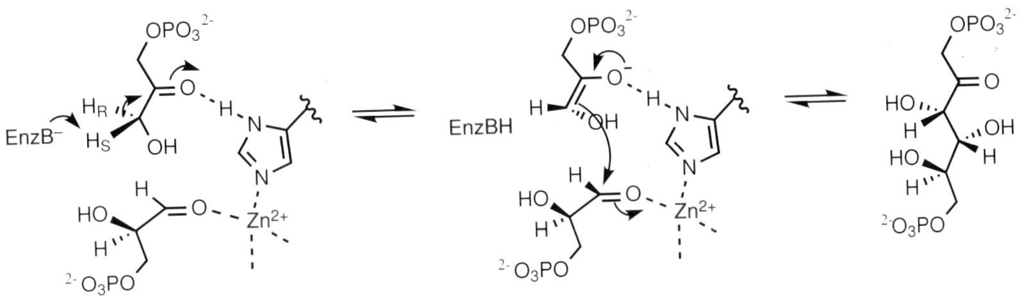

Fig. 7.5 Mechanism for yeast class II fructose-1,6-biphosphate aldolase.

for direct co-ordination. Studies of the enzyme–substrate complexes formed with the Zn^{2+}-containing enzyme by infrared spectroscopy have shown that carbonyl stretching frequency of DHAP (at $1730 \, cm^{-1}$) is unaffected by binding to the active site. However, the carbonyl stretching frequency of G3P is shifted from $1730 \, cm^{-1}$ in aqueous solution to $1706 \, cm^{-1}$ when bound to the enzyme. These data suggest that the aldehyde carbonyl group of G3P is activated (and hence further polarized) by co-ordination to Zn^{2+}. It has been proposed that an intervening histidine ligand stabilizes the enolate form of DHAP, as shown in Fig. 7.5.

Fructose-1,6-bisphosphate aldolase has been used for enantioselective carbon–carbon bond formation in organic synthesis. The class I enzyme is highly selective for the DHAP substrate, but can react with a wide range of aldehyde substrates. Reaction with racemic 2-hydroxy-3-azido-propion-aldehyde on a 1–10-mmol scale yields a single enantiomer of the aldol product containing three chiral centres. Biotransformation of the same substrates with aldolase enzymes of different stereospecificity yields dia-stereomeric products as shown in Fig. 7.6. Dephosphorylation of the prod-ucts followed by reduction of the azido group generates the corresponding amines, which undergo stereospecific reductive amination reactions with the free keto group. The cyclic amine products can be readily converted into a range of aza-sugar analogues which are of considerable interest as glycosid-ase inhibitors.

Claisen enzymes

In the presence of alkoxide ions, carboxylic esters react via an ester enolate intermediate to form β-keto esters. This reaction, illustrated in Fig. 7.7 for ethyl butyrate, is known as the Claisen ester condensation. The Claisen ester condensation requires more vigorous reaction conditions than the aldol condensation, since the C–H proton adjacent to an ester is significantly less acidic than the C–H proton adjacent to a ketone.

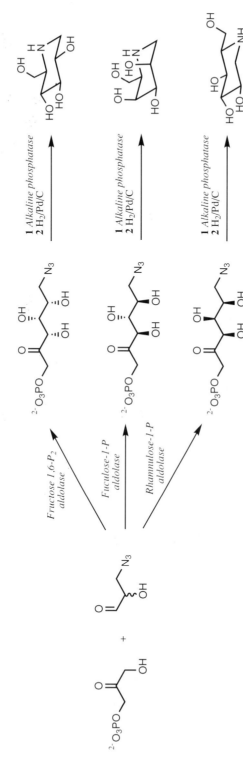

Fig. 7.6 Use of aldolase enzymes for organic synthesis. Pd, palladium.

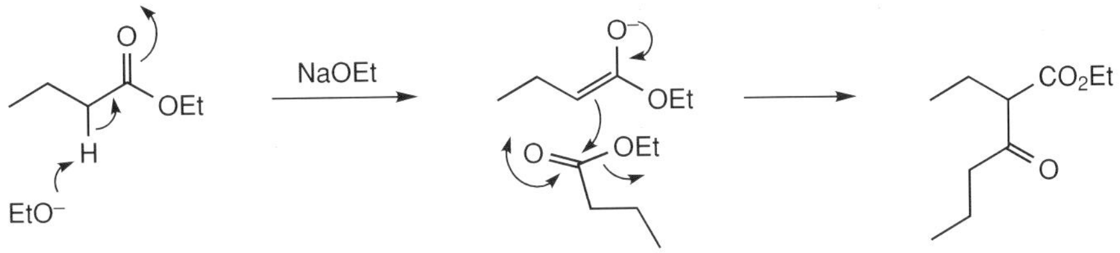

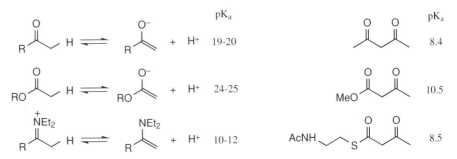

Fig. 7.7 Example of a Claisen reaction. NaOEt, sodium ethoxide.

Claisen reactions also occur in biological systems, using thioesters rather than oxygen esters. Since the sulphur atom of a thioester is larger than oxygen and utilizes $3p$-valence electrons rather than $2p$-electrons, there is consequently much less overlap between the sulphur lone pair of electrons and the carbonyl group than in the oxygen equivalent. Thus, the carbonyl group of a thioester behaves more like a ketone in terms of reactivity than an ester. Shown in Fig. 7.8 are pK_a values for formation of enolate or enolate equivalents for a selection of ketone and ester systems. The first set of data highlight the dramatic increase in acidity of an iminium salt compared to a ketone, which rationalizes the use of enamine intermediates in the class I aldolases. From the second set of data we can see that the pK_a of a β-keto-thioester is similar to that of a β-diketone, and significantly less than that for a β-keto-ester.

The thioester used for biological Claisen ester reactions is the coenzyme A (CoA) ester which we encountered in Chapter 5 as an acyl transfer coenzyme. Acetyl CoA is the substrate for several types of carbon–carbon forming reactions, shown in Fig. 7.9.

Mevalonic acid is an important cellular precursor to terpene and steroid natural products. The biosynthesis of mevalonic acid involves two reactions in which carbon–carbon bonds are formed from the α-position of acetyl CoA. The first reaction is a condensation reaction with a second molecule of acetyl CoA, to form 3-ketobutyryl CoA. Note that the good leaving group properties of the thiol group of CoA are also significant in this reaction.

Fig. 7.8 pK_a values for enolate formation.

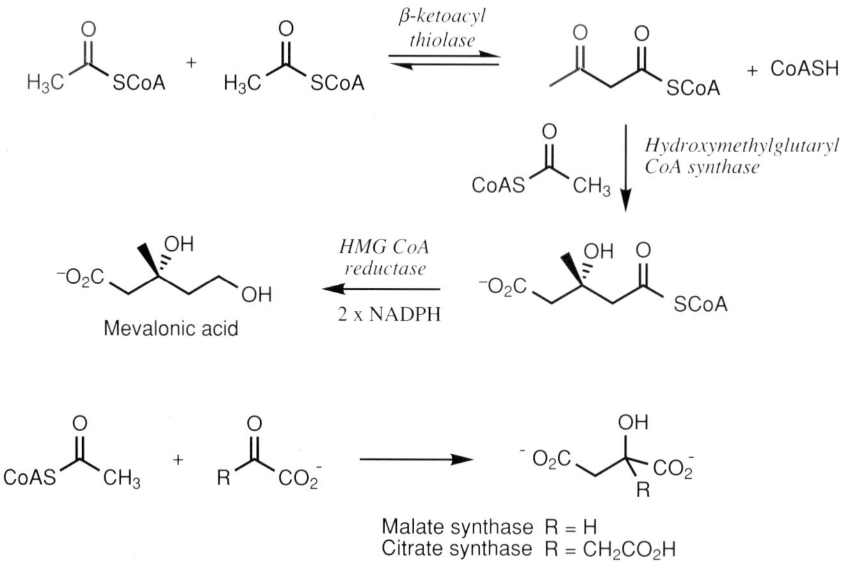

Fig. 7.9 Reactions catalysed by Claisen enzymes. HMG, hydroxymethylglutaryl; NADPH, nicotinamide adenine dinucleotide phosphate.

Reaction of the β-keto group with a further equivalent of acetyl CoA generates hydroxymethyl-glutaryl CoA, which is reduced by an NADPH-dependent reductase to give mevalonic acid. Two further examples shown in Fig. 7.9 are malate synthase and citrate synthase, which catalyse important metabolic reactions of acetyl CoA.

The mechanism of these 'Claisen enzymes' which react through the α-position of acetyl CoA could either proceed via formation of an α-carbanion intermediate, or by a concerted deprotonation/bond formation step. Incubation of acetyl CoA with malate synthase or citrate synthase in the presence of ^3H_2O followed by re-isolation of substrate gives no exchange of ^3H into acetyl CoA, ruling out the reversible formation of a carbanion intermediate. Stereochemical studies using chiral [2-^{2}H, ^{3}H]-labelled acetyl CoA, illustrated in Fig. 7.10, have established that the malate synthase reaction proceeds with overall inversion of stereochemistry, and with a kinetic isotope effect ($k_H/k_T = 2.7$).

These results suggest that there is a transient thioester enolate intermediate formed in the malate synthase reaction, which then reacts with the aldehyde carbonyl of glyoxylate. Formation of the thioester enolate intermediate would be thermodynamically unfavourable due to the high pK_a of the α-proton; however, the enzymatic reaction is made effectively irreversible by the subsequent hydrolysis of malyl CoA to malic acid. It is not known exactly how the formation of the thioester enolate (or enol) is made possible

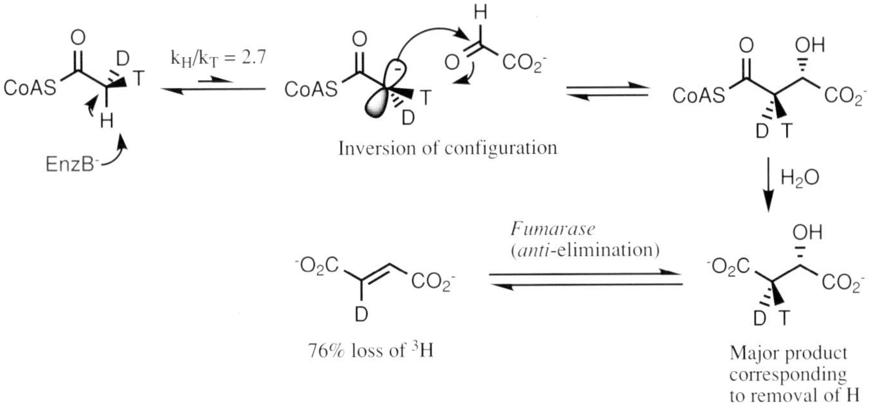

Fig. 7.10 Stereochemistry of the malate synthase reaction.

at the enzyme active site, but there must be significant stabilization of this intermediate *in situ* by the enzyme.

Assembly of fatty acids and polyketides

Acetyl CoA is also used in the formation of the carbon chains of fatty acids and polyketide natural products in biological systems. Fatty acids are long-chain carboxylic acids, such as stearic acid ($C_{17}H_{35}CO_2H$), which are found widely as triglyceride esters in the lipid component of living cells. Polyketide natural products are assembled from a polyketide precursor containing ketone functional groups on alternate carbon atoms. In many cases these polyketide precursors are cyclized to form aromatic compounds, such as orsellinic acid shown in Fig. 7.11.

There are many similarities between fatty acid biosynthesis and poly-ketide biosynthesis: they are both assembled from acyl CoA thioesters via a series of Claisen-type reactions by multi-enzyme synthase complexes (fatty acid synthase or polyketide synthase). The assembly is carried out via stepwise addition of two-carbon units onto a growing acyl chain which is attached to an acyl carrier protein (ACP) via a thioester linkage.

The first acetyl unit is transferred from acetyl CoA onto the ACP, but thereafter the substrate for carbon–carbon bond formation is malonyl CoA, formed by carboxylation of acetyl CoA (described on pp. 149–150). The malonyl group is transferred from malonyl CoA onto the ACP by an acyltransferase (AT) activity (Fig. 7.12). Carbon–carbon bond formation is achieved by attack of the α-carbon of the malonyl thioester onto the acetyl thioester intermediate, with decarboxylation of the malonyl group, by a ketosynthase (KS) activity. This reaction, producing a β-keto-thioester, is

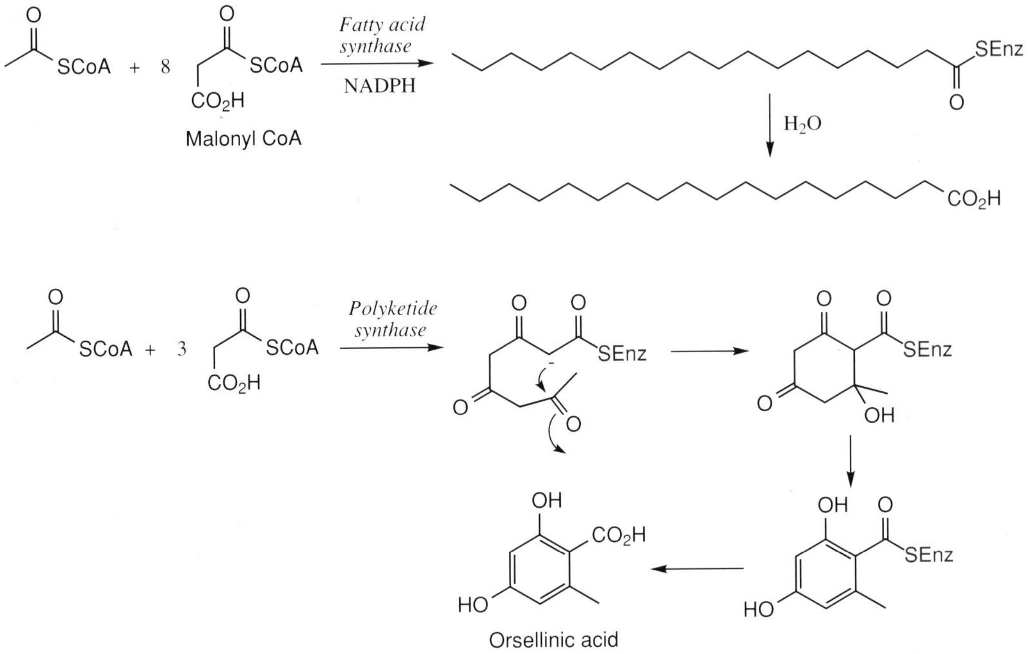

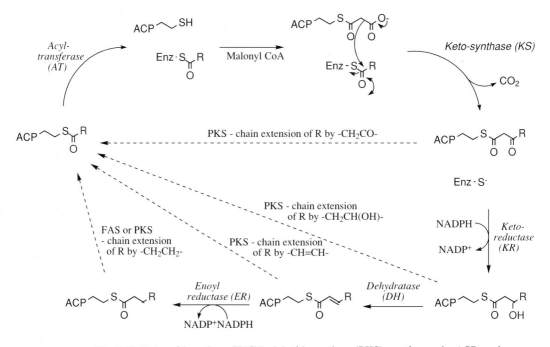

Fig. 7.11 Assembly of fatty acids and polyketides.

Fig. 7.12 Fatty acid synthase (FAS)/polyketide synthase (PKS) reaction cycle. ACP, acyl carrier protein.

similar to that of the Claisen enzymes above, except that decarboxylation occurs at the same point.

The question which then arises is whether decarboxylation occurs before, after or at the same time as carbon–carbon bond formation? Stereospecific labelling studies, shown in Fig. 7.13, have demonstrated that this reaction occurs with inversion of configuration at C-2 of the malonyl unit and with little or no hydrogen exchange at C-2. This implies that carbon–carbon bond formation and decarboxylation are, in fact, concerted.

In the case of fatty acid biosynthesis, the new β-keto-thioester is reduced to a β-hydroxy-thioester, eliminated to give an α,β-unsaturated thioester, and then further reduced to give a two-carbon-extended acyl chain, as shown in Fig. 7.12. In the case of polyketide biosynthesis, each two-carbon unit can be processed as either the β-keto-thioester, the β-hydroxy-thioester, the α,β-unsaturated thioester or as the fully reduced thioester. Assembly of each polyketide is therefore controlled by the arrangement of processing enzyme activities on the polyketide synthase multi-enzyme complex. How is this done?

Information regarding the molecular structure and organization of poly-ketide synthases is now emerging from the cloning and sequencing of genes which encode these enzymes. The genes responsible for the biosynthesis of the polyketide antibiotic erythromycin have been identified and their nucleo-tide sequences determined. They encode three huge multi-functional poly-peptides of size 300–500 kDa, illustrated in Fig. 7.14. The enzyme activities responsible for processing of the growing polyketide chain have been identi-fied by amino acid sequence alignments, and are found sequentially along the polypeptide chains. Remarkably, the arrangment of processing enzyme activities on the polyketide synthases matches the order of chemical steps required for biosynthesis of the polyketide precursor. It therefore appears that these multi-enzyme complexes function as molecular production lines built up of 'modules' of enzyme activities.

In the case of erythromycin the polyketide is assembled from the three-

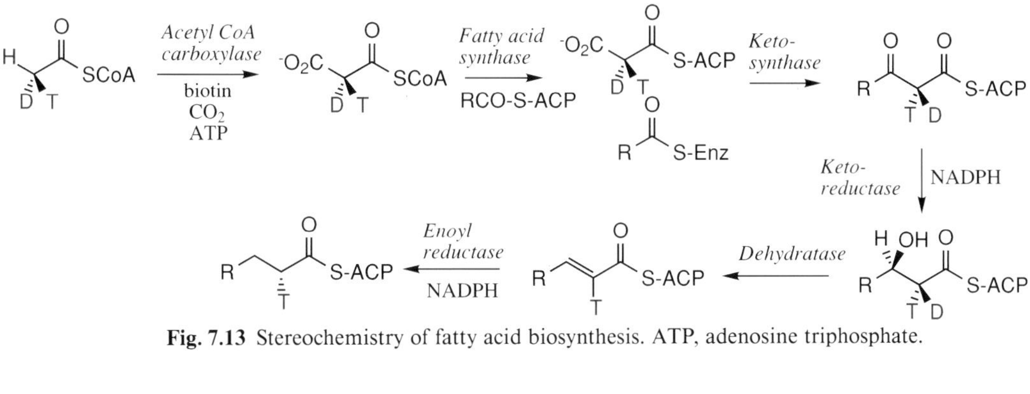

Fig. 7.13 Stereochemistry of fatty acid biosynthesis. ATP, adenosine triphosphate.

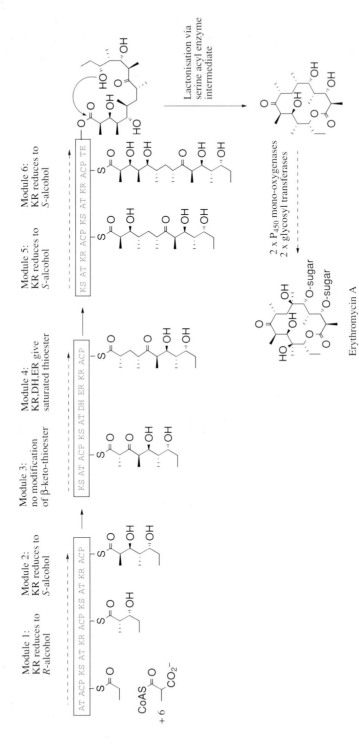

Fig. 7.14 Assembly of erythromycin A via multi-enzyme polyketide synthases. DH, dehydratase; ER, enoyl reductase; KR ketoreductase.

carbon unit of propionyl CoA, which is carboxylated to give methylmalonyl CoA. Acyl transfer and carbon–carbon bond formation takes place through the α-carbon of a methylmalonyl-thioester, in the same way as is shown in Fig. 7.12, giving α-methyl-β-keto-thioesters at each stage. Each 'module' of enzyme activities contains the enzymes required for the assembly of a new β-keto-thioester and its subsequent modification. For example, the first module contains KS and AT activities to make the new β-keto-thioester, and a ketoreductase (KR) activity to reduce the β-keto-thioester to a β-*R*-hydroxy-thioester, and so on. Each of the multifunctional polyketide synthases contains two such 'modules' of enzymatic activities. At the end of the third polyketide synthase is a thioesterase (TE) activity which catalyses an intramolecular lactonization via a serine acyl enzyme intermediate. Subsequent modification of the polyketide precursor to erythromycin A occurs by separate P_{450} mono-oxygenase and glycosyl transferase enzymes.

Carboxylases: use of biotin

We have already seen examples of nucleophilic attack of a carbanion equivalent onto aldehyde and ester electrophiles. There are finally a number of examples of nucleophilic attack of carbanion equivalents onto carbon dioxide to generate carboxylic acid products.

We have just seen that the carboxylation of acetyl CoA to give malonyl CoA is an important step in fatty acid and polyketide natural product biosynthesis. This step is catalysed by acetyl CoA carboxylase. This enzyme uses acetyl CoA and bicarbonate as substrates, but also requires adenosine triphosphate (ATP), which is converted to adenosine diphosphate (ADP) and inorganic phosphate (P_i), and the cofactor biotin. Biotin was first isolated from egg yolk in 1936, and was found to act as a vitamin whose deficiency causes dermatitis. Its structure is a bicyclic ring system containing a substituted urea functional group which is involved in its catalytic function. The biotin cofactor is covalently attached to the ε-amino side chain of an active site lysine residue.

How does such an apparently unreactive chemical structure serve to activate carbon dioxide for these carboxylation reactions, and what is the role of ATP in the reaction? These questions were answered by a series of experiments with isotopically labelled bicarbonate substrates. Bicarbonate is rapidly formed from carbon dioxide in aqueous solution and is the substrate for biotin-dependent carboxylases. Incubation of biotin-dependent β-methylcrotonyl-CoA carboxylase with ^{14}C-bicarbonate and ATP gave an intermediate ^{14}C-labelled enzyme species. Methylation with diazomethane followed by degradation of the enzyme structure revealed that the ^{14}C label was covalently attached to the biotin cofactor, in the form of a carbon

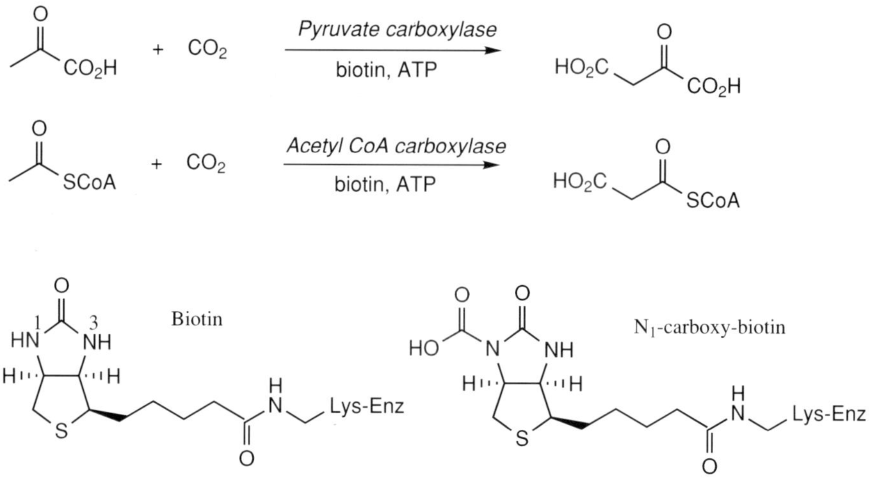

Fig. 7.15 Biotin-dependent enzymes.

dioxide adduct onto N-1 of the cofactor. This species, N_1-carboxy-biotin, illustrated in Fig. 7.15, has been shown to be chemically and kinetically competent as an intermediate in the carboxylation reaction.

Incubation of enzyme with ^{18}O-labelled bicarbonate led to the isolation of product containing two atoms of ^{18}O and P_i containing one atom of ^{18}O. The transfer of ^{18}O to P_i implies that ATP is used to activate bicarbonate by formation of an acyl phosphate intermediate, known as carboxyphosphate. It is thought that carboxyphosphate is then attacked by N-1 of the biotin cofactor. Since the N–H group of amides and urea is normally a very unreactive nucleophile, it is thought that N-1 is deprotonated prior to attack on carboxyphosphate. The carboxy-biotin intermediate thus formed is then attacked by a deprotonated substrate to form the carboxylated product and re-generate the biotin cofactor, as shown in Fig. 7.16. It has been shown that the carboxylation of pyruvate to oxaloacetate catalysed by pyruvate carboxylase proceeds with retention of configuration at C-3, as shown in Fig. 7.17. Other biotin-dependent carboxylases also proceed with retention of configuration.

Ribulose bisphosphate carboxylase/oxygenase (Rubisco)

All of the carbon-based molecules found in living systems are ultimately derived from the fixation of gaseous carbon dioxide by green plants during photosynthesis. Carbon dioxide is then re-generated from respiration of living organisms, from the decomposition of carbon-based material and dead organisms, and from the combustion of wood and fossil fuels. This global cycle of processes is known as the carbon cycle. The enzyme which is responsible for the fixation of carbon dioxide by green plants is an enzyme of

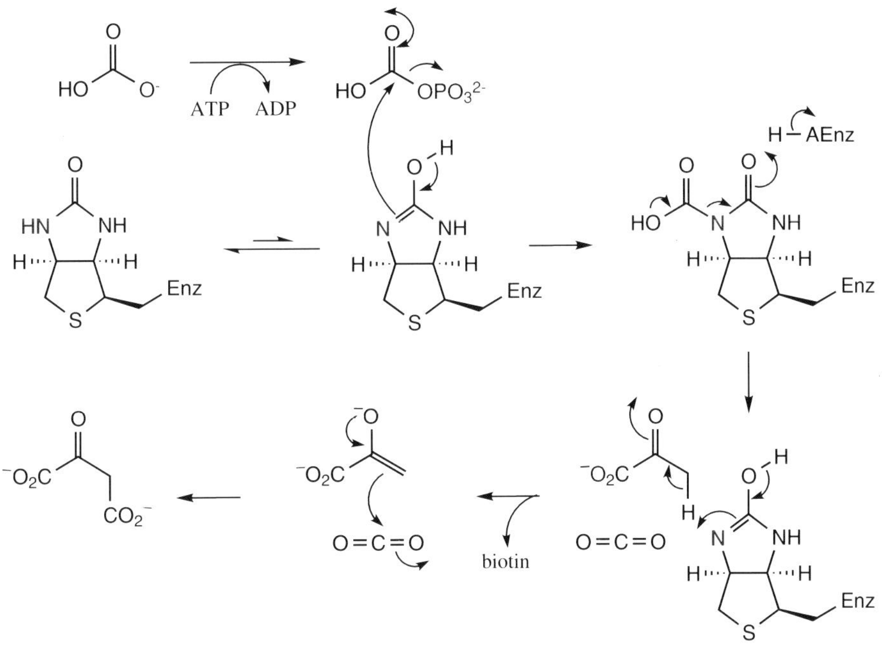

Fig. 7.16 Mechanism for biotin-dependent carboxylation.

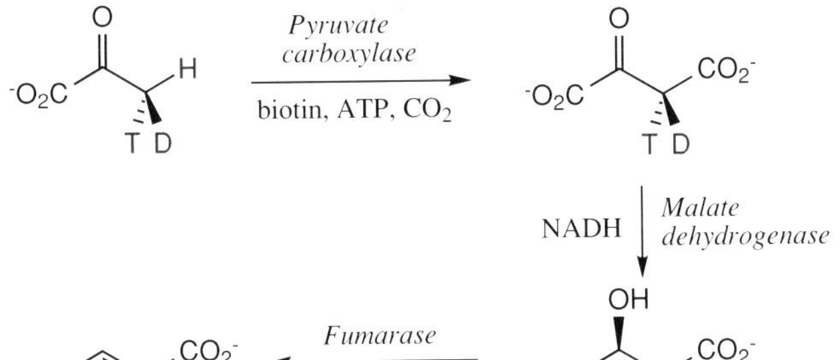

Fig. 7.17 Stereochemistry of the pyruvate carboxylase reaction. NADH, nicotinamide adenine dinucleotide.

primary importance to life on earth. This enzyme is ribulose bisphosphate carboxylate/oxygenase, commonly known as 'Rubisco'.

The reaction catalysed by Rubisco is the carboxylation and concomitant fragmentation of ribulose-1,5-bisphosphate to generate two molecules of 3-phosphoglycerate (Fig. 7.18). This is a step on the Calvin cycle of green plants, which transforms 3-phosphoglycerate via a series of aldolase and transketolase enzymes once again into ribulose-1,5-bisphosphate. Thus, each cycle incorporates one equivalent of carbon dioxide into cellular carbon.

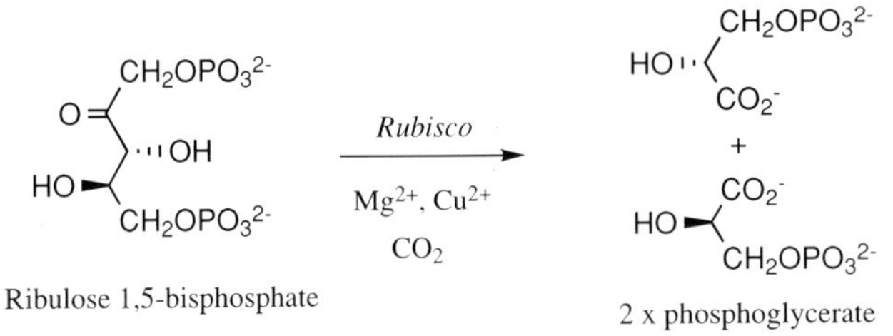

Fig. 7.18 Reaction catalysed by ribulose-1,5-bisphosphate carboxylase.

The enzymatic reaction has been shown to commence by deprotonation at C-3 of ribulose-1,5-bisphosphate, and enolization of the C-2 ketone, since [3-^{3}H]-ribulose-1,5-bisphosphate rapidly exchanges ^{3}H with solvent in the presence of enzyme. The resulting enediol intermediate reacts with carbon dioxide to generate a carboxylated intermediate. Carbon–carbon bond cleavage is thought to occur by hydration of the C-3 ketone, followed by fragmentation of the C-2–C-3 bond to generate a carbanion product, which protonates to give the second molecule of 3-phosphoglycerate. The proposed mechanism is shown in Fig. 7.19.

Vitamin K-dependent carboxylase

One other carboxylase enzyme worthy of note is the vitamin K-dependent carboxylase responsible for the activation of prothrombin via carboxylation

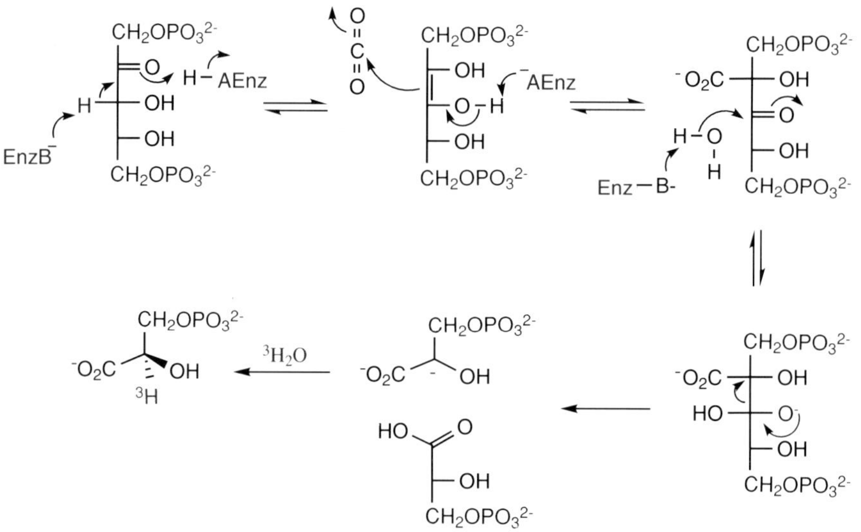

Fig. 7.19 Mechanism of the ribulose-1,5-bisphosphate carboxylase reaction.

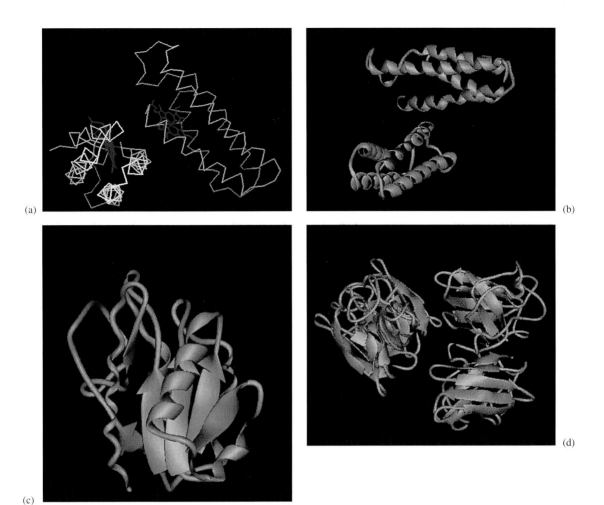

(a)

(b)

(c)

(d)

Plate 2.1 Examples of protein tertiary structures. (a) Structure of cytochrome b_{562} (α-carbon backbone), which consists of a 'bundle' of four α-helices, encompassing a heme cofactor (magenta). The unit cell determined by crystallography contains two molecules of protein. (b) Ribbon drawing of the cytochrome b_{562} structure, illustrating the 4 α-helices of the helix bundle. (c) Ribbon structure of flavodoxin, which is composed of 5 parallel β-sheets, with intervening α-helices. (d) Ribbon structure of superoxide dismutase. This metalloenzyme consists of a barrel of antiparallel β-sheets. The unit cell contains four such β-barrels.

[facing page 152]

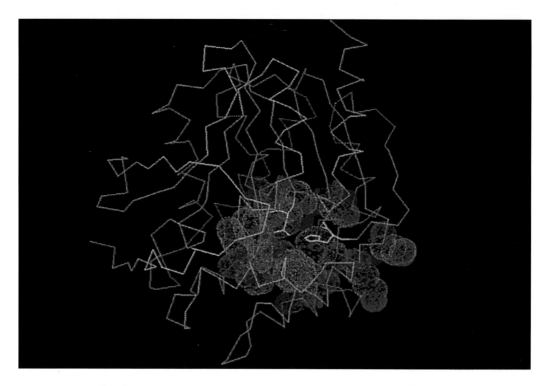

Plate 3.1 Structure of *Xanthobacter autotrophicus* haloalkane dehalogenase (α-carbon backbone), Active site residues Asp-124 and His-289 are drawn in blue. Residues lining the active site are highlighted through their Van der Waals surfaces (nitrogen atoms in red, oxygen atoms in dark blue).

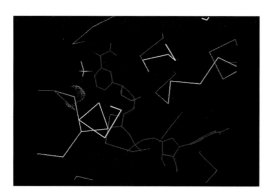

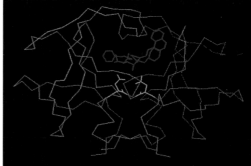

Plate 4.1 Active site view of alcohol dehydrogenase, showing the substrate ethanol (yellow, oxygen atom dark blue) bound to the active site zinc cofactor (blue, Van der Waals surface). The alignment of the *proR* hydrogen with respect to an NAD⁺ analogue (magenta) can be seen. Parts of the protein α-carbon backbone can be seen in grey.

Plate 5.1 Structure of HIV protease (α-carbon backbone), complexed with a statine-containing inhibitor (magenta). Active site residues Asp-25 and Asp-25′ are drawn in blue. Note the symmetrical structure of the homodimer, and the size of the substrate-binding cavity.

(a)

(b)

(c)

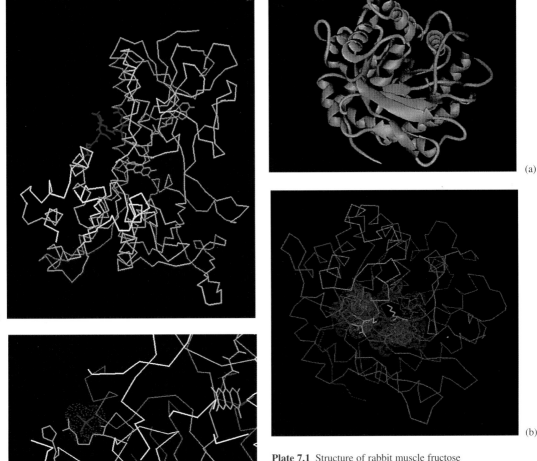

(a)

(b)

Plate 7.1 Structure of rabbit muscle fructose 1,6-bisphosphate aldolase. (a) Ribbon diagram of protein tertiary structure, showing αβ barrel structure composed of parallel β-sheets and intervening α-helices, encompassing a central active site cavity. (b) α-Carbon backbone of protein, with the Van der Waals surfaces of active site residues highlighed. Active site residues Lys-229 and Lys-146 are drawn in blue, with Lys-229 clearly visible in the centre of the active site cavity.

Plate 6.1 (a) Structure of human glutathione reductase (α-carbon backbone), showing the FAD cofactor (orange) and the glutathione substrate (magenta). The NADPH cofactor is not shown. (b) Binding site in glutathione reductase for 2′-phosphate of NADPH (magenta, Van der Waals surface). Residues His-219 and Arg-224 are drawn in blue (Arg-218 not shown). Note the neighbouring FAD cofactor (orange). (c) Binding site in human glutathione reductase for glutathione substrate (magenta). Residues Arg-37 (left), Ala-34 (centre) and Arg-347 (top) are drawn in blue. The proximity between the guanidinium sidechains of Arg-37 and Arg-347 and the substrate carboxylate groups can be seen. FAD cofactor is drawn in orange.

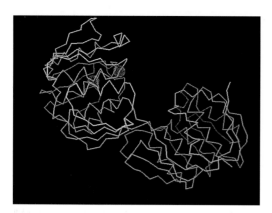

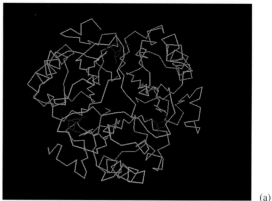

(a)

Plate 8.1 Structure of *Escherichia coli* EPSP synthase (α-carbon backbone). The two domains of the protein structure can clearly be seen. Residues Gly-96 and Pro-101, whose mutation causes resistance to glyphosate, are highlighted in blue.

(b)

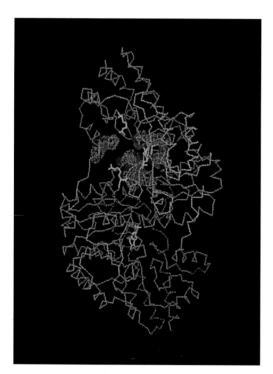

Plate 10.1 (a) Structure of *Bacillus subtilis* chorismate mutase trimer (α-carbon backbone), with bound transition state analogue (magenta). Active site residue Arg-90 is drawn in blue in each subunit. (b) Active site view, showing the proximity of the guanidinium sidechain of Arg-90 (light blue; nitrogen atoms in orange) to the transition state analogue (magenta; oxygen atoms in dark blue).

Plate 9.1 Structure of *Escherichia coli* aspartate amino-transferase dimer (α-carbon backbone), showing the PLP cofactors (orange) bound at each of the two active sites. Amino acid residues lining one of the two active sites are highlighted (Van der Waals surface). Active site residues Arg-386, Arg-292 and Lys-258 (close to PLP cofactor) are drawn in blue.

of a series of glutamate amino acid residues. This step is involved in the calcium-dependent activation of platelets during the blood clotting response. Vitamin K is a fat-soluble naphthoquinone which can exist in either oxidized (quinone form) or reduced (hydroquinone form) states. The carboxylase enzyme is an integral membrane protein which has only recently been purified. The carboxylation of prothrombin requires reduced vitamin K, carbon dioxide, oxygen and a carboxylation substrate (Fig. 7.20). Reduced vitamin K is converted stoichiometrically into vitamin K epoxide, which is re-cycled via separate reductase enzymes back to reduced vitamin K.

A major clue to the role of vitamin K in the enzymatic reaction came from a chemical model reaction, shown in Fig. 7.21. In the presence of molecular oxygen and a crown ether, a naphthol analogue of reduced vitamin K was found to act as a catalyst for the Dieckmann condensation of diethyl adipate, generating an epoxide byproduct. A mechanism was proposed for this reaction in which the naphthoxide salt and dioxygen react via a dioxetane intermediate to form a tertiary alkoxide species which acts as a base for the ester condensation reaction. Note that the naphthoxide salt itself is not a strong enough base to catalyse the reaction, but is converted into a much stronger base via reaction with oxygen.

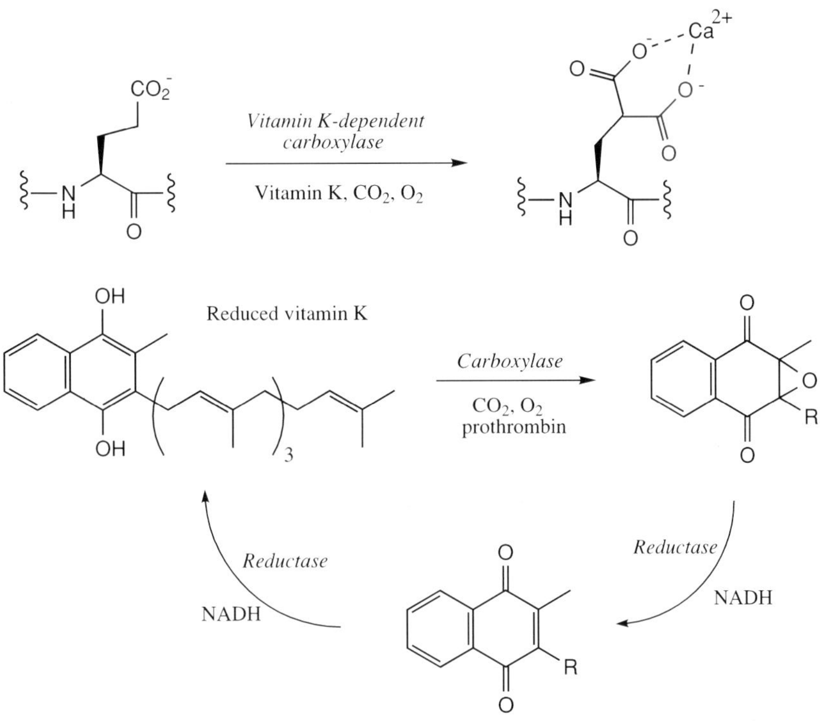

Fig. 7.20 The vitamin K-dependent carboxylase reaction. NADH, nicotinamide adenine dinucleotide.

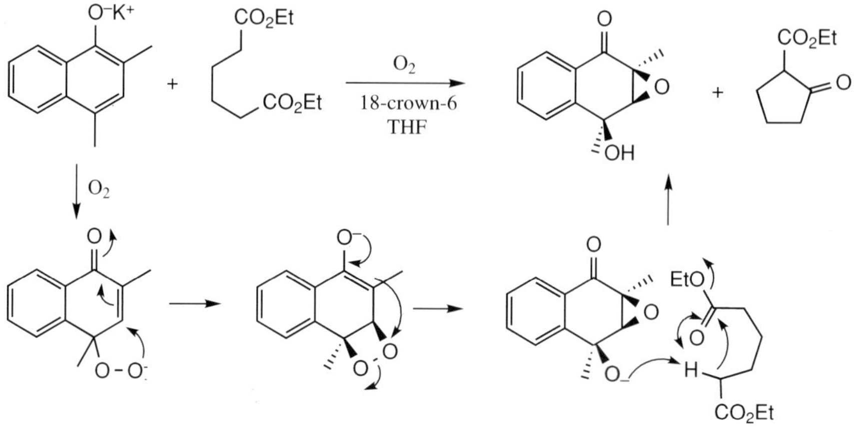

Fig. 7.21 Model reaction for role of vitamin K. THF, tetrahydrofuran.

In order to investigate the mechanism of the enzyme-catalysed carboxylation reaction, reduced vitamin K was incubated with carbon dioxide and enzyme in the presence of $^{18}O_2$. Complete incorporation of ^{18}O into the epoxide oxygen of vitamin K epoxide was observed, together with partial incorporation of ^{18}O (approximately 20%) into the ketone carbonyl, consistent with the dioxetane mechanism. It has been proposed that the hydrated alkoxide ion released from fragmentation of the dioxetane intermediate acts as a base to deprotonate the γ-H of a glutamyl substrate, followed by reaction with carbon dioxide, as shown in Fig. 7.22. It has been shown that the C-4

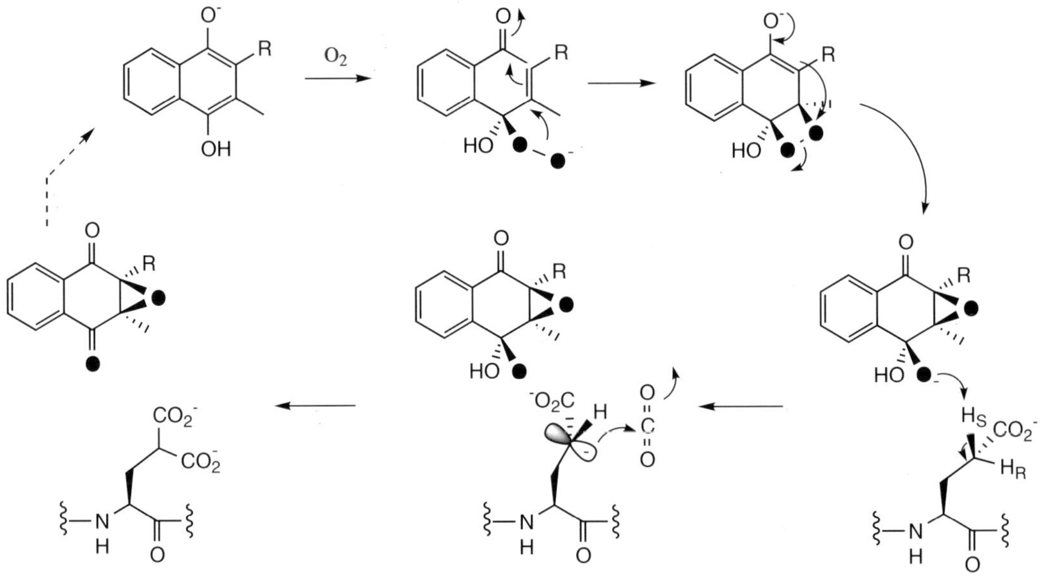

Fig. 7.22 Mechanism for the vitamin K-dependent carboxylase.

proS hydrogen is abstracted, and subsequent reaction with carbon dioxide occurs with inversion of configuration. Note that it is remarkable that the hydrated alkoxide ion does not simply collapse to form the corresponding ketone: presumably the micro-environment of the enzyme active site disfavours the loss of hydroxide ion from the lower face of vitamin K.

Thiamine pyrophosphate (TPP)-dependent enzymes

Decarboxylation reactions are also widely found in biological chemistry. In Chapter 3 we saw an example of the decarboxylation of a β-keto acid acetoacetate by the action of acetoacetate decarboxylase, via an imine linkage. In general terms β-keto acids can be fairly readily decarboxylated since the β-keto group provides an electron sink for the decarboxylation reaction. No such electron sink exists for α-keto acids; however, Nature has found a way of decarboxylating α-keto acids, and in many cases forming a carbon–carbon bond at the same time, using the coenzyme TPP.

TPP is formed by phosphorylation of the vitamin thiamine, lack of which causes the deficiency disease beriberi. The structure of TPP consists of two heterocyclic rings: a subsituted pyrimidine ring and a substituted thiazolium ring (Fig. 7.23). The thiazolium ring is responsible for the catalytic chemistry carried out by this coenzyme, due to two chemical properties:

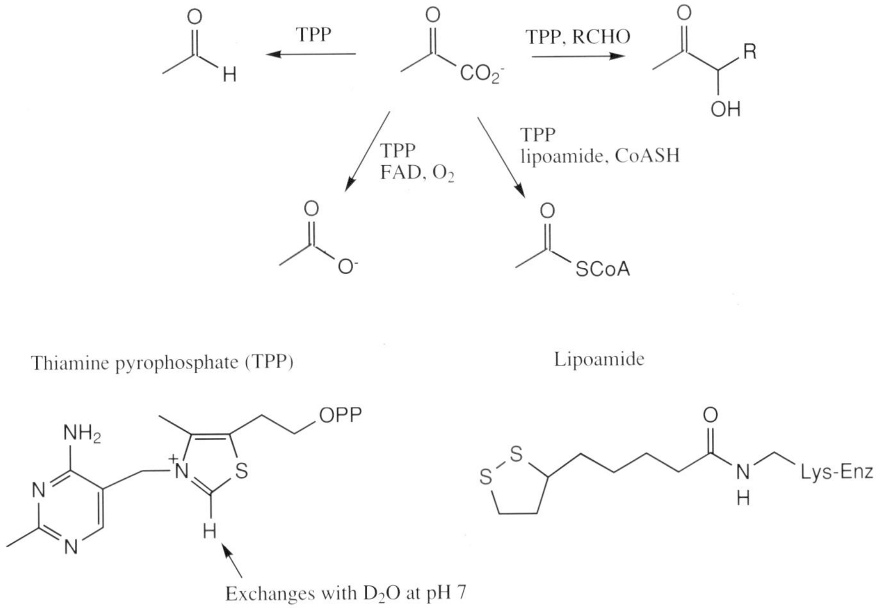

Fig. 7.23 TPP-dependent reactions of pyruvate. FAD, flavin adenine dinucleotide.

1 the acidity of the proton attached to the thiazolium ring, which will exchange with deuterium in ^2H_2O;

2 the presence of a carbon–nitrogen double bond which can act as an electron sink for decarboxylation.

The decarboxylation of pyruvic acid is carried out by several TPP-dependent enzymes, yielding acetaldehyde, acetic acid, acetyl CoA or α-hydroxyacetyl compounds. Each of these products is formed via closely related mechanisms.

The first step in each of the TPP-dependent reactions is deprotonation of the thiazolium ring to generate a carbanion ylid. This carbanion attacks the keto group of pyruvate to generate a covalently attached lactyl adduct (Fig. 7.24). Decarboxylation of this adduct can then take place, using the carbon–nitrogen double bond as an electron sink, forming an enamine intermediate. Protonation of this intermediate and fragmentation of the linkage with the coenzyme yield acetaldehyde and re-generates the coenzyme ylid. Alternatively, the enamine intermediate can react with an electrophile (E^+), generating after fragmentation of the linkage with the coenzyme an acetyl-E species.

An important reaction in the cellular breakdown of D-glucose is the conversion of pyruvate to acetyl CoA, which is catalysed by the pyruvate dehydrogenase complex. This reaction follows the mechanism shown in Fig. 7.24, but in this case the electrophile is a second cofactor present in the pyruvate dehydrogenase complex: lipoamide. The structure of lipoamide shown in Fig. 7.23 is simply a five-membered ring containing a disulphide linkage attached via an acyl chain to the ε-amino group of a lysine residue. Upon reaction of the TPP–enamine intermediate with oxidized lipoamide

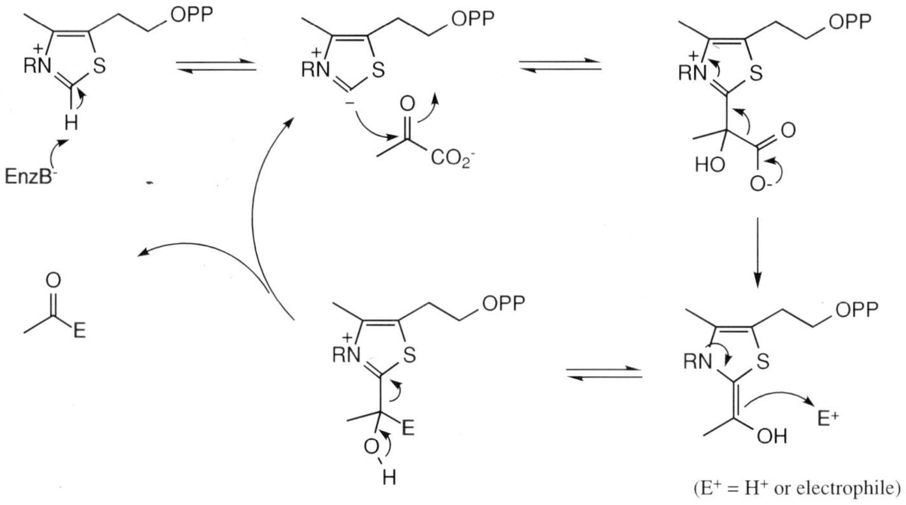

Fig. 7.24 Mechanism for TPP-dependent pyruvate decarboxylation.

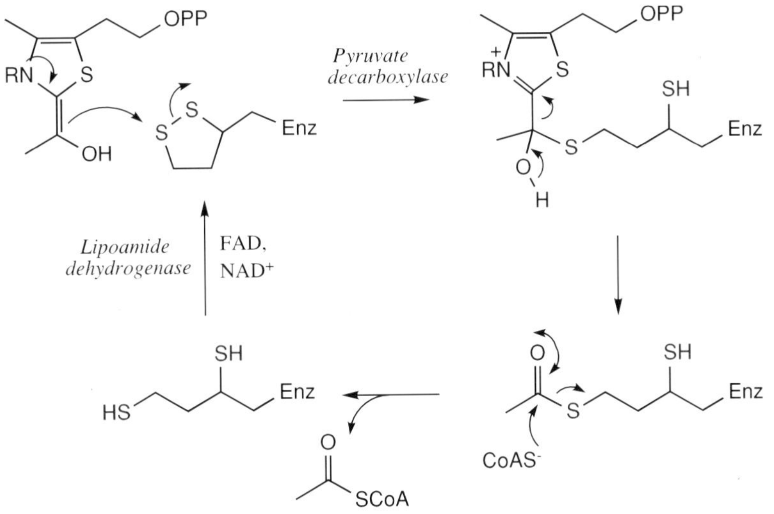

Fig. 7.25 TPP-dependent production of acetyl CoA using lipoamide. FAD, flavin adenine dinucleotide.

and fragmentation of the linkage with TPP, an acetyl-lipoamide thioester is formed, as shown in Fig. 7.25. Transfer of the acyl group to the thiol group of CoA generates acetyl CoA and reduced lipoamide, which is re-cycled to oxidized lipoamide by flavin-dependent lipoamide dehydrogenase (see Chapter 6, Section 6.5).

TPP is also used by several enzymes for carbon–carbon bond formation. Illustrated in Fig. 7.26 is one example of the family of TPP-dependent transketolase enzymes which carry out a range of carbohydrate 2-hydroxyacetyl transfer reactions. A similar mechanism can be written for these reactions initiated by attack of the thiazolium ylid on the keto group, followed in this case not by decarboxylation but by carbon–carbon bond cleavage. The ability of these enzymes to form carbon–carbon bonds enantioselectively is also being exploited for novel biotransformation reactions that are of use in organic synthesis.

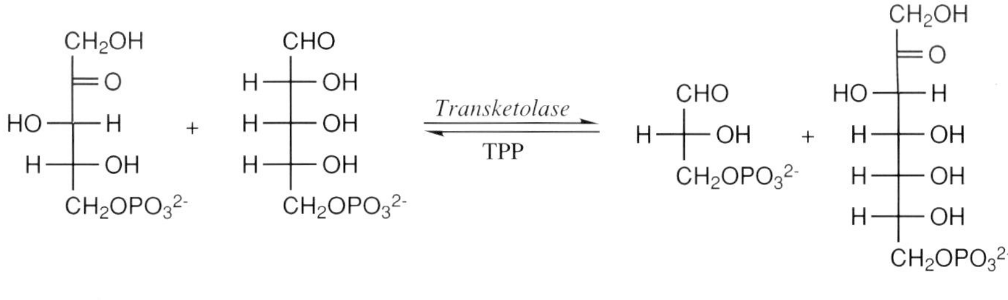

Fig. 7.26 Example of a transketolase reaction.

7.3   Carbon–carbon bond formation via carbonium ion intermediates

Carbon–carbon bond formation via carbonium intermediates is less widely found than via carbanion equivalents. However, there is one large class of biological reactions which involve highly stabilized carbonium ion intermediates: the conversion of allylic pyrophosphate metabolites into terpenoid natural products.

Terpene cyclases

Terpenes are a major class of natural products found widely in plants, but also including the steroid lipids and hormones found in animals. The common structural feature of the terpene natural products is the five-carbon isoprene unit. Some common examples shown in Fig. 7.27 are plant natural products menthol, camphor and geraniol.

The biosynthesis of terpene natural products proceeds from a family of allylic pyrophosphates containing multiples of five carbon atoms. Two five-carbon units are joined together by the enzyme geranyl pyrophosphate synthase, as shown in Fig. 7.28. Loss of pyrophosphate from dimethylallyl pyrophosphate (DMAPP) generates a stabilized allylic carbonium ion. Attack of the π-bond of isopentenyl pyrophosphate (IPP) generates a tertiary carbonium ion. Stereospecific loss of a proton generates the product geranyl pyrophosphate. Similar reactions with further units of IPP generate the C_{15} building block farnesyl pyrophosphate, the C_{20} geranylgeranyl pyrophosphate, and so on.

Geranyl pyrophosphate is converted to the family of C_{10} monoterpenes by the monoterpene cyclases. A 96-kDa cyclase enzyme responsible for the assembly of (+)-α-pinene has been purified from sage leaves, whilst a separate 55-kDa enzyme from the same source catalyses the production of the opposite enantiomer (–)-α-pinene. The 55-kDa cyclase II produces several other isomeric products. As well as (–)-α-pinene (26%) the enzyme produces (–)-β-pinene (21%), (–)-camphene (28%), myrcene (9%) and

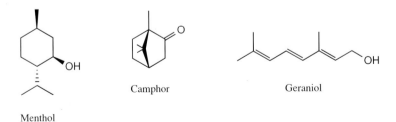

Menthol   Camphor   Geraniol

Fig. 7.27 Terpenoid natural products.

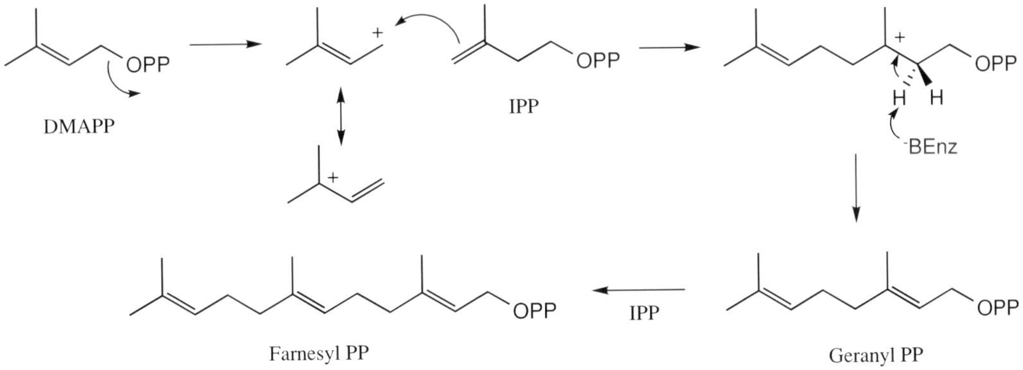

Fig. 7.28 Biosynthesis of farnesyl pyrophosphate. DMAPP, dimethylallyl pyrophosphate; IPP, isopentenyl pyrophosphate; PP, pyrophosphate.

(–)-limonene (8%). These products can be rationalized by the mechanism shown in Fig. 7.29. It has been shown that the first step in the assembly of these monoterpenes is a 1,3-migration of pyrophosphate to give linalyl pyrophosphate, which for cyclase II has the 3S stereochemistry. Loss of pyrophosphate gives an allylic cation, which is attacked by the second alkene to give a six-membered ring and a tertiary carbonium ion intermediate. Further ring closure to generate a four-membered ring gives another tertiary carbonium ion, which can deprotonate in either of two ways to give α-pinene or β-pinene. Limonene can be formed by elimination from the monocyclic

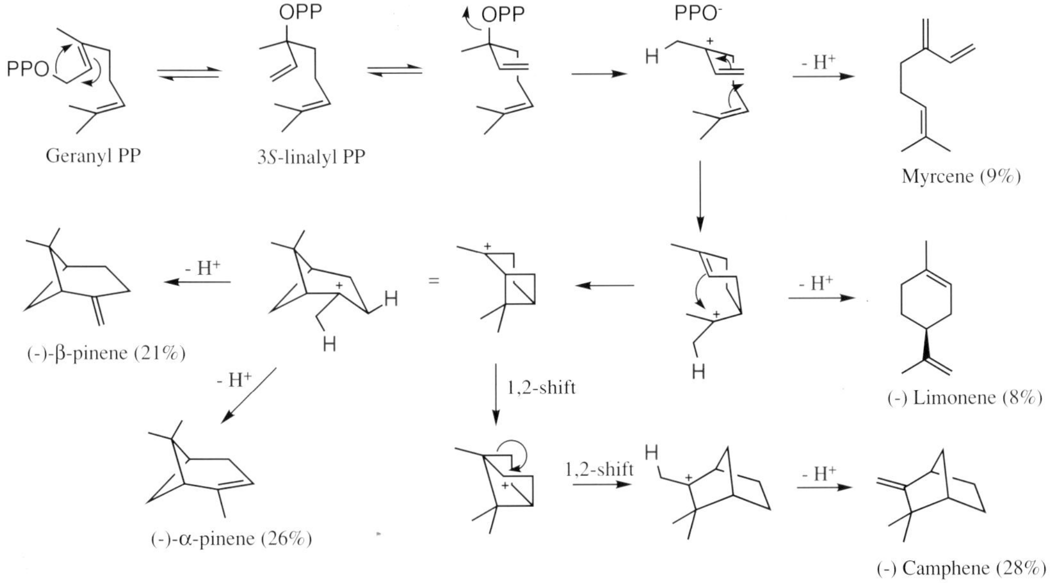

Fig. 7.29 Mechanism for (–)-α-pinene synthase.

carbonium ion, whilst camphene is formed by two consecutive 1,2-alkyl migrations, followed by elimination.

Most enzymes are specific for the production of a single isomeric product, so in this case how can we be sure that all of these products are from one enzyme? Apart from demonstrating that there is only one homogeneous protein present, a chemical test can be made. If there is a common precursor to two or more products, then by introducing deuterium atoms at specific points in the substrate, the ratio of products should be affected by the presence of a deuterium isotope effect which should favour one pathway over another. This test was applied to the (–)-pinene synthase reaction by incubation of the specifically deuterated substrate shown in Fig. 7.30. At the point of divergence to α-pinene or β-pinene the presence of a terminal $-CD_3$ group would be expected to favour formation of α-pinene. This was indeed observed: the proportion of α-pinene in the product mixture rose from 26 to 38%, whilst the proportion of β-pinene dropped from 21 to 13%. The proportion of myrcene product also dropped significantly from 9 to 4%, also consistent with an isotope effect operating on the allylic carbonium ion intermediate.

Farnesyl pyrophosphate is cyclized to form the large family of sesquiterpene natural products by a further class of terpene cyclase enzymes, some of which have been purified to homogeneity. Pentalenene synthase catalyses the cyclization of farnesyl pyrophosphate to give pentalenene, which is further processed in *Streptomyces* UC5319 to give the pentalenolactone family of antibiotics, as shown in Fig. 7.30. The enzyme is a 41-kDa protein with a K_M of 0.3 μM for farnesyl pyrophosphate and a k_{cat} of 0.3 s^{-1}, and an absolute requirement for Mg^{2+} ions. A series of isotopic labelling experiments have been carried out to support the mechanism of cyclization shown in Fig. 7.31. Cyclization of farnesyl pyrophosphate is proposed to form an

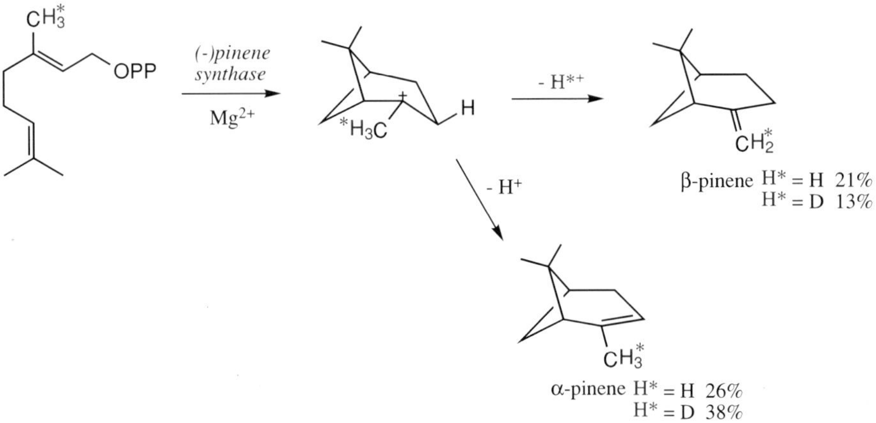

Fig. 7.30 Kinetic isotope effect in the (–)-pinene synthase reaction.

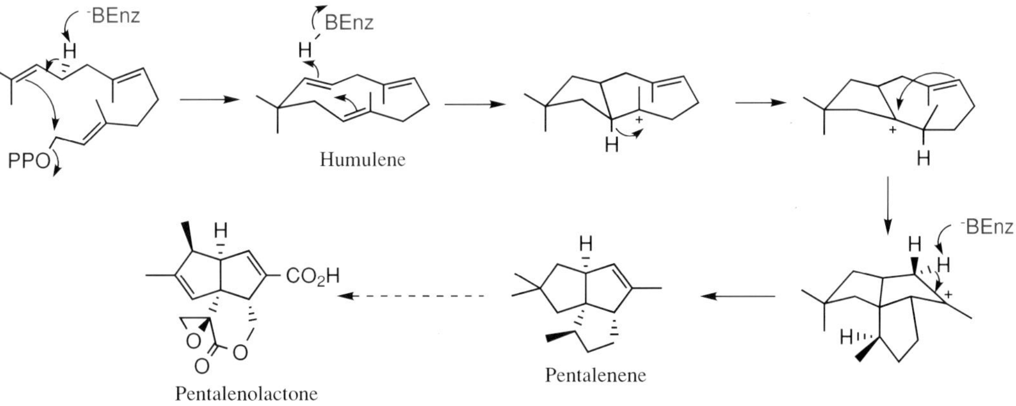

Fig. 7.31 Mechanism for pentalenene synthase.

11-membered intermediate, humulene, which is followed by a five-membered ring closure to form a bicyclic tertiary carbonium ion. 1,2-Hydride migration followed by a further five-membered ring closure gives a tricyclic carbonium ion, which upon elimination gives pentalenene.

Examination of the inferred amino acid sequence of pentalenene synthase revealed an amino acid sequence motif Asp–Asp–X–X–Asp found in other pyrophosphate-utilizing enzymes. It is believed that the Asp–Asp pair are involved in chelating the essential Mg^{2+} ion, which in turn chelates the pyrophosphate ion, as shown in Fig. 7.32. An indication of the tight binding

Pentalenene synthase	F	L	D D	F	L	D
Trichodiene synthase	V	L	D D	S	K	D
Aristolochene synthase	L	I	D D	V	L	E
Casbene synthase	L	I	D D	T	I	D
Limonene synthase	V	I	D D	I	Y	D
Farnesyl PP synthase	V	Q	D D	I	L	D
Geranylgeranyl PP synthase	I	A	D D	Y	H	N

Consensus   D D   X   X(D)

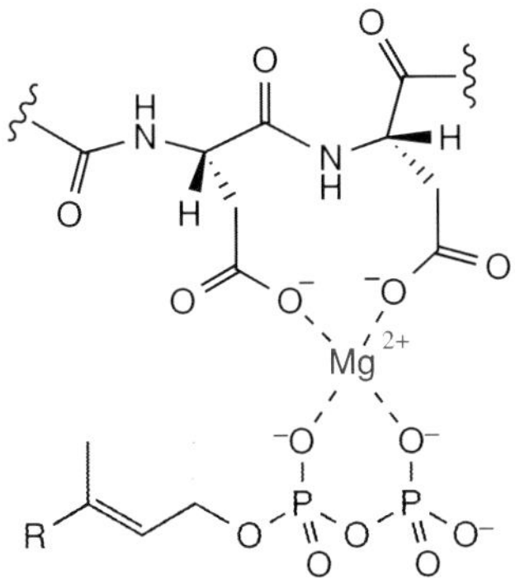

Fig. 7.32 Mg^{2+}–pyrophosphate binding motif.

of pyrophosphate is that this product inhibits the enzymatic reaction potently (K_i 3 μM).

How do these cyclase enzymes control the precise regiochemistry and stereochemistry of these multi-step cyclizations? Presumably the aliphatic chain of the precursor is bound in a specific orientation at the enzyme active site, but in addition it is likely that there is specific stabilization of the carbonium ion intermediates by nearby active site groups. There is evidence that the pyrophosphate anion remains tightly associated with the enzyme throughout these cyclizations. Thus, the intermediate carbonium ions are, in practice, stabilized ion pairs. The regiochemistry of the final deprotonation step will be controlled by the location of an active site base.

7.4   Carbon–carbon bond formation via radical intermediates

In Chapter 6 we have seen how redox enzymes are able to generate radical intermediates in enzyme-catalysed reactions. In certain instances these radical intermediates are used for carbon–carbon bond formation reactions. The two examples which we shall examine involve phenol radical couplings, used in natural product biosynthesis and in lignin biosynthesis.

Phenolic radical couplings

Certain aromatic natural products are formed by the radical coupling of two phenol precursors in an enzyme-catalysed process. Information about these enzymes is scarce, but in the cases where enzymes have been purified they are found to be blue copper (Cu^{2+}) proteins (the characteristic colour due to the metal cofactor). For example, the phenol radical coupling of sulochrin shown in Fig. 7.33 is catalysed by a 157-kDa oxidase enzyme containing six atoms of Cu^{2+}. The enzyme exhibits a deep blue colour (λ_{max} 605 nm) which disappears upon reduction with ascorbate under nitrogen. These observations suggest that there are active site Cu^{2+} metal ions which accept one electron from the phenol substrates to generate Cu^+ intermediates. The phenoxy radicals thus generated are relatively stable, and react together as shown in Fig. 7.33.

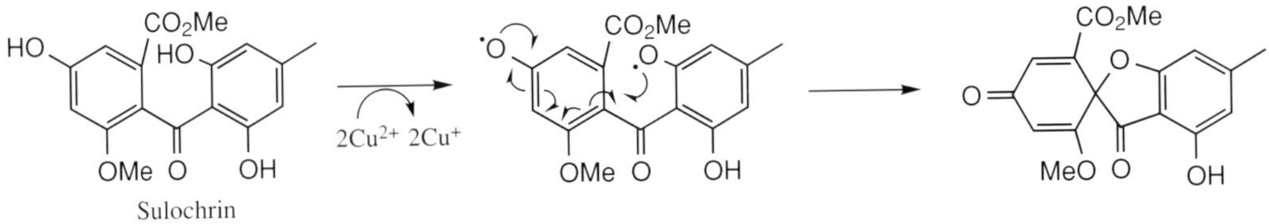

Fig. 7.33 Enzymatic phenolic radical coupling.

Lignin biosynthesis

Lignin is a complex aromatic polymeric material which is a major structural component of woody tissues in plants. In most trees lignin constitutes 30–40% of the dry weight of the wood. It is highly cross-linked and exceptionally resistant to chemical degradation, making elucidation of its chemical structure very difficult. However, it is now known to be heterogeneous in structure, composed of phenylpropanoid (aromatic ring plus three-carbon alkyl chain) units linked together by a variety of types of carbon–

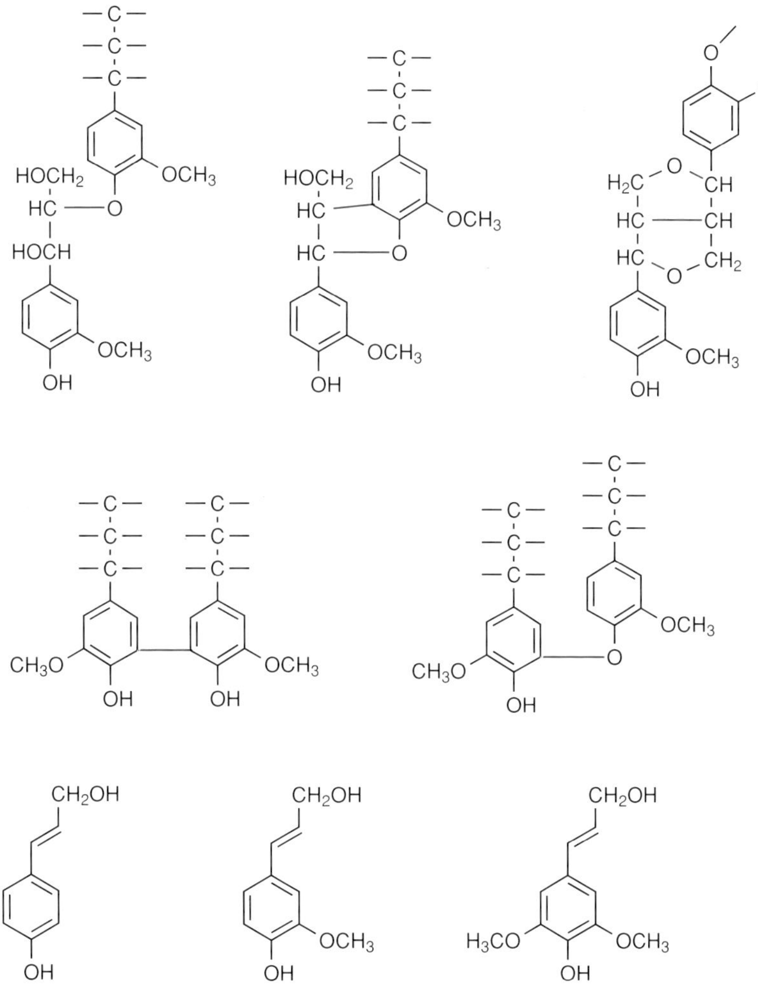

p-hydroxycinnamyl alcohol   Coniferyl alcohol   Sinapyl alcohol

Fig. 7.34 Five common structural components of lignin, and the three biosynthetic precursors. Bonds formed by radical coupling are highlighted.

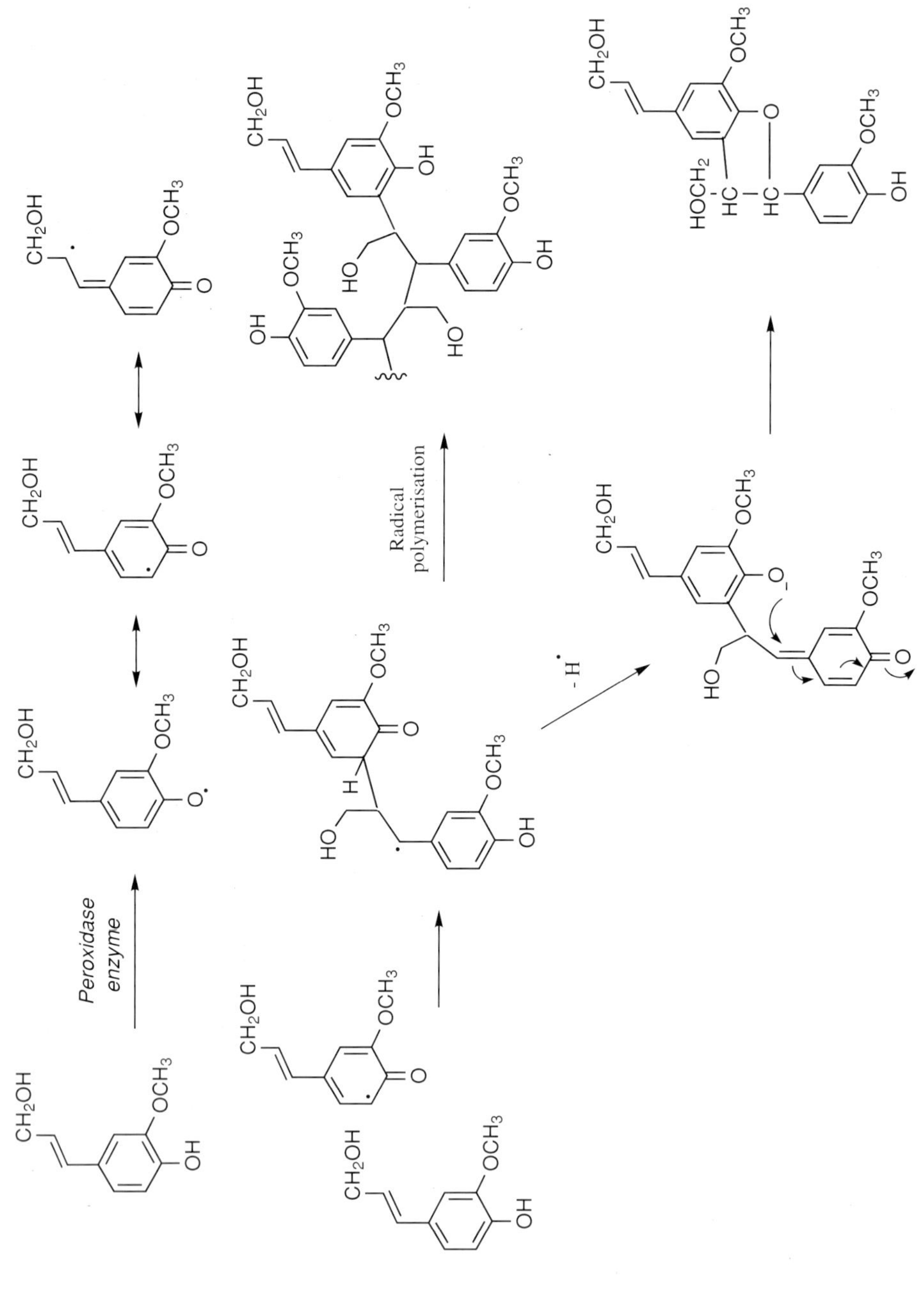

Fig. 7.35 Lignin formation via radical coupling.

carbon linkage. Shown in Fig. 7.34 are a selection of structural components found in lignin. Although structurally diverse, the common feature of these components is that they can all be formed via radical couplings.

The biosynthesis of lignin starts from three cinnamyl alcohol precursors: p-hydroxycinnamyl alcohol, coniferyl alcohol and sinapyl alcohol, shown in Fig. 7.34. The ratio of these precursors used for lignin assembly is species-dependent. They are each activated by formation of a phenoxy radical at the C-4 hydroxyl group by peroxidase, phenolase and tyrosinase enzymes widely found in plants. As shown in Fig. 7.35, the phenoxy radical can exist in a number of resonance structures in which radical character is found, for example, on oxygen at the C-5 position and at the β-position of the side chain. Carbon–carbon bond formation can then take place with the α,β-double bond of another molecule, generating a new radical species which can form a further carbon–carbon bond. Thus, the formation of a phenoxy radical initiates a radical polymerization reaction which forms a highly heterogeneous polymer. Evidence that the polymerization is a chemical reaction and not an enzyme-catalysed reaction comes from the observation that lignin is not optically active, despite containing many chiral centres.

There are also examples of carbon–carbon forming reactions involving the coenzyme vitamin B_{12} which proceed via radical mechanisms. Since these enzymes are classified as isomerases, we shall meet them later in Chapter 10, Section 10.6.

Problems

1 N-Acetylneuraminic acid aldolase catalyses the reaction shown below. Given that the enzyme requires no cofactors, suggest a mechanism. (Hint: use the open chain forms of the monosaccharides.)

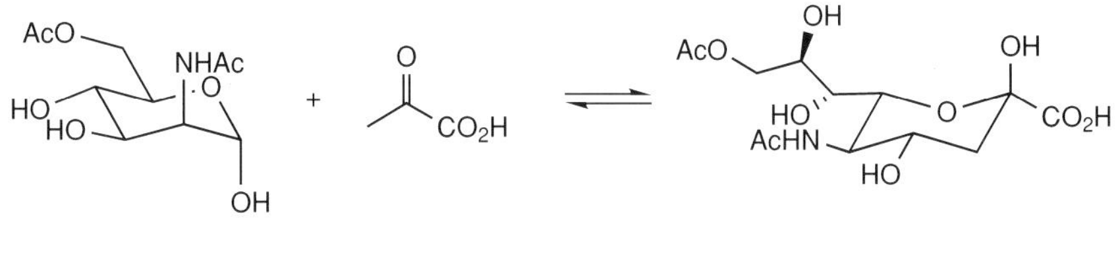

2 The two plant enzymes chalcone synthase and resveratrol synthase catalyse the reactions shown on p. 166. Suggest mechanisms for the two enzymatic reactions. Comment on the observation that when these enzymes were sequenced, they were found to share 70–75% amino acid sequence identity.

3 Phosphoenolpyruvate (PEP) carboxylase catalyses the carboxylation of PEP using Mn^{2+} as a cofactor. When incubated with $HC^{18}O_3^{-}$ two atoms

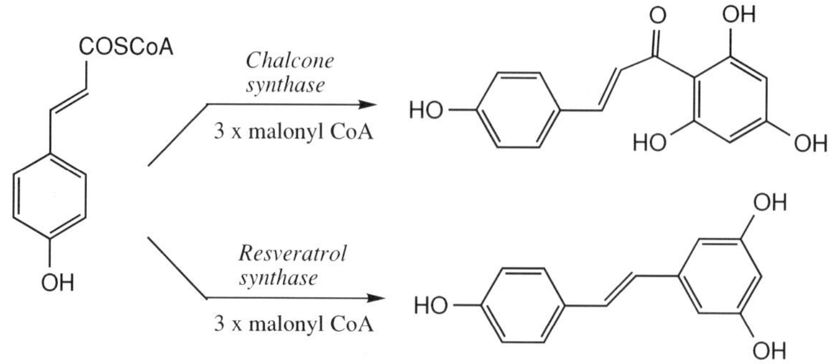

of ^{18}O were found in the oxaloacetate product, and one atom of ^{18}O in inorganic phosphate. Suggest a mechanism.

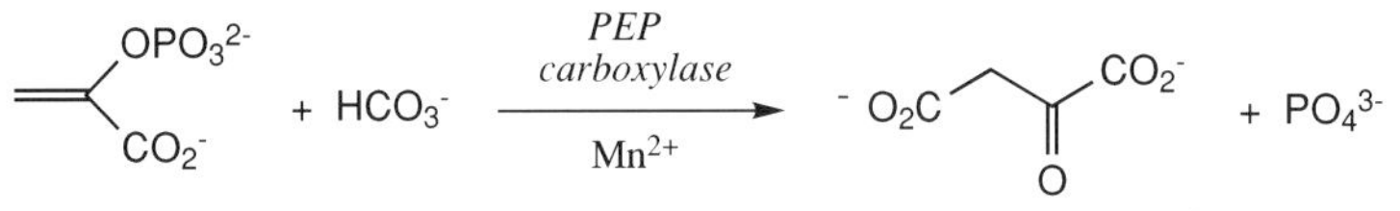

4 Transcarboxylase catalyses the simultaneous carboxylation of propionyl CoA to methylmalonyl CoA and decarboxylation of oxaloacetate to pyruvate, as shown below. The enzyme requires biotin as a cofactor, but does *not* require ATP. Suggest a mechanism.

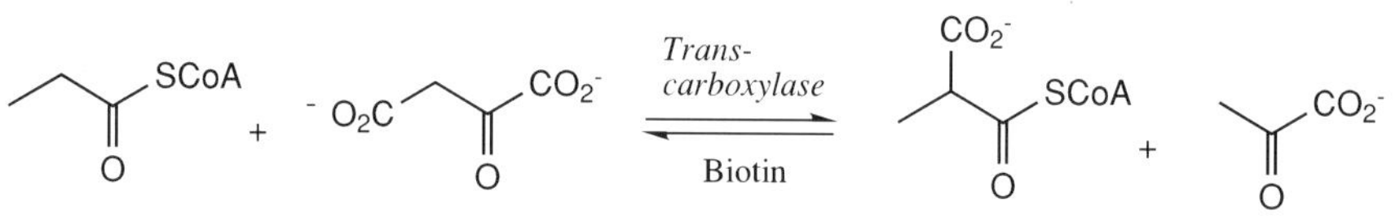

5 Suggest a mechanism for the TPP-dependent transketolase reaction illustrated in Fig. 7.26.

6 Pyruvate oxidase catalyses the oxidative decarboxylation of pyruvate to acetate (not acetyl CoA). The enzyme requires TPP and oxidized flavin as cofactors, and utilizes oxygen as an electron acceptor (i.e. it is converted to hydrogen peroxide). Suggest a mechanism.

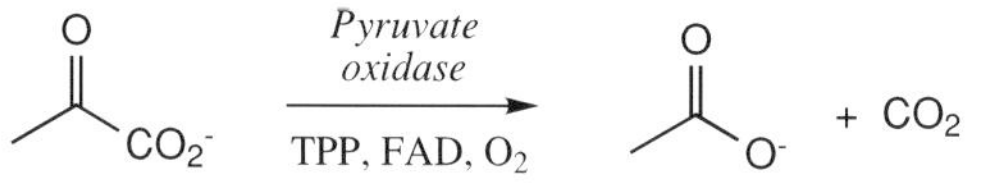

7 The conversion of geranyl pyrophosphate (PP) into bornyl PP is catalysed by bornyl PP synthetase. Write a mechanism for the enzymatic reaction. When the enzyme was incubated with geranyl PP specifically labelled with ^{18}O attached to C-1, the product was found to contain ^{18}O only in the oxygen attached to carbon, and not in any of the other six oxygens. Comment on this remarkable result.

8 Kaurene synthetase catalyses the cyclization of geranylgeranyl PP to kaurene. An intermediate in the reaction is thought to be copalyl PP.

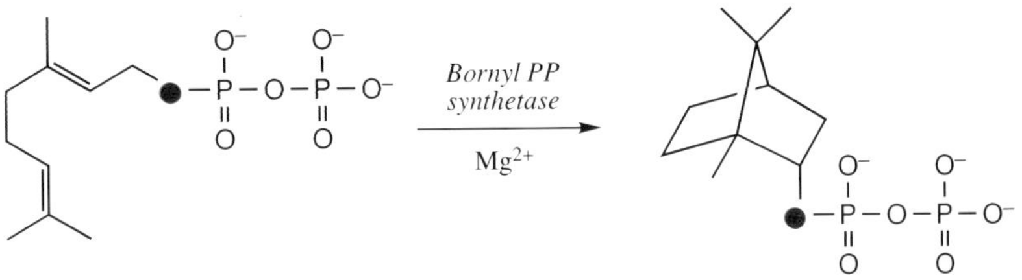

Suggest mechanisms for conversion of geranylgeranyl PP into copalyl PP and thence to kaurene. How would you attempt to prove that copalyl PP is an intermediate in this reaction?

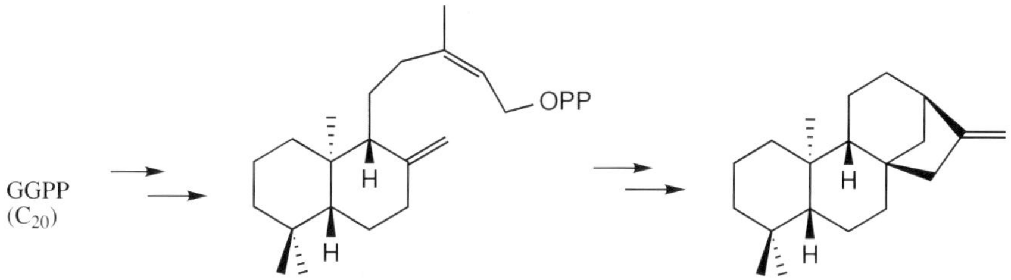

GGPP
(C_{20})

Copalyl PP   Kaurene

9 Usnic acid is a natural product thought to be biosynthesized from two molecules of an aromatic precursor. Suggest a biosynthetic pathway to the aromatic precursor from acetyl CoA, and a mechanism for the conversion to usnic acid.

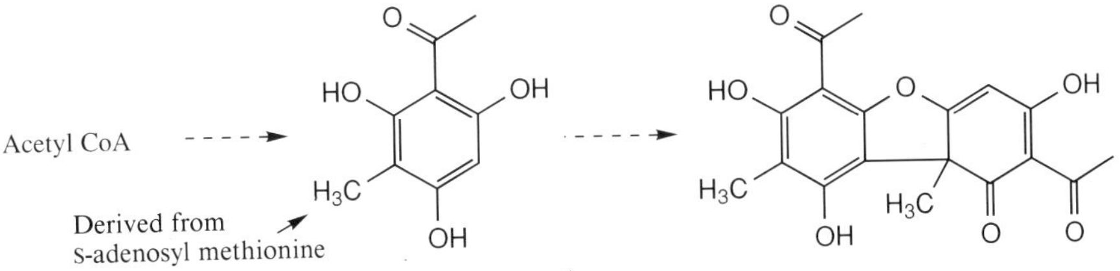

Acetyl CoA

Derived from
s-adenosyl methionine

Usnic acid

Further reading

General

Abeles, R.H., Frey, P.A. & Jencks, W.P. (1992) *Biochemistry*. Jones & Bartlett, Boston.
Mann, J. (1987) *Secondary Metabolism*, 2nd edn. Clarendon Press, Oxford.
Staunton, J.S. (1978) *Primary Metabolism: a Mechanistic Approach*. Clarendon Press, Oxford.
Walsh, C.T. (1979) *Enzymatic Reaction Mechanisms*. Freeman, San Francisco.

Aldolases

Class I

Sygusch, J., Beaudry, D. & Allaire, M. (1987) *Proc Nat Acad Sci USA*, **84**, 7846–50.
Morris, A.J. & Tolan, D.R. (1994) *Biochemistry,* **33**, 12 291–7.

Class II

Belasco, J.G. & Knowles, J.R. (1983) *Biochemistry*, **22**, 122–9.
Smith, G.M., Mildvan, A.S. & Harper, E.T. (1980) *Biochemistry*, **19**, 1248–55.

Use in organic synthesis

Jones, J.B. (1986) *Tetrahedron*, **42**, 3341–3403.
Toone, E.J., Simon, E.S., Bednarski, M.D. & Whitesides, G.M. (1989) *Tetrahedron*, **45**, 5365–422.
Whitesides, G.M. & Wong, C.H. (1985) *Angew Chem Int Ed Engl*, **24**, 617.
Wong, C.H. & Whitesides, G.M. (1994) *Enzymes in Synthetic Organic Chemistry.* Pergamon, Oxford.

Claisen enzymes

Malate synthase

Clark, J.D., O'Keefe, S.J. & Knowles, J.R. (1988) *Biochemistry*, **27**, 5961–71.

Fatty acid synthase

Chang, S.I. & Hammes, G.G. (1990) *Acc Chem Res*, **23**, 363–9.

Erythromycin synthase

Staunton, J.S. (1991) *Angew Chem Int Ed Engl*, **30**, 1302–6.

Carboxylases

Biotin-dependent

Knowles, J.R. (1989) *Annu Rev Biochem*, **58**, 195–222.
Phillips, N.F.B., Shoswell, M.A., Chapman–Smith, A., Keech, D.B. & Wallace, J.C. (1992) *Biochemistry*, **31**, 9445–50.

Rubisco

Hartman, F.C. & Harpel, M.R. (1994) *Adv Enzymol*, **66**, 1–76.

Vitamin K-dependent

Dowd, K.P., Hershline, R., Ham, S.W. & Naganathan, S. (1994) *Nat Prod Rep*, **11**, 251–64.
Suttie, J.W. (1985) *Annu Rev Biochem*, **54**, 459–78.

Thiamine pyrophosphate

Kluger, R. (1990) *Chem Rev*, **90**, 1151–69.
Kluger, R., Chin, J. & Smyth, T. (1981) *J Am Chem Soc*, **103**, 884–8.

Terpene cyclases

Cane, D.E. (1990) *Chem Rev*, **90**, 1089–103.

Pinene cyclase

Croteau, R.B., Wheeler, C.J., Cane, D.E., Ebert, R. & Ha, H.J. (1987) *Biochemistry*, **26**, 5383–9.

Pentalenene synthase

Cane, D.E., Sohng, J.K., Lamberson, S.M. *et al.* (1994) *Biochemistry*, **33**, 5846–57.

Radical couplings

Sulochrin oxidase

Nordlov, H. & Gatenbeck, S. (1982) *Arch Microbiol*, **131**, 208–11.

Lignin biosynthesis

Freudenberg, K. (1965) *Science*, **148**, 595–600.
Higuchi, T. (1971) *Adv Enzymol*, **34**, 207–84.

8 Enzymatic Addition/Elimination Reactions

8.1 Introduction

The addition and elimination of the elements of water is a common process in biochemical pathways. Particularly common is the dehydration of β-hydroxy-ketones or β-hydroxy-carboxylic acids. In this chapter we shall examine the mechanisms employed by hydratase/dehydratase enzymes, and the related ammonia lyases.

In particular we shall discuss several enzymes on the shikimate pathway, which is depicted in Fig. 8.1. This pathway is responsible for the biosynthesis of the aromatic amino acids L-phenylalanine, L-tyrosine and L-tryptophan in plants and micro-organisms. Nature synthesizes the aromatic amino acids starting from D-glucose via a pathway involving a series of interesting elimination reactions. This is an important pathway for plants, since it is also responsible for the biosynthesis of the precursors to the structural polymer lignin, encountered in Chapter 7, Section 7.4. Animals, which do not utilize this pathway, have to consume the aromatic amino acids as part of their diet.

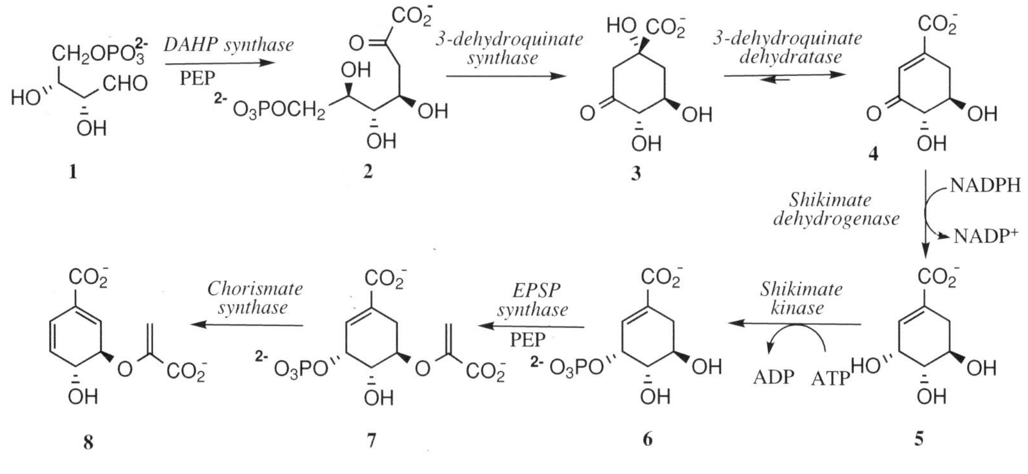

Fig. 8.1 The shikimate pathway. 1, erythrose-4-phosphate; 2, 3-deoxy-D-arabinoheptulosonic acid 7-phosphate (DAHP); 3, 3-dehydroquinate; 4, 3-dehydroshikimate; 5, shikimate; 6, shikimate-3-phosphate; 7, 5-enolpyruvyl shikimate-3-phosphate (EPSP); 8, chorismate. NADP, nicotinamide adenine dinucleotide phosphate; PEP, phosphoenol pyruvate.

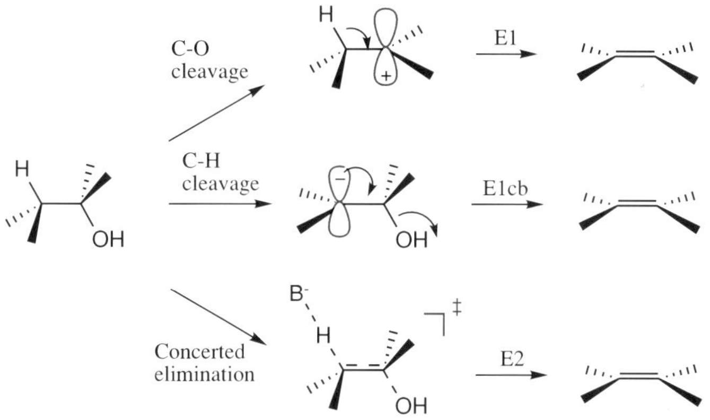

Fig. 8.2 Mechanisms for elimination of water.

The elimination of the elements of water involves cleavage of a C–H bond and cleavage of a C–O bond. The timing of C–H versus C–O cleavage determines the type of elimination mechanism involved, as illustrated in Fig. 8.2. If C–O cleavage takes place first, a carbonium ion is generated which eliminates via cleavage of an adjacent C–H bond. This E1 mechanism is found in organic reactions, but is rare in biological systems. If C–H bond cleavage occurs first a carbanion intermediate is generated, which eliminates via subsequent C–O bond cleavage. This E1cb mechanism is quite common in biological chemistry in cases where the intermediate carbanion is stabilized, as in the elimination of β-hydroxy-ketones or β-hydroxy-thioesters. If C–O and C–H bond cleavage are concerted then a single-step E2 mechanism is followed.

The stereochemical course of enzymatic elimination reactions is strongly dependent upon the mechanism of elimination. If an E2 mechanism is operating then an *anti*-elimination will ensue. This is generally observed in the enzyme-catalysed dehydrations of β-hydroxy-carboxylic acids. However, if an E1cb elimination mechanism is operating then the stereochemistry of the reaction can either be *syn* or *anti* depending upon the positioning of the catalytic groups at the enzyme active site. Commonly, the *syn* stereochemistry is observed for enzyme-catalysed elimination of β-hydroxy-ketones, although this is not true in all cases.

8.2   Hydratases and dehydratases

Enzymes which catalyse the elimination of water are usually known as dehydratases. However, since addition/elimination reactions are reversible sometimes the biologically relevant reaction is the hydration of an alkene, in which case the enzyme would be known as a hydratase. We shall consider

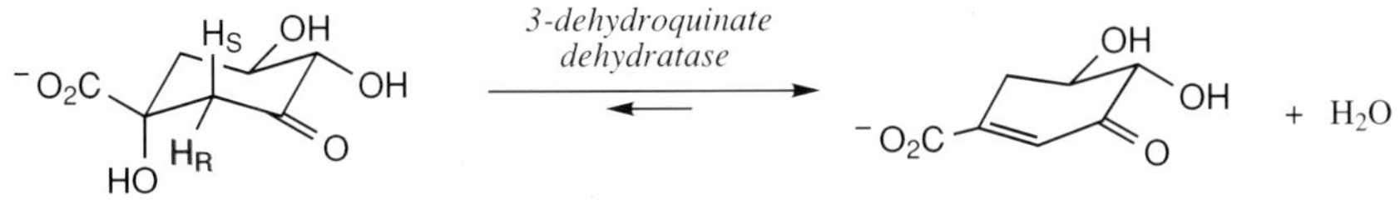

Fig. 8.3 Reaction catalysed by 3-dehydroquinate dehydratase.

in turn enzymes which catalyse the elimination of β-hydroxy-ketones, β-hydroxy-thioesters and β-hydroxy-carboxylic acids.

The dehydration of 3-dehydroquinic acid to 3-dehydroshikimic acid, shown in Fig. 8.3, is a well-studied example of an enzyme-catalysed elimination of a β-hydroxy-ketone. This reaction is catalysed by 3-dehydroquinate dehydratase, the third enzyme on the shikimate pathway.

The enzymatic reaction is reversible, although the equilibrium constant of 16 lies in favour of the forward reaction. The stereochemical course of the enzymatic reaction in *Escherichia coli* was shown, using stereospecifically labelled substrates, to be a *syn*-elimination of the equatorial C-2 *proR* hydrogen and the C-1 hydroxyl group. The stereochemistry of the reaction is remarkable, since the axial C-2 *proS* hydrogen is much more acidic in basic solution than the *proR* hydrogen, due to favourable overlap with the C–O π-bond.

A major clue to the enzyme mechanism is that the enzyme is inactivated irreversibly by treatment with substrate and sodium borohydride. This indicates that an imine linkage is formed between the C-3 ketone and the ε-amino group of an active site lysine (Lys) residue. Peptide mapping studies have established that this residue is Lys-170.

A mechanism for the enzymatic reaction is shown in Fig. 8.4. Upon formation of the imine linkage at C-3, a conformational change is thought to take place, giving a twist boat structure. In this conformation the C–H bond of the *proR* hydrogen lies parallel to the orbital axes of the adjacent C–N π-bond, favouring removal of this hydrogen. Proton abstraction by an active site base leads to the formation of a planar enamine intermediate. This intermediate acts as a stabilized carbanion intermediate for an E1cb elimination mechanism. Extrusion of water, presumably assisted by general acid catalysis, is followed by hydrolysis of the iminium salt linkage to give 3-dehydroshikimic acid.

The elimination of β-hydroxyacyl coenzyme A (CoA) thioesters is one of the reactions involved in the fatty acid synthase cycle described in Chapter 7, Section 7.2. β-Hydroxydecanoyl thioester dehydratase catalyses the reversible dehydration of β-hydroxydecanoyl thioesters to give *trans*-2-decenoyl thioesters, which are subsequently isomerized to give *cis*-3-decenoyl thioesters. This reaction sequence represents a branch point in the biosynthesis of saturated and unsaturated fatty acids, as shown in Fig. 8.5.

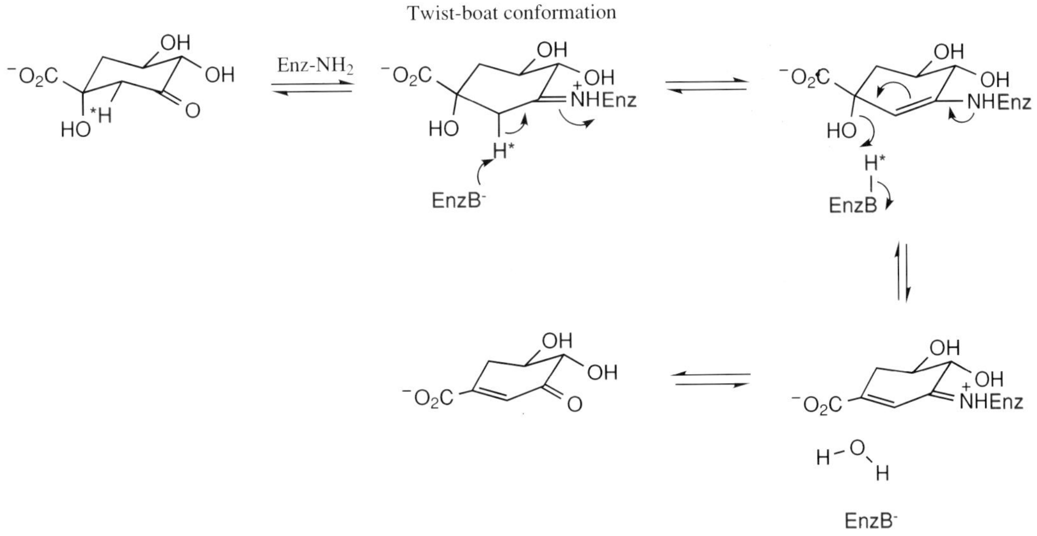

Twist-boat conformation

Fig. 8.4 Mechanism of reaction catalysed by *E. coli* 3-dehydroquinate dehydratase.

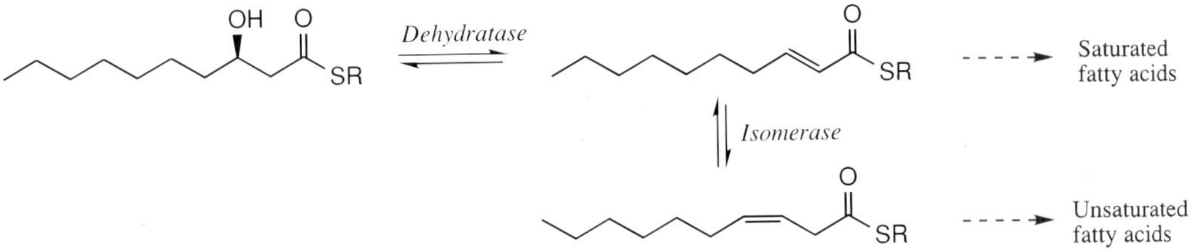

Fig. 8.5 Reactions catalysed by β-hydroxydecanoyl thioester dehydratase.

β-Hydroxydecanoyl thioester dehydratase from *E. coli* is a dimer of subunit molecular weight 18 kDa and requires no cofactors for activity. Stereochemical studies have revealed that the *proS* hydrogen is abstracted at C-2, so elimination of the 3*R* alcohol results in an overall *syn*-elimination of water. The subsequent isomerization involves abstraction of the *proR* hydrogen at C-4. There is considerable evidence for a histidine (His) active site base, which has been identified by peptide mapping studies as His-70. It is thought that His-70 also mediates the isomerization reaction, as shown in Fig. 8.6. However, there is no intramolecular transfer of deuterium from C-4 to C-2, suggesting that the protonated His-70 exchanges readily with solvent water.

This enzyme is specifically inactivated by a substrate analogue 3-decynoyl-*N*-acetyl-cysteamine. The α-proton of the inhibitor is abstracted by His-70 in the normal fashion, but protonation of the alkyne at the γ-position generates a highly reactive allene intermediate. This intermediate is then attacked by the nearby His-70 resulting in irreversible inactivation,

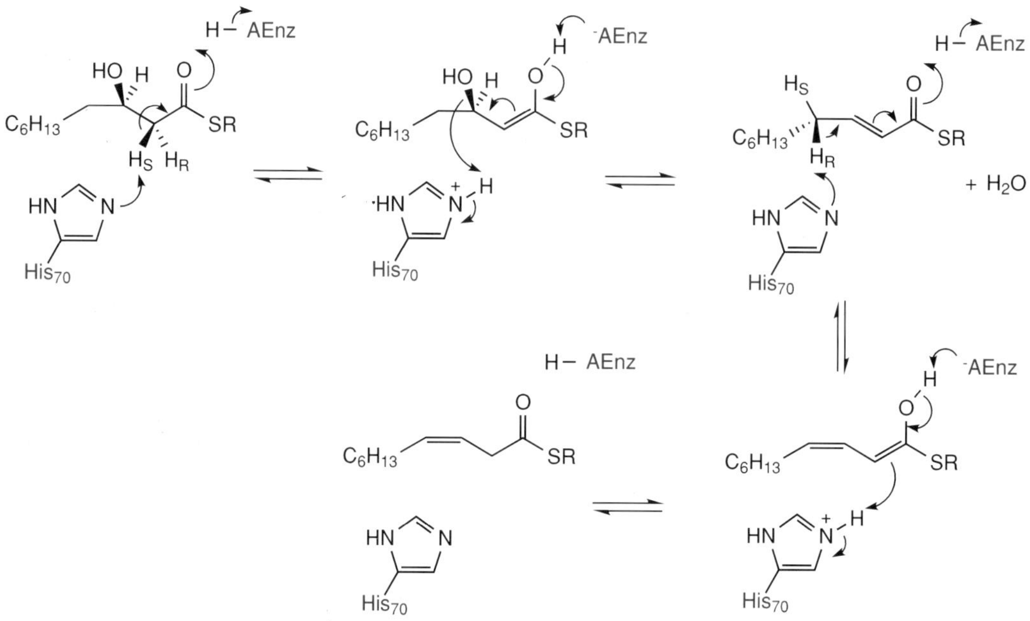

Fig. 8.6 Mechanism for β-hydroxydecanoyl thioester dehydratase.

as shown in Fig. 8.7. Since the inhibitor is activated by deprotonation at C-2 the enzyme brings about its own inactivation by way of the enzyme mechanism. This class of mechanism-based inhibitors are therefore often known as 'suicide inhibitors'.

The dehydration of β-hydroxycarboxylic acids is a reaction which occurs

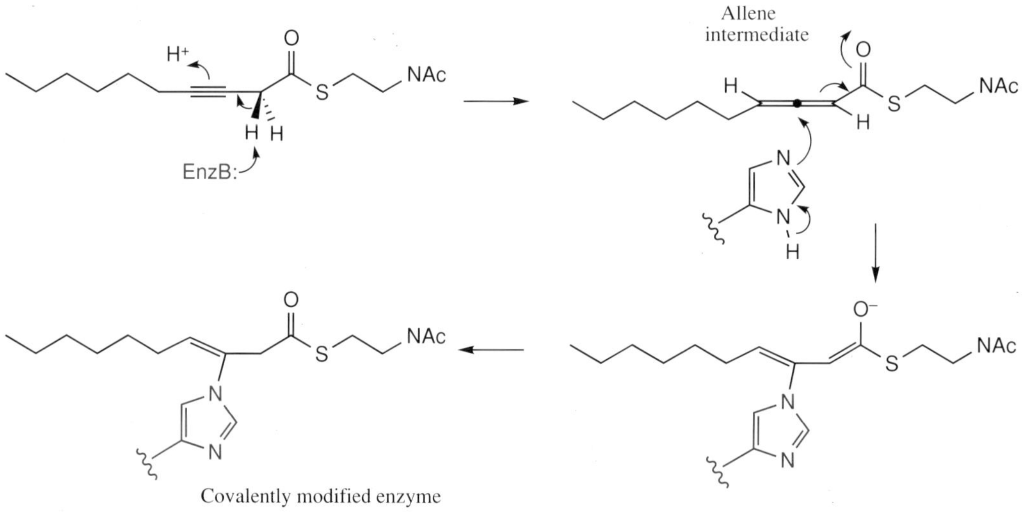

Fig. 8.7 Irreversible inhibition of β-hydroxydecanoyl thioester dehydratase.

quite frequently in biochemical pathways. Two dehydrations which occur on the citric acid cycle are shown in Fig. 8.8: (i) the dehydration of citrate and re-hydration of *cis*-aconitate to give isocitrate, catalysed by the single enzyme aconitase; and (ii) the hydration of fumarate to malate catalysed by the enzyme fumarase. Both these enzymes were found at an early stage to be dependent upon iron for activity.When the enzymes were purified and characterized, both enzymes were found to contain Fe_4S_4 iron–sulphur clusters at their active sites. The usual biological function of iron–sulphur clusters, as explained in Chapter 6, Section 6.7, is single electron transport. What then is the role of an iron–sulphur cluster in a hydratase enzyme?

Determination of an X-ray crystal structure for mitochondrial aconitase revealed that one of the iron atoms in the cluster co-ordinates the hydroxyl group of isocitrate. This observation suggests that the role of the iron–sulphur cluster is to function as a Lewis acid group to facilitate C–O cleavage during the catalytic mechanism. Subsequently, the activation of water as iron(II) hydroxide could provide a reactive nucleophile for water addition. Examination of the enzyme crystal structure indicated that serine (Ser)-642 is well positioned as the base for proton abstraction of isocitrate. The stereochemical course of both citrate and isocitrate dehydrations has been shown to be *anti*, thus an E2 mechanism is likely, as shown in Fig. 8.9. It has also been shown that the proton abstracted by Ser-642 is returned upon hydration of *cis*-aconitate (although the hydroxyl group removed is exchanged). This implies that Ser-642 is somehow shielded from solvent water and also that the intermediate *cis*-aconitate is flipped over in the active site before re-hydration.

This brief discussion covers by no means all of the types of hydratase enzymes found in biological systems, but illustrates a few of the most

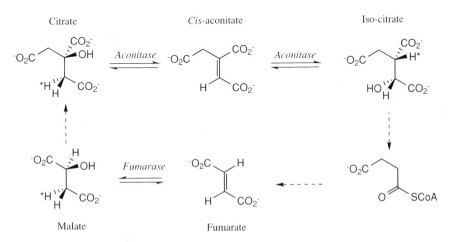

Fig. 8.8 Reactions catalysed by aconitase and fumarase.

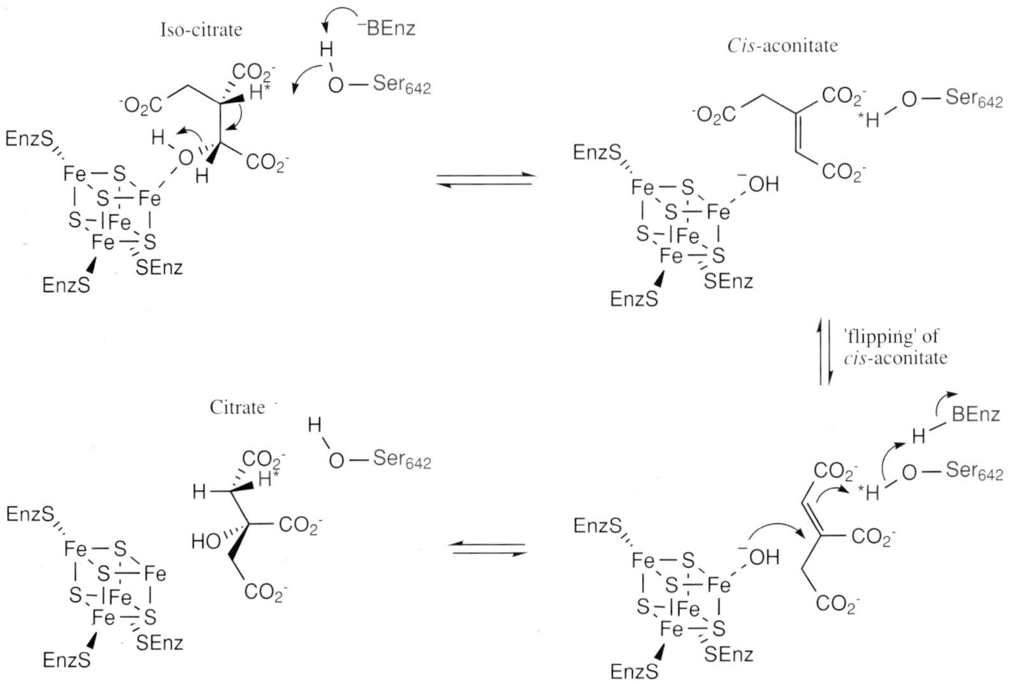

Fig. 8.9 Mechanism for aconitase.

common examples, and the types of mechanistic and stereochemical methods used in other cases.

8.3   Ammonia lyases

The enzymatic elimination of ammonia is carried out on several L-amino acids in biological systems: L-phenylalanine, L-tyrosine, L-histidine and L-aspartic acid, as shown in Fig. 8.10. In each of these cases the C_β–H bond being broken lies adjacent to an activating group in the β-position of the amino acid side chain.

The histidine and phenylalanine ammonia lyases utilize a novel mechanism to assist the leaving group properties of ammonia, which would nor-

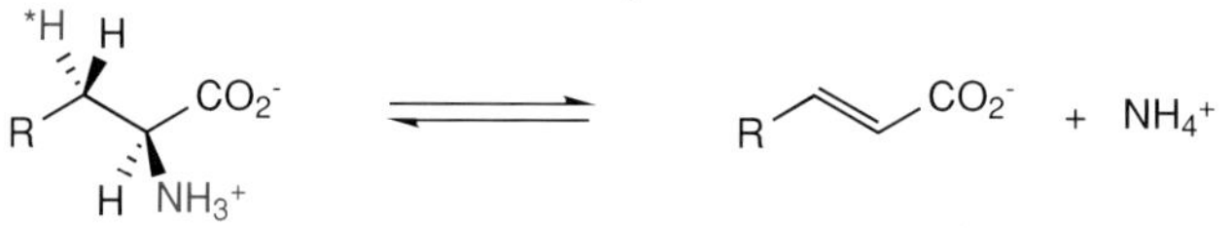

R = Ph	phenylalanine ammonia lyase
R = imidazole	histidine ammonia lyase
R = CO$_2$H	aspartase

Fig. 8.10 Reactions catalysed by some ammonia lyases.

mally be a poor leaving group. These enzymes are readily inactivated by treatment with nucleophilic reagents, implying that there is an electrophilic group present at the enzyme active site. Examination of purified enzyme revealed that the electrophilic cofactor has a characteristic ultraviolet absorption at 340 nm.

Inactivation of phenylalanine ammonia lyase with NaB^3H_4 followed by acidic hydrolysis of the protein gave ^3H-alanine. This implies that the electrophilic group is a dehydroalanine amino acid residue, formed by dehydration of a serine residue. In the *Pseudomonas* histidine ammonia lyase the site of the dehydroalanine residue has been identified as position 143, derived from Ser-143. Substitution for threonine (Thr)-143 using site-directed mutagenesis gives an inactive enzyme; however, a cysteine-(Cys)-143 mutant enzyme is able to form the dehydroalanine cofactor by elimination of hydrogen sulphide.

A mechanism for the phenylalanine ammonia lyase reaction is illustrated in Fig. 8.11. Attack of the α-amino group of the substrate upon the dehydroalanine residue generates a secondary amine intermediate. *Anti-*

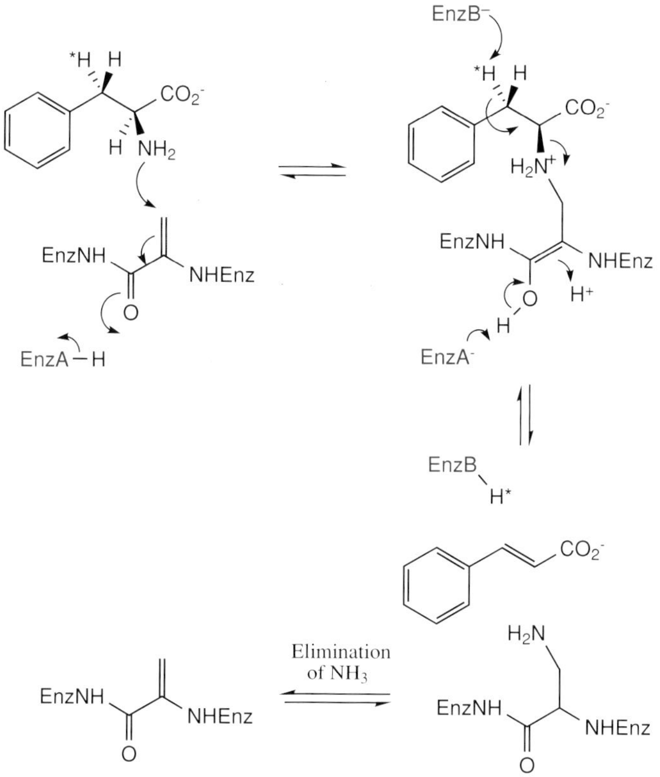

Fig. 8.11 Mechanism for phenylalanine ammonia lyase.

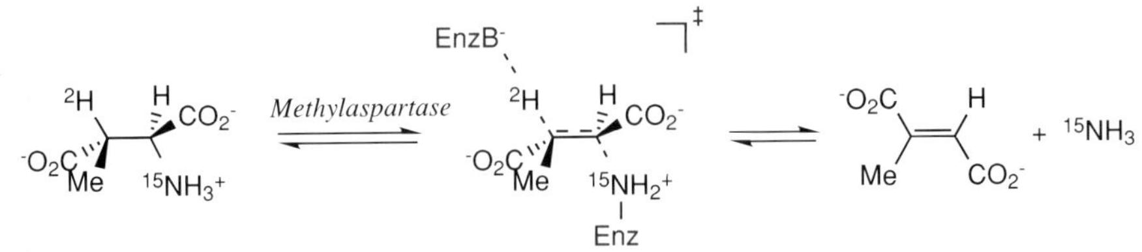

Fig. 8.12 Concerted C–H and C–N cleavage in the methylaspartase reaction.

elimination of this intermediate leaves an aminomethyl–glycine inter-mediate, which is re-cycled to the dehydroalanine residue by elimination of ammonia.

The question then arises: is the elimination of the substrate concerted or stepwise? This question has been addressed using kinetic isotope effects in the reaction of methylaspartase, shown in Fig. 8.12. This enzyme catalyses the *anti*-elimination of *threo*-β-methyl-aspartic acid, also utilizing a dehydroalanine cofactor. Measurement of the rate of the enzymatic reaction using the [3-^{2}H] substrate revealed a primary kinetic isotope effect of 1.7, indicating that C–H bond cleavage is partially rate-determining. However, there was also found to be a ^{15}N isotope effect of 1.025 upon the [2-^{15}N]-labelled substrate. If the reaction is concerted, then both of these kinetic isotope effects are operating on the same step, in which case the effects should be additive. So, a [3-^{2}H,2-^{15}N] substrate was prepared and a kinetic isotope effect of 1.042 was observed, indicating that the isotope effects are additive and that the elimination is indeed concerted.

8.4   Elimination of phosphate and pyrophosphate

In all elimination reactions an important determinant of reaction rate and mechanism is whether a good leaving group is available. Elimination of water is hindered by the fact that the hydroxyl group is a poor leaving group, since the pK$_a$ of the conjugate acid water is 15.7. In the dehydratase enzymes this problem is alleviated by acid or Lewis acid catalysis. However, another strategy found in biochemical pathways for provision of an efficient leaving group is phosphorylation of the leaving group.

The pK$_a$ values for the three dissociation equilibria of phosphoric acid, as shown in Fig. 8.13, are 2.1, 7.2 and 12.3. At neutral pH a phosphate monoester would be singly deprotonated, thus for departure of a phosphate leaving group the pK$_a$ of its conjugate acid (H$_2$PO$_4^-$) would be 7.2—thus, phosphate is a much better leaving group than hydroxide ion.

Two important examples of the elimination of phosphate are shown in Fig. 8.14. The first is the formation of isopentenyl pyrophosphate by an eliminative decarboxylation. The enzyme catalysing this reaction has been

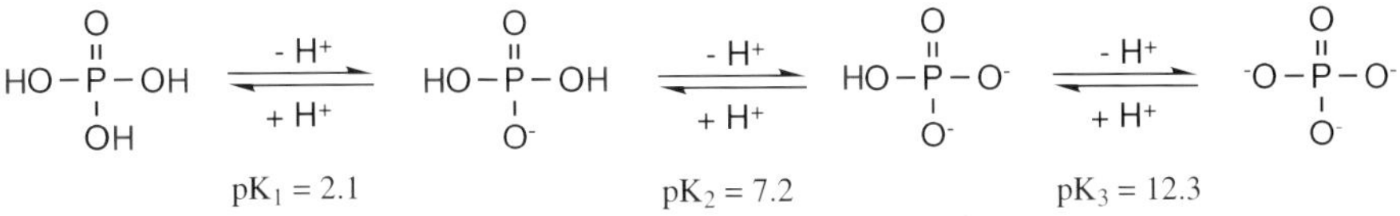

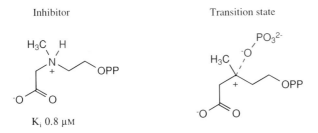

Fig. 8.13 pK_a values for phosphate anions.

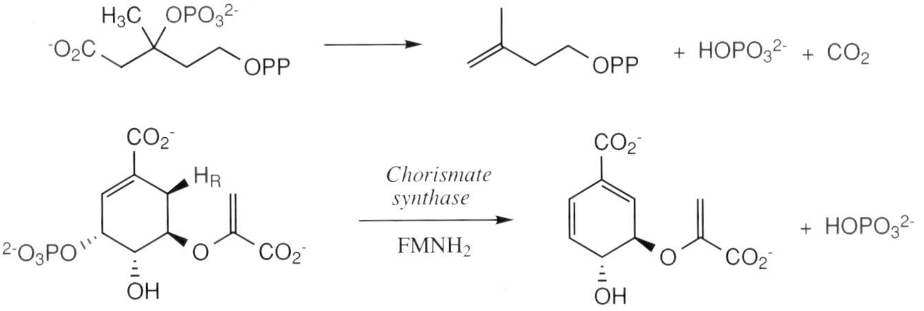

Fig. 8.14 Enzymatic eliminations of phosphate. $FMNH_2$, Flavin mononucleotide (reduced).

purified from baker's yeast, and has been shown to be strongly inhibited by a tertiary amine substrate analogue ($K_i = 0.8$ μM). At neutral pH the tertiary amine will be protonated, thus the potent inhibition can be explained by this analogue mimicking a tertiary carbonium ion intermediate in the mechanism of this enzyme, as shown in Fig. 8.15.

Chorismate synthase catalyses the 1,4-elimination of phosphate from 5-enolpyruvyl-shikimate-3-phosphate (EPSP) to give chorismic acid (see Fig. 8.1). The stereochemistry of the enzymatic elimination in *E. coli* has been shown to be *anti*. The highly unusual feature of this enzymatic reaction is that the enzyme requires reduced flavin as a cofactor, although the overall reaction involves no change in redox level. Recent experiments have indicated that the flavin cofactor is involved somehow in the elimination reaction, since incubation of the 6R-6-fluoro analogue (in which the proton removed is substituted with fluorine) with enzyme leads to the observation of a flavin semiquinone species. It therefore appears that one-electron transfers may be involved in the mechanism employed by this enzyme.

Fig. 8.15 Transition state inhibitor for mevalonate pyrophosphate decarboxylase.

8.5   **Case study: EPSP synthase**

There are a number of enzymes which catalyse multi-step addition–
elimination reactions, of which the best characterized is the enzyme EPSP
synthase. This enzyme catalyses the sixth step on the shikimate pathway,
namely the transfer of an enolpyruvyl moeity from phosphoenol pyruvate
(PEP) to shikimate-3-phosphate (see Fig. 8.1).

Early experiments on the mechanism of this enzyme examined ^{3}H
exchange processes. Incubation of [3-^{3}H]-PEP with enzyme and shikimate-
3-phosphate led to the release of ^{3}H label into solvent, consistent with the
existence of an intermediate containing a rotatable methyl group, as shown
in Fig. 8.16.

Subsequent overexpression of this enzyme allowed a detailed examination of the reaction using stopped flow kinetics and rapid quench methods.
Analysis of a single turnover of the enzymatic reaction by stopped flow
methods revealed the existence of an intermediate formed from shikimate-
3-phosphate and PEP after 5 ms and eventually consumed after 50 ms.
Incubation of large amounts of enzyme with shikimate-3-phosphate and
PEP followed by rapid quench of the emzymatic reaction into 100% triethylamine led to the isolation of small amounts of the desired intermediate. This
substance was a good substrate for both the forward and reverse reactions of
EPSP synthase, satisfying the criteria for establishment of an intermediate in
an enzymatic reaction.

Large-scale rapid quench studies using 500 mg of pure enzyme gave
300 µg of intermediate, which was characterized by ^{1}H, ^{13}C and ^{31}P nuclear
magnetic resonance (NMR) spectroscopy, verifying the structure of the
tetrahedral intermediate previously suspected. This intermediate was subsequently observed transiently at the active site of the enzyme by NMR
spectroscopy. Synthetic phosphonate analogues of the tetrahedral intermediate have been synthesized and were found to be potent inhibitors of the
enzyme as shown in Fig. 8.17.

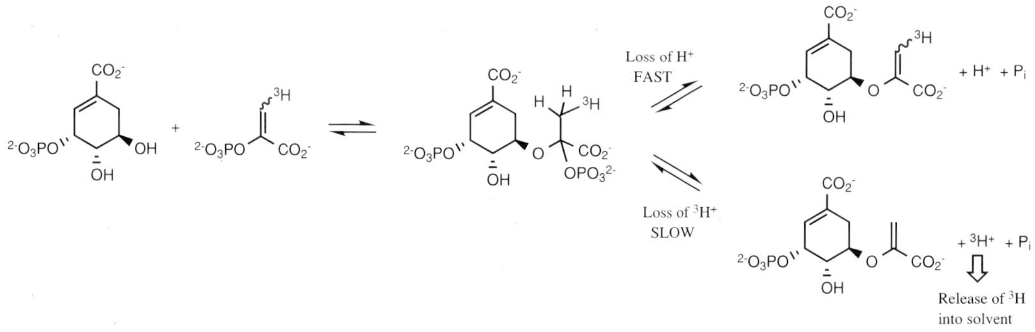

Fig. 8.16 ^{3}H exchange in the EPSP synthase reaction.

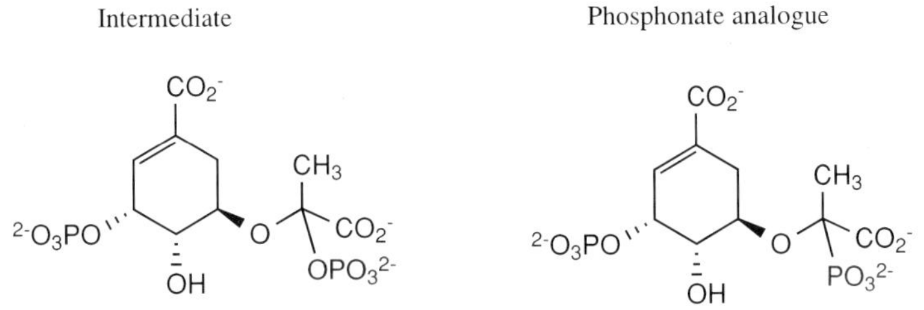

Intermediate   Phosphonate analogue

Fig. 8.17 Tetrahedral intermediate in the EPSP synthase reaction.

The X-ray crystal structure of EPSP synthase is shown in Plate 8.1 (facing p. 152). Examination of the structure reveals that the enzyme contains two domains connected by a flexible hinge, suggesting that there is a conformational change of the protein occurring during catalytic turnover. As yet there are no firm indications of where the enzyme active site is located.

So why all the interest in this enzyme? The reason is that EPSP synthase is the target for a herbicide called glyphosate, or phosphonomethyl-glycine. Inhibition of the shikimate pathway in plants is catastrophic, since the plant can no longer synthesize the aromatic amino acids required for metabolism and for the construction of the structural polymer lignin. However, glyphosate is non-toxic to animals as they do not utilize the shikimate pathway and hence have to consume the aromatic amino acids as part of their diet. Since glyphosate is non-toxic and remarkably biodegradable (by enzymes which cleave carbon–phosphorus bonds!), • it represents an 'environmentally friendly' herbicide. How does it inhibit EPSP synthase? It is thought that the protonated nitrogen of glyphosate mimics the transition state for attack of shikimate-3-phosphate upon PEP. It is likely that significant positive charge accumulates in this transition state on C-2 of PEP, as shown in Fig. 8.18.

There is only one other example of an enolpyruvyl transfer reaction, the conversion of UDP-*N*-acetyl-glucosamine to enolpyruvyl-UDP-*N*-acetyl-glucosamine. The transferase enzyme responsible for this transformation has been purified, overexpressed and analysed by similar approaches. A tetrahedral intermediate has been found for this enzyme also, although there is evidence that PEP forms a reversible adduct with an active site cysteine residue. A mechanism for this enzyme is shown in Fig. 8.19.

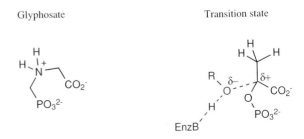

Glyphosate   Transition state

Fig. 8.18 Inhibition of EPSP synthase by glyphosate.

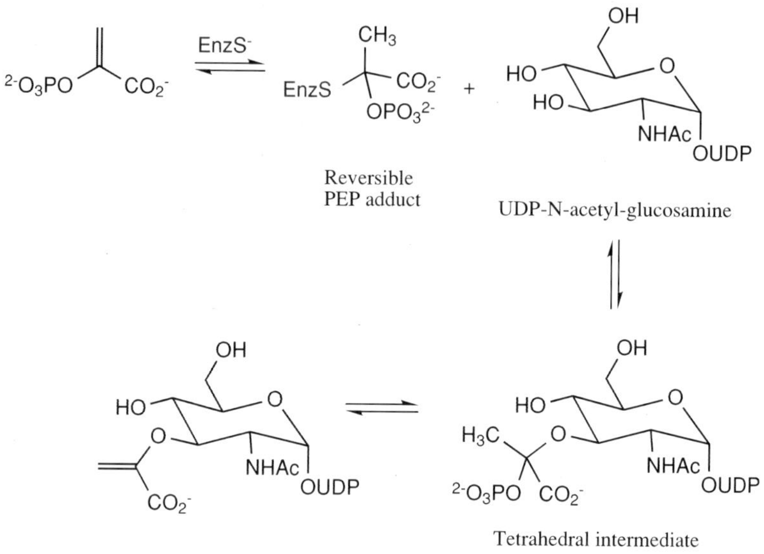

Fig. 8.19 Reaction catalysed by UDP-*N*-acetyl-glucosamine enolpyruvyl transferase.

Problems

1 In the 3-dehydroquinate dehydratase reaction, how would you identify the active site lysine residue which forms an imine linkage with the substrate?

2 The following intramolecular addition reaction is catalysed by a cyclo-isomerase enzyme from *Pseudomonas putida*. When the cycloisomerase reaction was carried out in ^3H_2O, product was found to contain tritium at C-5 and was found to have the 4*S*, 5*R* stereochemistry. Deduce whether the addition reaction occurs with *syn* or *anti* stereochemistry, and suggest a mechanism for the reaction.

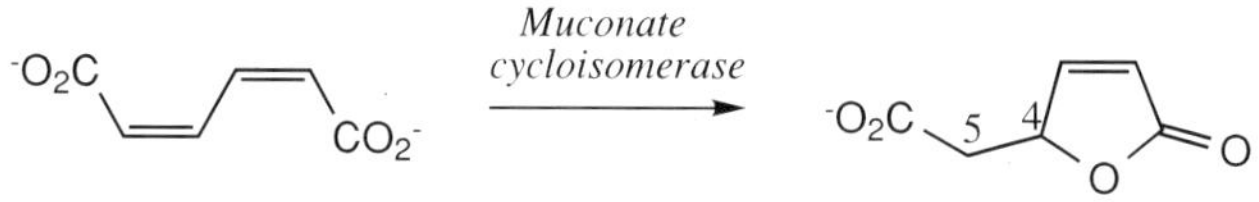

3 Dihydroxy acid dehydratase is an iron–sulphur cluster-containing hydratase enzyme which catalyses the interconversion of α,β-dihydroxy acids with α-keto acids. Suggest possible mechanisms for this enzyme, and predict what would be observed for each mechanism if an α-^2H-dihydroxy acid substrate was incubated with the enzyme.

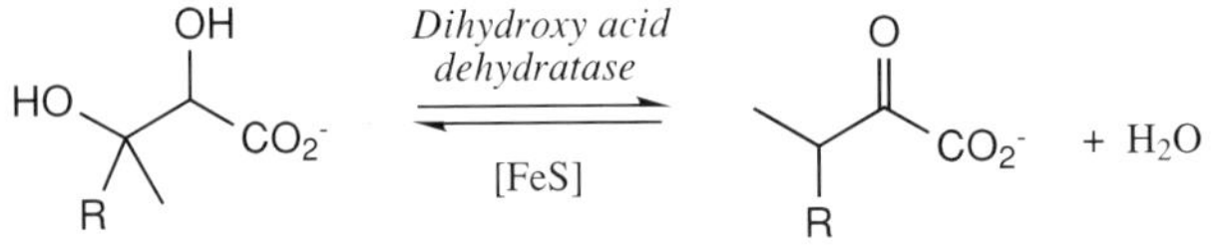

4 S-Adenosyl homocysteine hydrolase catalyses the conversion of S-adenosyl homocysteine to adenosine and homocysteine. Given the following data, suggest a mechanism for this enzyme: (i) the enzyme contains a catalytic amount (1 mol mol^{-1} enzyme subunit) of tightly bound nicotinamide adenine dinucleotide (NAD$^+$); (ii) the enzyme catalyses the exchange of the C–4′ hydrogen (H*) with ^{2}H$_2$O.

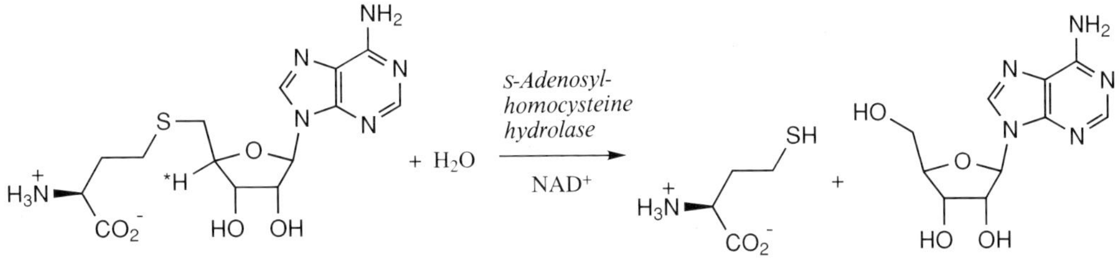

S-Adenosylhomocysteine

5 Chorismic acid (see Fig. 8.1) is a substrate for the three enzymes shown below: anthranilate synthase, p-aminobenzoate synthase and isochorismate synthase. The amino acid sequences of the three enzymes are similar, suggesting that they may follow a similar mechanistic course. Suggest intermediates and possible mechanisms for these three reactions.

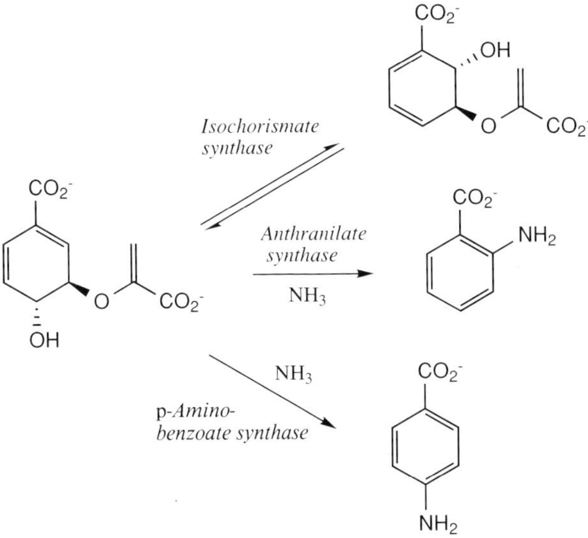

Further reading

General

Abeles, R.H., Frey, P.A. & Jencks, W.P. (1992) *Biochemistry*. Jones & Bartlett, Boston.
Walsh, C.T. (1979) *Enzymatic Reaction Mechanisms*. Freeman, San Francisco.

Dehydratases

3-Dehydroquinate dehydratase

Butler, J.R., Alworth, W.L. & Nugent, M.J. (1974) *J Am Chem Soc*, **96**, 1617–18.
Turner, M.J., Smith, B.W. & Haslam, E. (1975) *J Chem Soc Perkin Trans*, **1**, 52–5.

Thiol ester dehydratase

Schwab, J.M. & Henderson, B.S. (1990) *Chem Rev*, **90**, 1203–45.

Aconitase

Lauble, H., Kennedy, M.C., Beinert, H. & Stout, C.D. (1992) *Biochemistry*, **31**, 2735–48.

Ammonia lyases

Histidine ammonia lyase

Langer, M., Reck, G., Reed, J. & Retey, J. (1994) *Biochemistry*, **33**, 6462–7.
Langer, M., Lieber, A. & Retey, J. (1994) *Biochemistry*, **33**, 14 034–8.

Methylaspartase

Botting, N.P., Jackson, A.A. & Gani, D. (1989) *J Chem Soc Chem Commun*, 1583–5.

Elimination of phosphates

Mevalonate pyrophosphate

Dhe-Paganon, S., Magrath, J. & Abeles, R.H. (1994) *Biochemistry*, **33**, 13 355–62.

Chorismate synthase

Ramjee, M.N., Coggins, J.R., Hawkes, T.R., Lowe, D.J. & Thorneley, R.N.F. (1991) *J Am Chem Soc*, **113**, 8566–7.
Ramjee, M.N. Balasubramanian, S., Abell, C., *et al.* (1992) *J Am Chem Soc*, **114**, 3151–3.
Welch, D.R., Cole, K.W. & Gaertner, F.H. (1974) *Arch Biochem Biophys*, **165**, 505–18.

EPSP synthase

Anderson, K.S. & Johnson, K.A. (1990) *Chem Rev*, **90**, 1131–49.
Stallings, W.C., Abdel-Meguid, S.S., Lim, L.W. *et al.* (1991) *Proc Natl Acad Sci*, **88**, 5046–50.

9 Enzymatic Transformations of Amino Acids

9.1 Introduction

α-Amino acids are primary cellular metabolites which are required for the assembly of proteins, as discussed in Chapter 2. They are also used for the biosynthesis of alkaloids—a class of nitrogen-containing natural products found in many plants. There is, therefore, a sizeable group of enzymatic reactions involved in the biosynthesis, breakdown and transformation of α-amino acids. In this chapter we shall discuss the common enzymatic transformations of α-amino acids, focusing in particular on the role of the coenzyme pyridoxal-5′-phosphate (PLP).

Illustrated in Fig. 9.1 are the general types of transformations found for α-amino acids. As discussed in Chapter 2, the amino acids used for protein

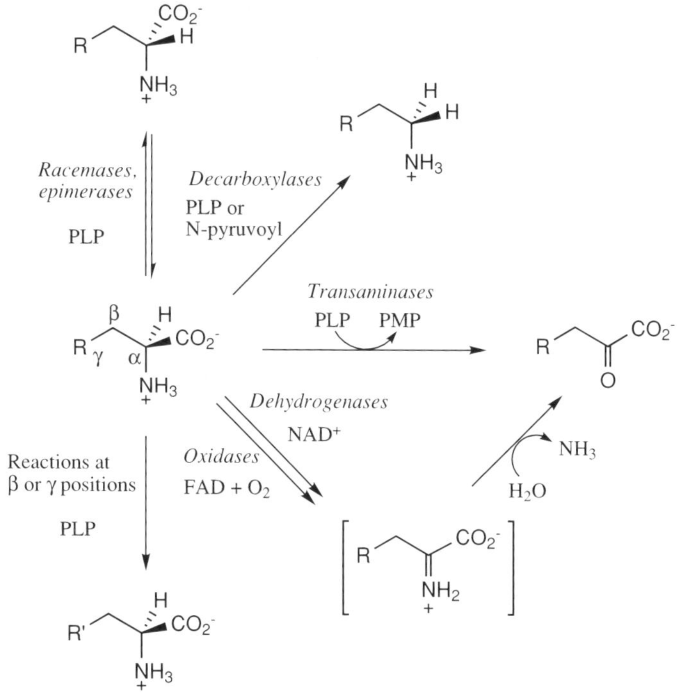

Fig. 9.1 Types of enzymatic transformations of α-amino acids. FAD, flavin adenine dinucleotide; NAD, nicotinamide adenine dinucleotide; PLP, pyridoxal 5′-phosphate; PMP, pyridoxamine 5′-phosphate.

185

biosynthesis exclusively have the L-configuration. However, there are a number of racemase and epimerase enzymes producing D-amino acids which are used for a small number of specific purposes in biological systems. α-Amino acid decarboxylases produce the corresponding primary amines, some of which have important bodily functions in mammals, and others which are used in the biosynthesis of alkaloids in plants. Oxidation of α-amino acids has been mentioned in Chapter 6: there are several nicotinamide adenine dinuclectide (NAD$^+$)- and flavin-dependent dehydrogenase and oxidase enzymes which oxidize amino acids via the corresponding iminium salt to the α-keto acid. Imine intermediates then make possible a number of further transformations as we shall see later in the chapter.

9.2   PLP-dependent reactions at the α-position

Pyridoxal 5′-phosphate is a coenzyme derived from vitamin B$_6$ (pyridoxine). Pyridoxine is oxidized and phosphorylated in the body to give the active form of the coenzyme, as shown in Fig. 9.2. This vitamin was first isolated from rice bran in 1938, and was found to be active against the deficiency disease pellagra.

A wide range of reaction types are catalysed by PLP-dependent enzymes; however, in general the substrates for these enzymes are α-amino acids. The structural features of the coenzyme which make this chemistry possible are a pyridine ring which acts as an electron sink, and an aldehyde substituent at the C-4 position through which the coenzyme becomes covalently attached to the amino acid substrate.

Enzymes which utilize PLP bind the cofactor through an imine linkage between the aldehyde group of PLP and the ε-amino group of an active site lysine (Lys) residue. At neutral pH this imine linkage is protonated to form a more electrophilic iminium ion. Upon binding of the α-amino acid substrate, the α-amino group attacks the iminium ion, displacing the lysine residue and forming an imine linkage itself with the pyridoxal cofactor. This aldimine intermediate, shown in Fig. 9.2, is the starting point for each of the mechanisms which we shall meet in the following sections. Although for sake of simplicity I shall write the phenolic hydroxyl group of PLP in protonated form, there is evidence that it is deprotonated when bound to the enzyme, and that the phenolate anion forms a hydrogen bond to the protonated iminium ion, as shown in Fig. 9.2.

Formation of the aldimine adduct dramatically increases the acidity of the amino acid α-proton. This activation is utilized by a family of racemase and epimerase enzymes which utilize PLP as a cofactor. In these enzymes formation of the aldimine intermediate is followed by abstraction of the α-proton of the amino acid utilizing the pyridine ring as an electron sink and

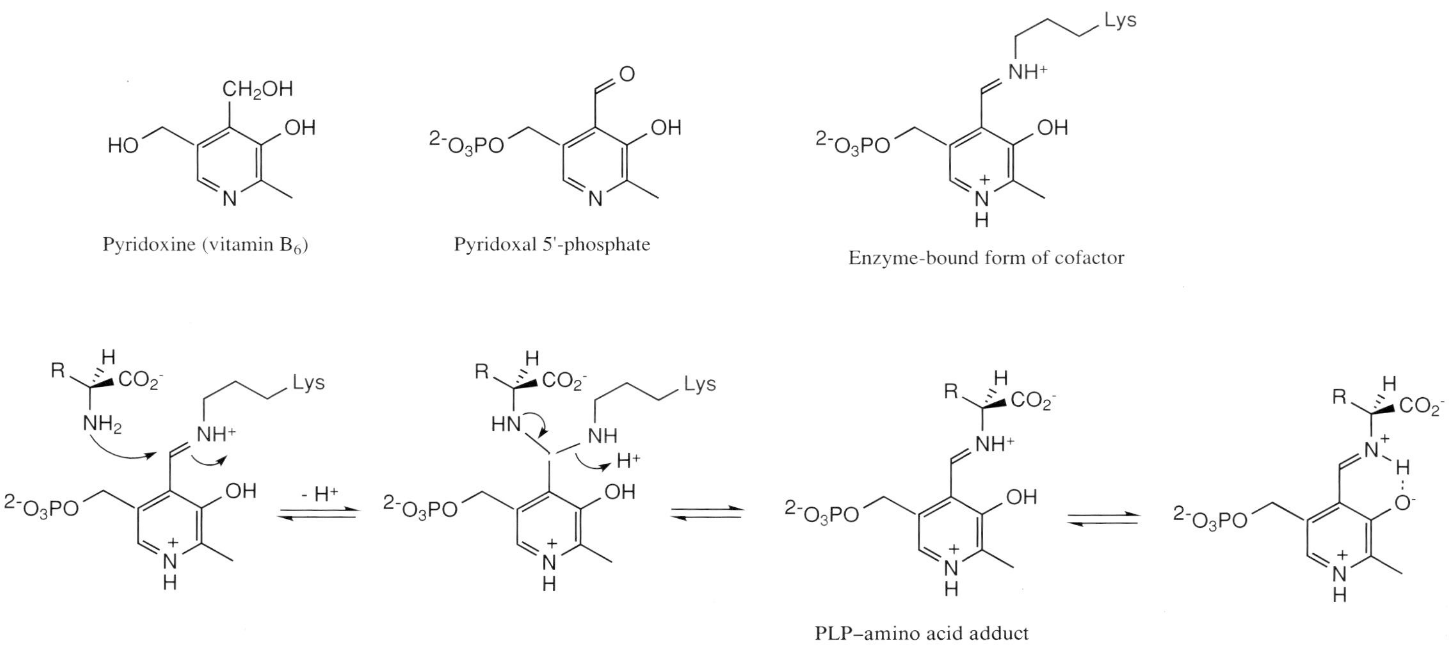

Fig. 9.2 PLP and its attachment to PLP-dependent enzymes.

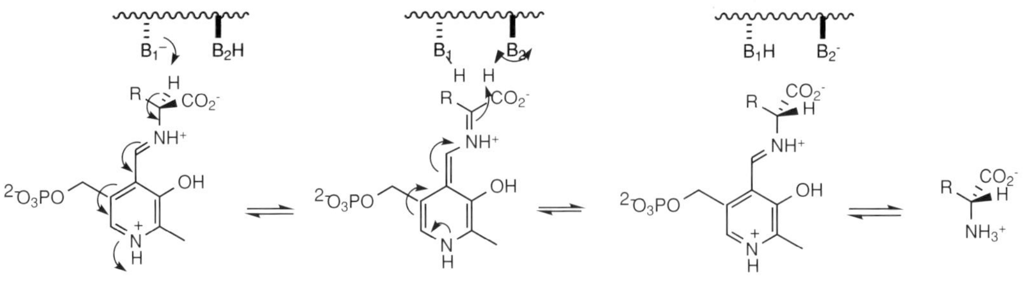

Fig. 9.3 Mechanism of PLP-dependent racemases (two-base mechanism illustrated).

generating a quinonoid species. Delivery of a proton from the opposite face of the molecule results in inversion of configuration at the α-centre, as shown in Fig. 9.3. Detachment of the product from the coenzyme is carried out by attack of the active site lysine residue.

In some racemases re-protonation is carried out by a second active site residue (the 'two-base' mechanism). In other cases deprotonation and re-protonation are carried out by a single active site base which is able to access both faces of the ketimine adduct. The latter 'one-base' mechanism can be demonstrated in a single turnover experiment by incubating a 2-^{2}H-L-amino acid substrate with a stoichiometric amount of enzyme. Isolation of D-amino acid product containing deuterium at the α-position implies intramolecular atom transfer by a single active site base.

One important example of a PLP-dependent racemase is alanine (Ala) racemase, which is used by bacteria to produce D-alanine. D-Alanine is then incorporated into the peptidoglycan layer of bacterial cell walls in the form of a D-Ala-D-Ala dipeptide. Inhibition of alanine racemase is therefore lethal to bacteria, since without peptidoglycan the cell walls are too weak to withstand the high osmotic pressure, and the bacteria lyse. One inhibitor of alanine racemase which has antibacterial properties is β-chloro-D-alanine, which inhibits the enzyme via an interesting mechanism shown in Fig. 9.4. β-Chloro-D-alanine is accepted as a substrate by the enzyme, which proceeds to bind the inhibitor covalently to its PLP cofactor. However, once in 800 turnovers deprotonation at the α-position is followed by loss of chloride, generating a PLP-bound enamine intermediate. This is detached from the PLP cofactor by attack of the lysine ε-amino group; however, the liberated free enamine reacts with the carbon centre of the PLP–enzyme imine, generating an irreversibly inactivated species. Further examples of such mechanism-based inhibitors will be given in the Problems section. Note also that there is a family of cofactor-independent racemase/epimerase enzymes which will be discussed in Chapter 10, Section 10.2.

Amino acid decarboxylases proceed from the PLP–amino acid adduct, shown in Fig. 9.2, this time using the PLP structure as an electron sink for

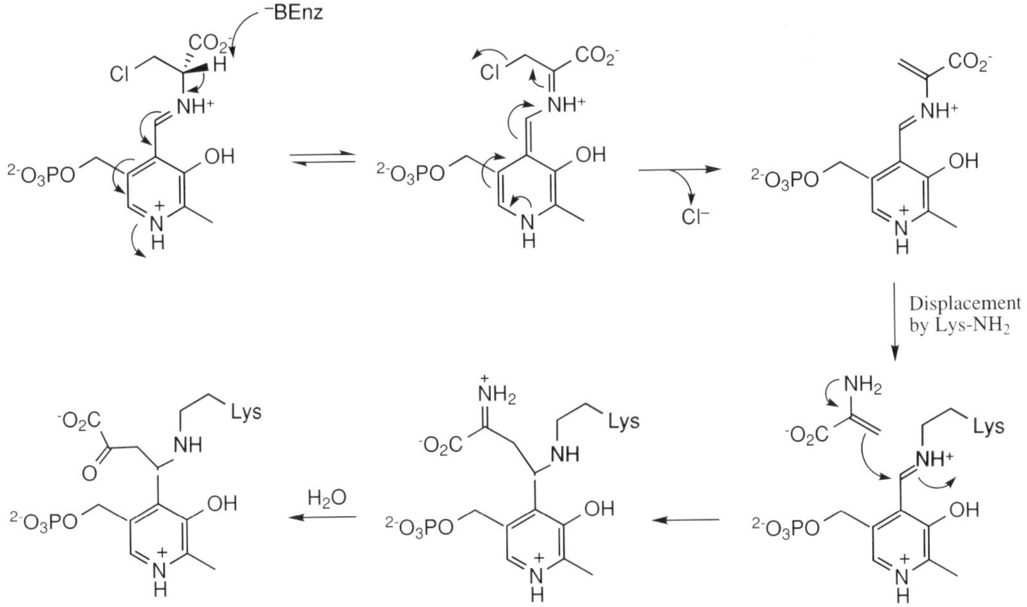

Fig. 9.4 Mechanism of inhibition of alanine racemase by β-chloro-D-alanine.

decarboxylation of this adduct. Re-protonation at (what was) the α-position, followed by detachment of the product from the PLP cofactor, generates the corresponding primary amine. Re-protonation usually takes place with retention of configuration at the α-position, as shown in Fig. 9.5. One important example of a PLP-dependent decarboxylase is the mammalian 3,4-dihydroxyphenylalanine (dopa) decarboxylase, which catalyses the decarboxylation of 3,4-dihydroxyphenylalanine to dopamine. Dopamine is a precursor to the neurotransmitters epinephrine and norepinephrine. This enzyme also catalyses the decarboxylation of 5-hydroxytryptophan to give another neurotransmitter serotonin, as shown in Fig. 9.6.

The third transformation carried out at the α-position of amino acids by PLP-dependent enzymes is transamination: conversion of the α-amino acid to an α-keto acid. This class of enzymes will be illustrated by the case study of aspartate aminotransferase.

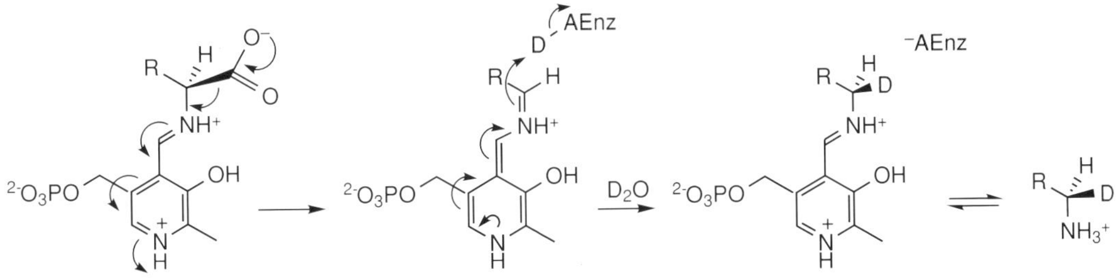

Fig. 9.5 Mechanism for PLP-dependent α-amino acid decarboxylases.

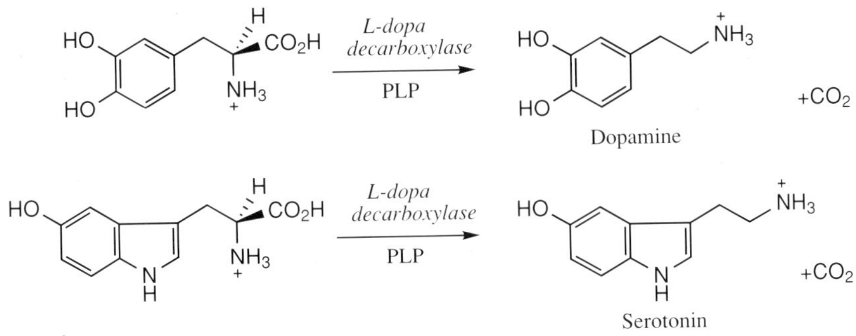

Fig. 9.6 Reactions catalysed by L-dopa decarboxylase.

9.3   Case study: aspartate aminotransferase

Aspartate aminotransferase catalyses the transamination of L-aspartic acid into oxaloacetate, at the same time converting α-ketoglutarate into L-glutamic acid. The enzymatic reaction therefore consists of two half-reactions, shown in Fig. 9.7.

Mammals contain two forms of aspartate aminotransferase, a cytosolic form and a mitochondrial form, both of which have been purified and studied extensively. The bacterial enzyme from *Escherichia coli* has also been purified, overexpressed and crystallized, allowing a detailed study of its mechanism of action, which is depicted in Fig. 9.8. The resting state of the enzyme in the absence of substrate contains the Lys-258–PLP–aldimine adduct, which absorbs at 430 nm. Upon binding of L-aspartate, the PLP–aldimine intermediate is formed, which also absorbs at 430 nm. The ε-amino group of Lys-258, released from binding the PLP cofactor, acts as a base for deprotonation of the α-hydrogen. This forms the quinonoid intermediate (visible at 490 nm by stopped flow kinetics) also found in the PLP-dependent racemases. However, in this case the quinonoid intermediate is re-protonated adjacent to the heterocyclic ring, generating a ketimine intermediate which can be observed at 340 nm by stopped flow kinetics. Hydrolysis of the ketimine intermediate releases the product oxaloacetate,

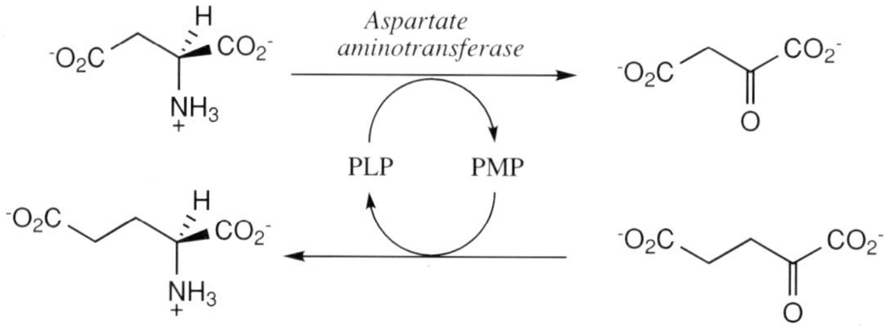

Fig. 9.7 Two half-reactions catalysed by aspartate aminotransferase. PMP, pyridoxamine-5′-phosphate.

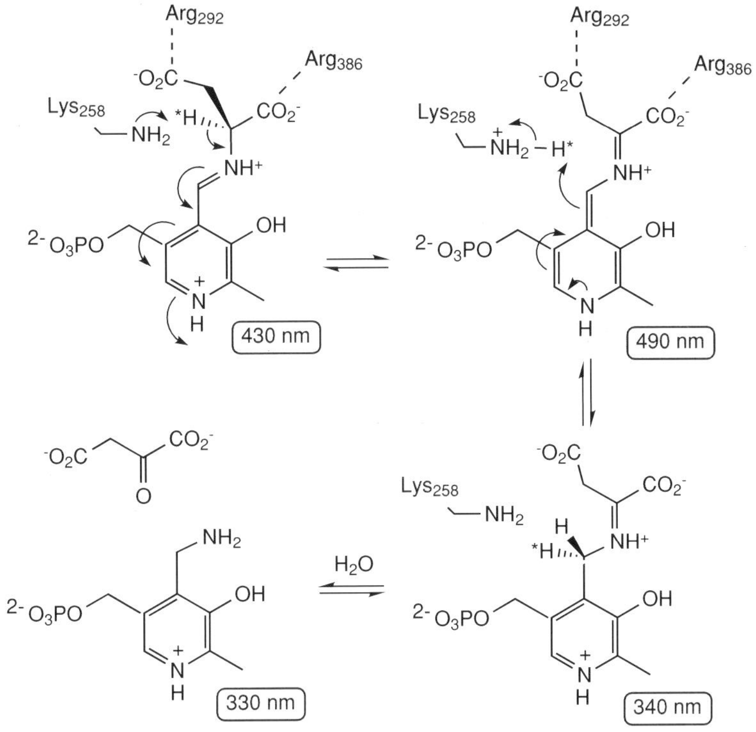

Fig. 9.8 Mechanism of aspartate aminotransferase half-reaction. Arg, arginine.

and generates a modified form of the cofactor known as pyridoxamine 5′-phosphate (PMP), visible at 330 nm.

The reaction is completed by carrying out the reverse transamination on the other α-keto acid substrate for this enzyme. α-Ketoglutarate is bound via a ketimine linkage, which is isomerized as before to the aldimine intermediate. Displacement of the aldimine linkage by Lys-258 releases L-glutamic acid and re-generates the PLP form of the cofactor.

Examination of the X-ray crystal structure of aspartate aminotransferase reveals that Lys-258 is suitably positioned to act as an intramolecular base for proton transfer, as indicated in Fig. 9.8. Replacement of Lys-258 for alanine by site-directed mutagenesis gives a completely inactive mutant enzyme, as expected, since there is no point of attachment or active site base. A cysteine (Cys)-258 mustant enzyme is similarly inactive. However, if this mutant is alkylated with 2-bromoethylamine an active enzyme is obtained which contains a thioether analogue of lysine at its active site, as shown in Fig. 9.9. This enzyme has 7% of the activity of the wild-type enzyme with a slightly shifted pH/rate profile of enzymatic activity, since the thioether-containing lysine analogue is slightly less basic than lysine.

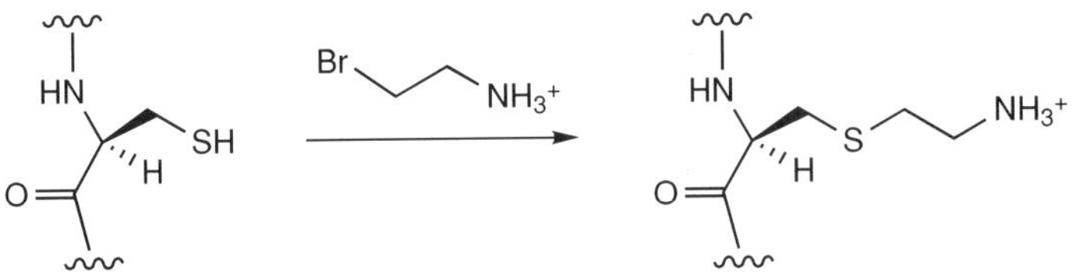

Inactive   7% activity

Fig. 9.9 Cys-258 mutant of aspartate aminotransferase.

A view of the aspartate aminotransferase X-ray crystal structure is shown in Plate 9.1 (facing p. 152), and an active site view is shown in Fig. 9.10. It is clear that both carboxylate groups of L-aspartate are bound by electrostatic interactions to active site arginine (Arg) residues: the α-carboxylate by Arg-386 and the β-carboxylate by Arg-292. In principle, the substrate specificity of this enzyme could be changed by replacing Arg-292 by other amino acids. Mutation of Arg-292 to an aspartate (Asp) residue gave an enzyme whose catalytic efficiency for L-aspartate had dropped from $34\,500\ M^{-1}s^{-1}$ to $0.07\ M^{-1}s^{-1}$. However, the mutant enzyme was found to be capable of processing L-amino acid substrates containing positively-charged side chains which could interact favourably with Asp-292. So, L-arginine, L-lysine and L-ornithine (one carbon shorter side chain than lysine) were all processed by the mutant enzyme, the best substrate being L-arginine with a k_{cat}/K_M of $0.43\ M^{-1}s^{-1}$. Figure 9.10 shows the interaction of L-aspartate with Arg-292, and the interaction of L-arginine with the Asp-292 mutant.

9.4   Reactions at the β- and γ-positions of amino acids

There is a smaller group of enzymatic reactions which take place at the β- and γ-positions of α-amino acids which are also dependent upon PLP. These reactions also make use of the PLP cofactor as an electron sink, but we shall see that there are examples in this class in which PLP acts as a four-electron sink rather than a two-electron sink. I shall illustrate one example of a reaction at the β-position, threonine dehydratase, and one at the γ-position, methionine-γ-lyase (Fig. 9.11).

Threonine dehydratase catalyses the conversion of L-threonine into α-ketobutyrate and ammonia. The enzymatic reaction starts from the aldimine adduct of PLP with L-threonine, which is deprotonated to generate the familiar quinonoid species. In this case the hydroxyl substituent at the β-position acts as a leaving group, presumably with acid catalysis, and an α,β-elimination reaction ensues. Attack at the imine linkage by the active site lysine residue releases the enamine equivalent of α-ketobutyrate. Hydrolysis of the enamine generates α-ketobutyrate and ammonia, and re-

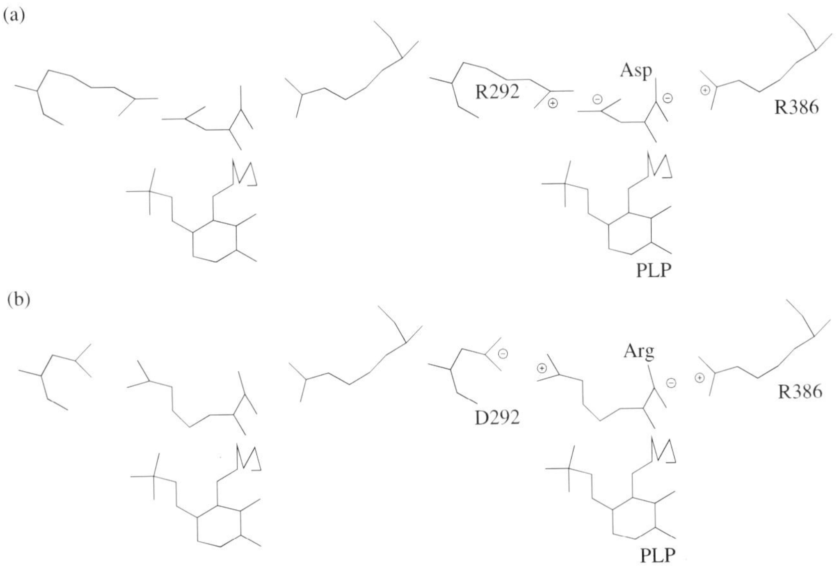

(a)

(b)

Fig. 9.10 (a) Stereoview of the active site of aspartate aminotransferase with substrate L-aspartate bound by Arg-292 and Arg-386. (b) A computer-generated model of the R292D mutant enzyme with L-arginine bound. (Reproduced with permission from C.N. Cronin & J.F. Kirsch (1988) *Biochemistry*, **27**, 4572–9.)

generates the PLP cofactor. The mechanism is depicted in Fig. 9.12.

Methionine-γ-lyase catalyses the conversion of L-methionine into α-keto-butyrate, ammonia and methanethiol (a particularly smelly enzyme to work with!). The mechanism of this enzyme starts from the aldimine adduct of PLP with L-methionine, which is deprotonated to generate the quinonoid intermediate. However, at this point a *second* deprotonation takes place at the β-position, utilizing the adjacent iminium species as a second electron sink to stabilize the β-carbanion. Elimination of the γ-substituent can then take place, followed by re-protonation at the γ-position, to generate the enamine intermediate seen above. The enamine equivalent of α-ketobutyrate

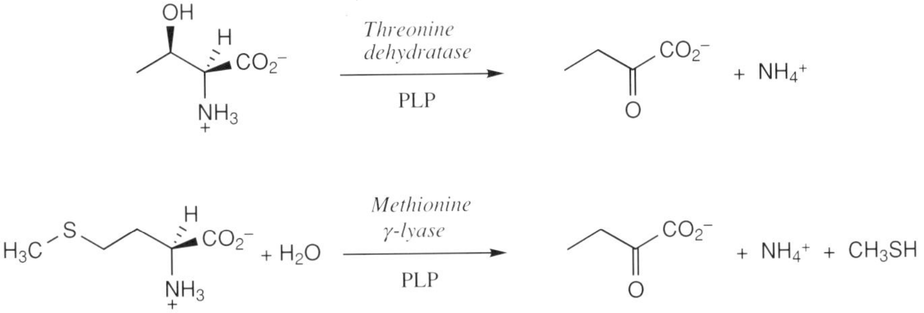

Fig. 9.11 Reactions at the β- and γ-positions of α-amino acids.

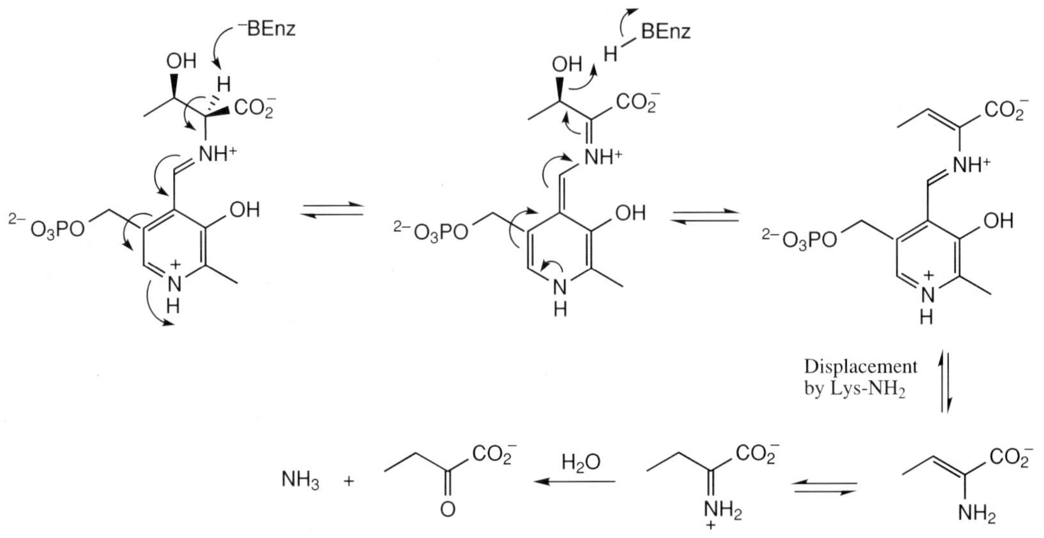

Fig. 9.12 Mechanism for threonine dehydratase.

is released, which after hydrolysis generates α-ketobutyrate and ammonia. The mechanism is depicted in Fig. 9.13.

There are a number of other enzymes in this class, most of which also employ a second deprotonation step, effectively utilizing the PLP–amino

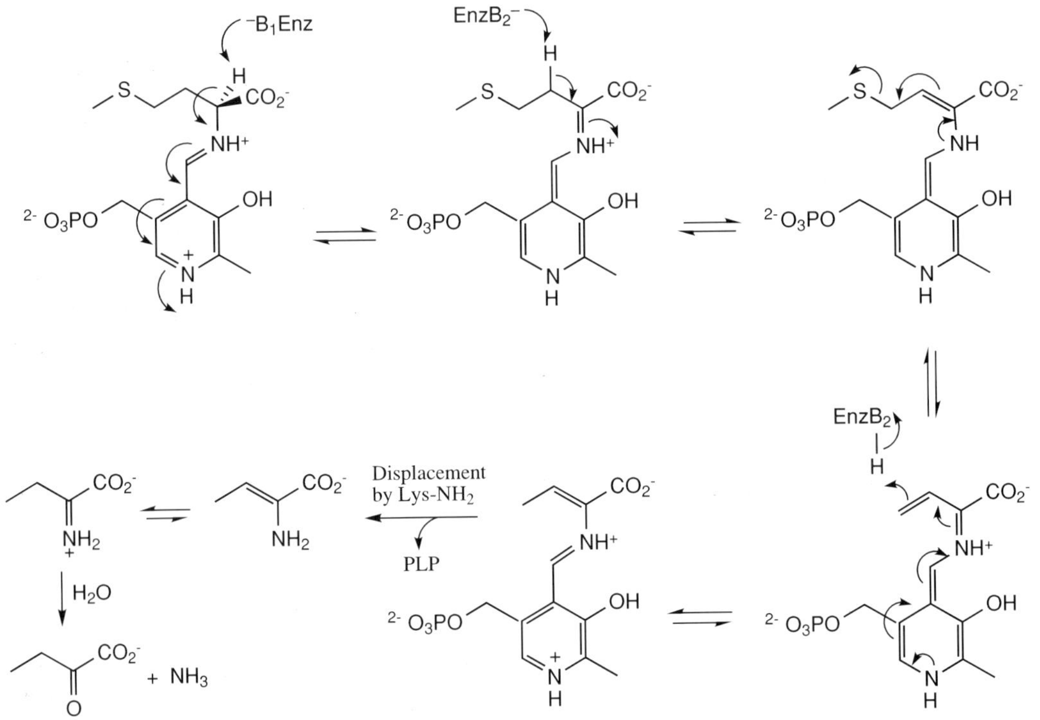

Fig. 9.13 Mechanism for methionine-γ-lyase.

acid iminium salt as a four-electron sink. The cellular role for such enzymes is often for the degradation of the respective amino acids and re-cycling of their nitrogen content.

9.5   Serine hydroxymethyltransferase

Serine hydroxymethyltransferase catalyses the interconversion of glycine and L-serine, using PLP and tetrahydrofolate as cofactors. This enzyme is unusual in that it utilizes the PLP coenzyme for carbon–carbon bond formation.

The mechanism for this reaction is depicted in Fig. 9.14, in the glycine-to-serine direction (the reaction is freely reversible). Following attachment of glycine to the PLP cofactor, deprotonation generates a quinonoid intermediate as seen above. However, this intermediate now reacts with N_5-methylene tetrahydrofolate, forming the carbon–carbon bond. A second deprotonation at the α-position allows the elimination of the tetrahydrofolate cofactor, generating a PLP–enamine adduct. This intermediate is attacked by water to generate the hydroxymethyl side chain of L-serine.

This is an important cellular reaction, since in the reverse direction it can be used to generate N_5-methylene tetrahydrofolate from L-serine, and this enzyme is largely responsible for the provision of cellular one-carbon methylene equivalents (see also Chapter 5 Section 5.8).

9.6   N-Pyruvoyl-dependent amino acid decarboxylases

A small number of amino acid decarboxylases have been found which show no requirement for PLP. Historically, the first of these enzymes to be discovered was histidine decarboxylase from *Lactobacillus*. The enzyme can, however, be inactivated by treatment with either sodium borohydride or phenylhydrazine, suggesting the presence of an electrophilic carbonyl cofactor. This cofactor has been shown to be an N-terminal pyruvoyl group, which forms an imine linkage with the α-amino group of L-histidine. The amide carbonyl group then acts as an electron sink for decarboxylation, as shown in Fig. 9.15.

Other examples of N-pyruvoyl-dependent decarboxylases include an S-adenosylmethionine decarboxylase and an aspartate decarboxylase from *E. coli*.

9.7   Imines and enamines in alkaloid biosynthesis

Certain α-amino acids are used as biosynthetic precursors for the assembly of alkaloid natural products. Examples of alkaloid natural products include nicotine, morphine and strychnine, shown in Fig. 9.16.

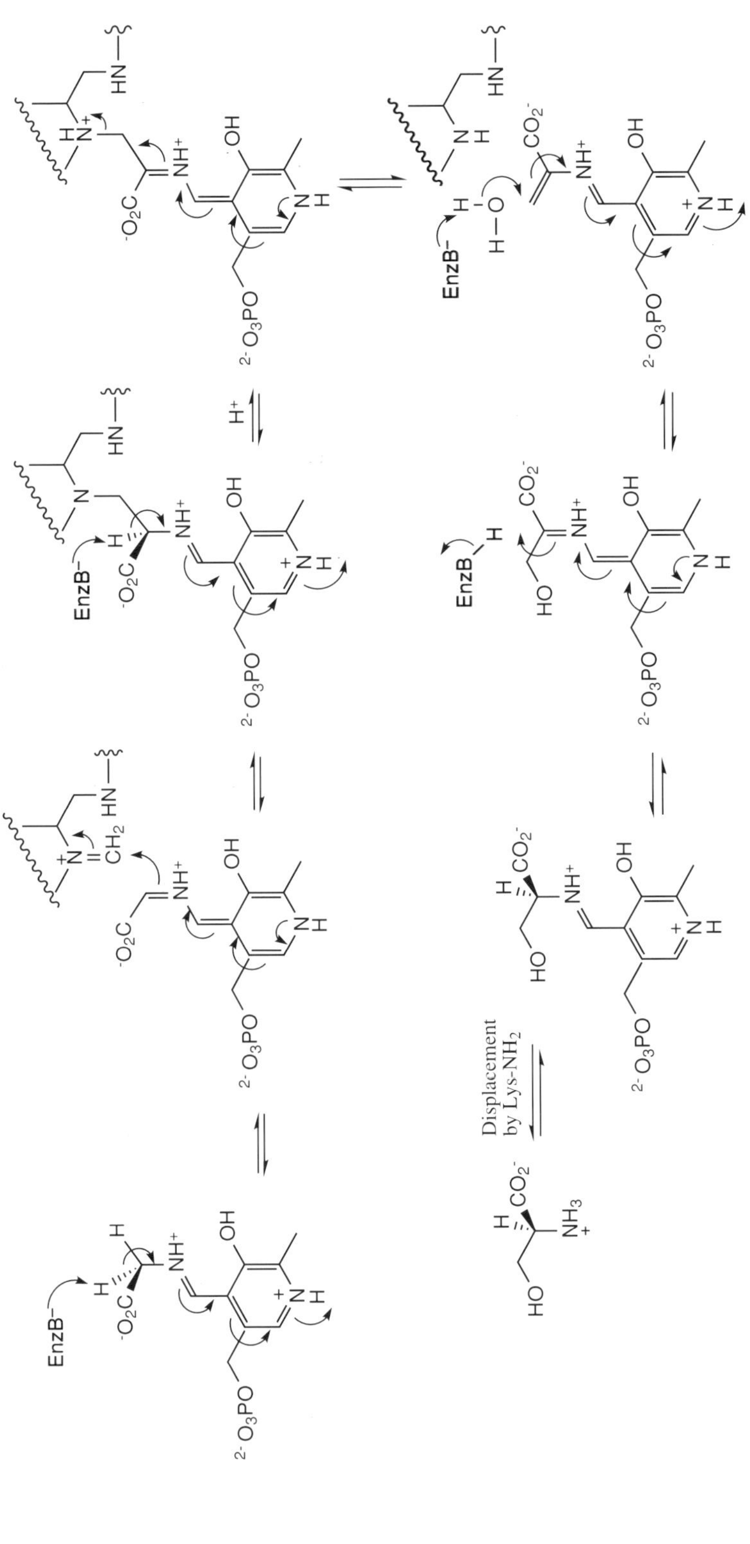

Fig. 9.14 Mechanism for serine hydroxymethyltransferase.

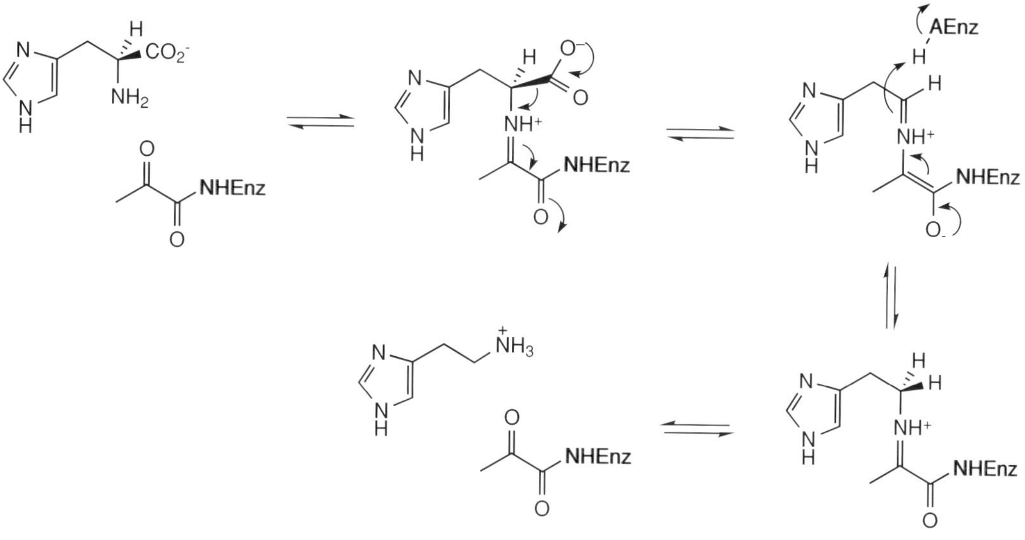

Fig. 9.15 Mechanism for N-pyruvoyl-dependent histidine decarboxylase.

The complex carbocyclic skeletons of these alkaloid natural products are assembled by intricate biosynthetic pathways. Many of the carbon–carbon bond forming steps involve imines and enamines derived from α-amino acids. For example, lupinine is a quinolizidine alkaloid found in *Lupinus* plants which is assembled from two molecules of L-lysine, as shown in Fig. 9.17. The amino acid is first decarboxylated by a PLP-dependent decarboxylase enzyme to give a symmetrical diamine. Oxidative deamination of one amino group, probably by a flavin-dependent oxidase, generates an amino-aldehyde. This condenses to give a cyclic imine, which in turn isomerizes to give a cyclic enamine. Reaction of one equivalent of the enamine with another equivalent of the imine forms a carbon–carbon bond between the two lysine-derived species. Hydrolysis of the resulting imine, followed by formation of a second imine linkage, leads to the natural product lupinine as shown in Fig. 9.17.

The biosyntheses of nicotine (from L-lysine and L-ornithine), morphine (from two molecules of L-tyrosine) and strychnine (from L-tryptophan) also

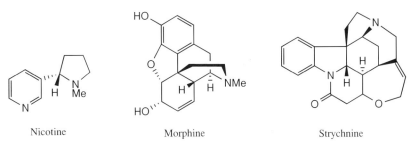

Nicotine   Morphine   Strychnine

Fig. 9.16 Alkaloid natural products.

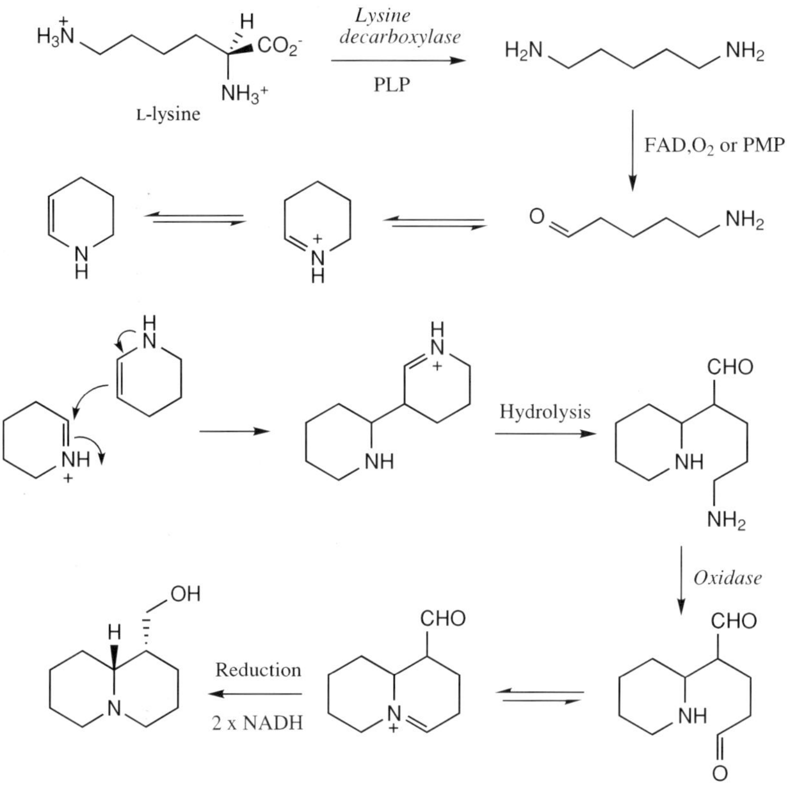

Fig. 9.17 Biosynthesis of lupinine. FAD, flavine adenine dinucleotide.

involve intermediate imine and enamine species, as well as examples of phenolic radical couplings encountered in Chapter 7, Section 7.4. Readers interested further in alkaloid biosynthesis should consult the references below.

Problems

1 Suggest a mechanism for the reaction catalysed by threonine-β-epimerase, a PLP-dependent enzyme.

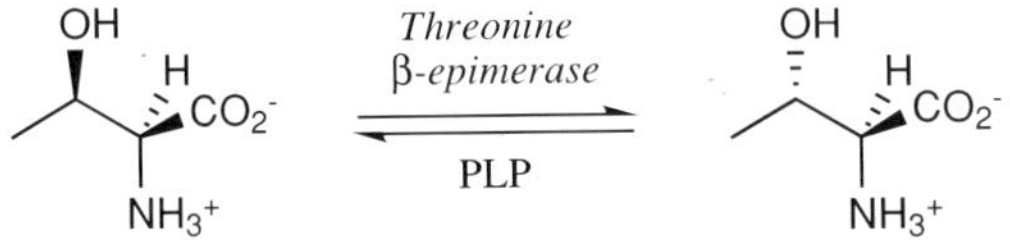

2 Aspartate aminotransferase is inactivated in time-dependent fashion by L-vinyl-glycine. Suggest a mechanism of inactivation.

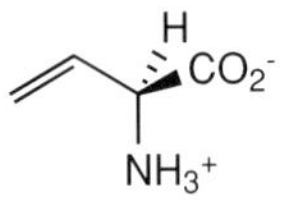

L-Vinyl-glycine

3 γ-Aminobutyrate transaminase has an important function in the human brain, where it catalyses the conversion of γ-aminobutyrate (a neurotransmitter) to succinate semialdehyde. This enzyme is inactivated (K_i 0.6 μM) by gabaculine. Suggest a mechanism of inactivation.

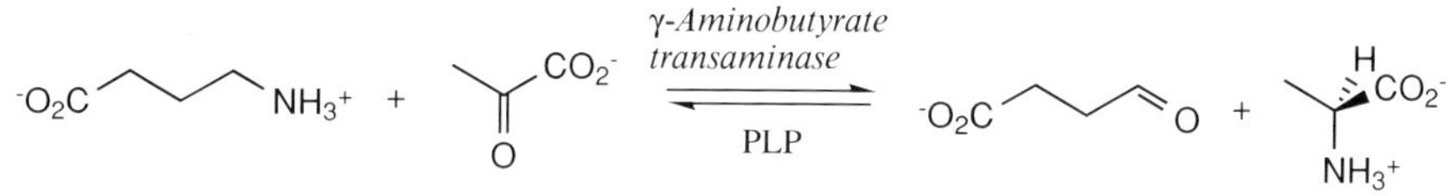

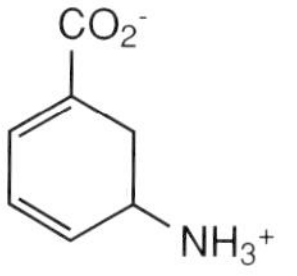

Gabaculine

4 The enzyme threonine synthase catalyses the converson of L-homoserine phosphate to L-threonine using PLP as a cofactor. Write a mechanism for the enzymatic reaction that is consistent with the following observations: (i) carrying out the enzymatic reaction in $H_2{}^{18}O$ leads to the incorporation of ^{18}O into the β-hydroxyl group of L-threonine; (ii) incubation of $3S$-[3-^3H]-homoserine phosphate with the enzyme leads to loss of the ^3H label in the product, whilst incubation of the $3R$-[3-^3H] isomer leads to retention of the ^3H label.

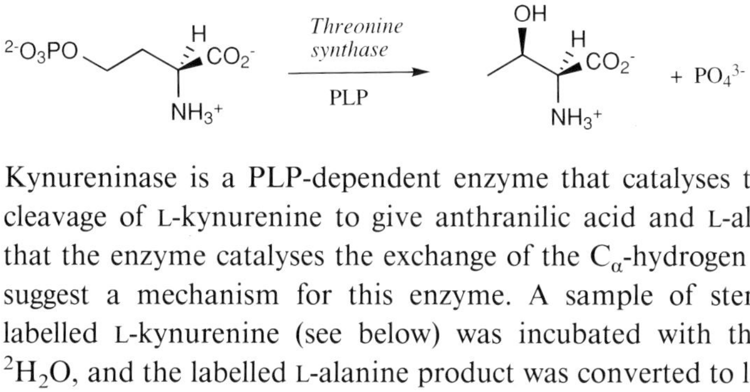

5 Kynureninase is a PLP-dependent enzyme that catalyses the hydrolytic cleavage of L-kynurenine to give anthranilic acid and L-alanine. Given that the enzyme catalyses the exchange of the C_α-hydrogen with solvent, suggest a mechanism for this enzyme. A sample of stereospecifically labelled L-kynurenine (see below) was incubated with the enzyme in ^2H_2O, and the labelled L-alanine product was converted to labelled acetic acid. The resulting chiral methyl group was found to have the S configu-

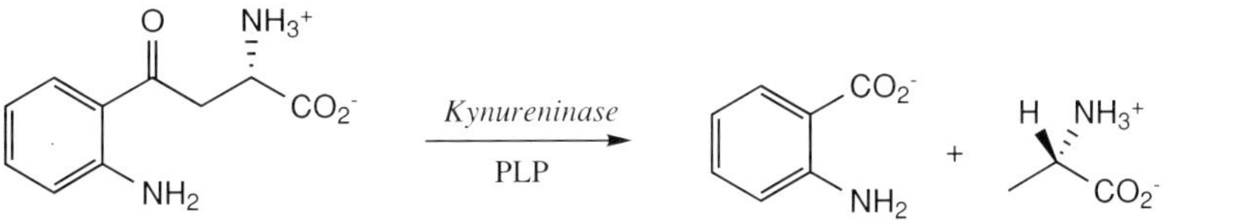

L-Kynurenine

Anthranilic acid

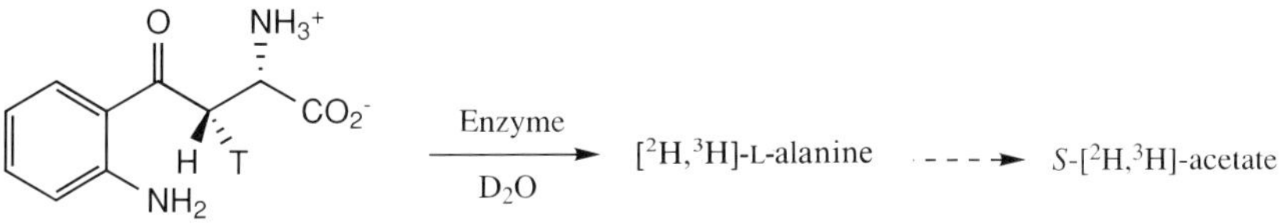

$[^2H,^3H]$-L-alanine ----> S-$[^2H,^3H]$-acetate

ration. Deduce whether the enzymatic reaction proceeds with retention or inversion of configuration at the labelled centre.

Further reading

General

Walsh, C.T. (1979) *Enzymatic Reaction Mechanisms.* Freeman, San Francisco.

PLP-dependent enzymes

Hayashi, H., Wada, H., Yoshimura, T., Esaki, N. & Soda, K. (1990) *Annu Rev Biochem*, **59**, 87–110.
Martell, A.E. (1982) *Adv Enzymol*, **53**, 163–200.
Vederas, J.C. & Floss, H.G. (1980) *Acc Chem Res*, **13**, 455–63.

Aspartate aminotransferase

Cronin, C.N. & Kirsch, J.F. (1988) *Biochemistry*, **27**, 4572–9.
Jäger, J., Moser, M., Sauder, U. & Jansonius, J.N. (1994) *J Mol Biol*, **239**, 285–305.
Planas, A. & Kirsch, J.F. (1991) *Biochemistry*, **30**, 8268–76.

Reactions at β-position

Braunstein, A.E. & Goryachenkova, E.V. (1984) *Adv Enzymol*, **56**, 1–90.

Serine hydroxymethyltransferase

Schirch, L. (1982) *Adv Enzymol*, **53**, 83–112.

Mechanism-based inhibitors

Walsh, C.T. (1982) *Tetrahedron*, **38**, 871–909.

Pyruvoyl-dependent enzymes

Recsei, P.A. & Snell, E.E. (1984) *Annu Rev Biochem*, **53**, 357–88.
van Poelje, P.D. & Snell, E.E. (1990) *Annu Rev Biochem*, **59**, 29–60.

Alkaloid biosynthesis

Hashimoto, T. & Yamada, Y. (1994) *Annu Rev Plant Physiol Plant Mol Biol*, **45**, 257–85.
Herbert, R.B. (1989) *The Biosynthesis of Secondary Metabolites*, 2nd edn. Chapman & Hall. London.
Mann, J. (1987) *Secondary Metabolism*, 2nd edn. Clarendon, Press, Oxford.

10 Isomerases

10.1 Introduction

The final group of enzymatic transformations that we shall meet are the isomerization reactions: the interconversion of two isomeric forms of a molecule. The interconversion of enantiomers is catalysed by racemase enzymes, as discussed in Chapter 9, Section 9.2. In this chapter we shall meet a second family of racemases which require no cofactors. We shall also meet enzymatic proton transfer reactions which involve the interconversion of tautomeric forms of ketones, and the interconversion of positional isomers of allylic compounds. These isomerization reactions are summarized in Fig. 10.1. Finally, we shall meet some examples of enzymatic skeletal rearrangements: an enzyme-catalysed Claisen re-arrangement and the radical-mediated vitamin B_{12} re-arrangements.

10.2 Cofactor-independent racemases and epimerases

In Chapter 9, Section 9.2 we met the family of α-amino acid racemase enzymes which utilize the coenzyme pyridoxal-5′-phosphate (PLP). In addition there is a family of racemase and epimerase enzymes which require no cofactors at all. A separate section is being devoted to these enzymes because of a single important mechanistic issue: how do these enzymes achieve the deprotonation of the α-proton of an α-amino acid? In the PLP-dependent racemases the formation of an imine linkage with the α-amino group dramatically increases the acidity of the α-proton. If these enzymes contain no PLP then no such assistance is possible, so they must have some

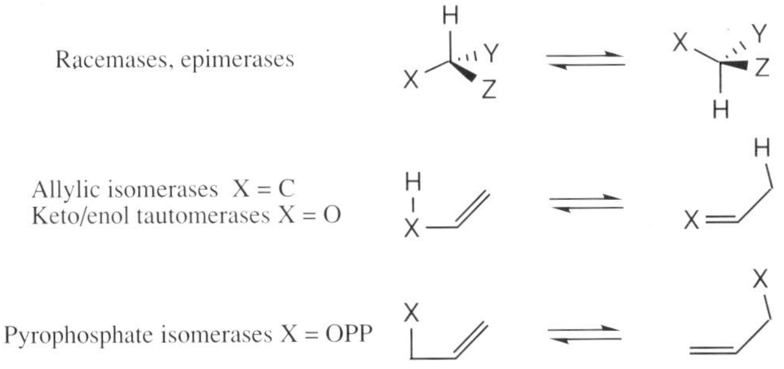

Fig. 10.1 Summary of enzyme-catalysed isomerization reactions.

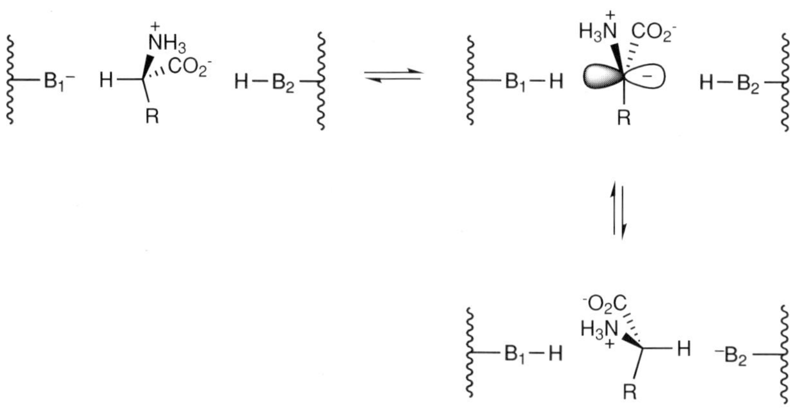

Fig. 10.2 Two-base mechanism for cofactor-independent racemases.

alternative way of carrying out this deprotonation. This question is of wider significance, since there is evidence for enzymatic deprotonation adjacent to carboxylic acids in other enzymes such as the flavin-dependent amino acid oxidase enzymes (see Chapter 6, Section 6.3).

Examples of cofactor-independent α-amino acid racemases are glutamate racemase from *Lactobacillus fermenti* and aspartate racemase from *Streptococcus thermophilus*. There is evidence in both these enzyme-catalysed reactions for a two-base mechanism of racemization, as shown in Fig. 10.2. In this mechanism an α-proton is removed from one face of the amino acid by an active site base and a proton delivered onto the other face by a protonated base on the opposite side of the enzyme active site. In this mechanism the unstable α-carbanion would exist only fleetingly.

How is experimental evidence obtained for such a two-base mechanism? One such method is illustrated in Fig. 10.3. In this experiment a stoichiometric amount of enzyme is incubated with substrate for a short period of time in tritiated water, and the distribution of ^3H label examined in the products. In a one-base active site the α-hydrogen is removed from the L-enantiomer and returned to the opposite face by the same base, giving the D-enantiomer. Hence, any ^3H label incorporated by exchange of the active

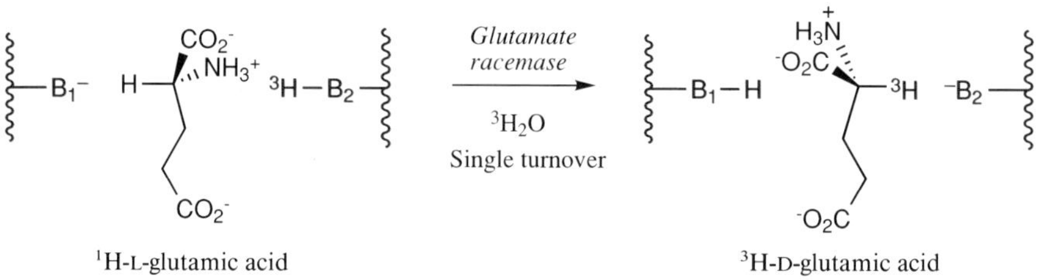

Fig. 10.3 Experimental evidence in favour of a two-base mechanism.

site base with ^3H_2O would be delivered equally to both L- and D-enantiomers. However, in a single catalytic cycle of a two-base active site one would expect the α-proton of the L-enantiomer to be abstracted by one base and a ^3H label to be delivered from the opposite face by the other base, resulting in increased incorporation of ^3H label into the D-enantiomer. Thus, starting with L-glutamate, glutamate racemase catalyses the preferential incorporation of ^3H from ^3H_2O into D-glutamic acid.

This approach has established the likelihood of a two-base mechanism for a number of cofactor-independent racemases and epimerases. However, this does not explain how the intermediate carbanion, however fleeting, is stabilized. Recent studies on mandelate racemase have provided an insight into this problem. Mandelate racemase is a cofactor-independent enzyme which catalyses the interconversion of R and S-mandelic acid. A two-base mechanism has been implicated for this enzyme, and is supported by a recent X-ray crystal structure of the enzyme in which active site residues lysine (Lys)-166 and histidine (His)-297 are suitably positioned to act as the two bases. Replacement of His-297 by a glutamine (Gln) residue using site-directed mutagenesis gives a mutant enzyme which is unable to catalyse the racemization reaction. This mutant enzyme (still containing Lys-166) is able to catalyse the exchange of the α-proton of S-mandelic acid with ^2H_2O to give ^2H-S-mandelic acid, but does not catalyse exchange with R-mandelic acid. This result implies that Lys-166 deprotonates the α-proton of S-mandelic acid, forming a carbanion intermediate. This intermediate is sufficiently stable to exchange with the ε-NHD_2^+ of Lys-166.

How is the carbanion intermediate stabilized? It is evident from the X-ray crystal structure that the carboxylate group of the substrate forms hydrogen bonds with a protonated Lys-164 residue and a protonated Glu-317 residue. Replacement of Glu-317 by a glutamine residue using site-directed mutagenesis gives a mutant enzyme with 10^4-fold reduced catalytic efficiency. It has been proposed that Glu-317 forms a strong 'low-barrier' hydrogen bond with the substrate carboxylate which stabilizes the formation of the α-carbanion, as shown in Fig. 10.4. It may be that similar methods of stabilization are employed in other cases of enzymatic deprotonation adjacent to carboxylate groups.

10.3   Keto–enol tautomerases

The enolization of ketones is a well-known reaction in organic chemistry that is utilized as an intermediate process in many enzyme-catalysed reactions, notably the aldolases encountered in Chapter 7, Section 7.2. In most cases enols and enolate anions are thermodynamically unstable species which are not isolable. However, in a few cases enol tautomers of ketones are

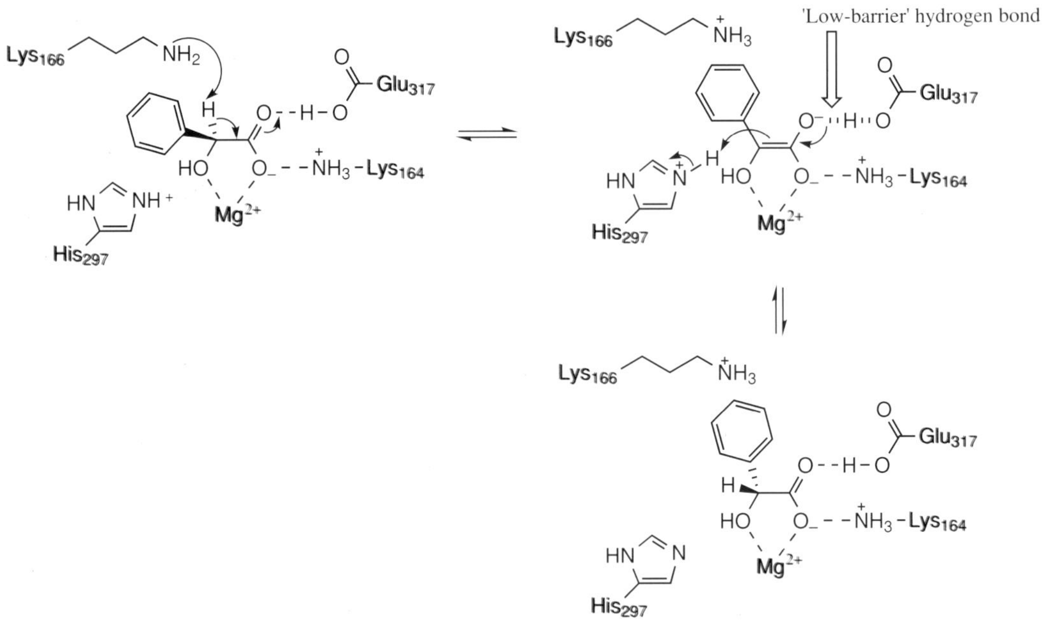

Fig. 10.4 Mechanism for mandelate racemase.

sufficiently stabilized to be isolable, and there are several enzymes which catalyse the interconversion of keto and enol tautomers.

One simple example is that of phenylpyruvate tautomerase, which catalyses the interconversion of phenylpyruvic acid with its stabilized enol form, as shown in Fig. 10.5. The enzyme catalyses the stereospecific exchange of the *proR* hydrogen with ^2H_2O via the *E* enol isomer, using acid/base active site chemistry.

Two-step keto–enol isomerizations are involved in the mechanisms of a number of isomerase enzymes. Ketosteroid isomerase (see Chapter 3, Section 3.5) catalyses the isomerization of Δ-2-ketosteroid to Δ-3-ketosteroid through a dienol intermediate, as shown in Fig. 10.6. This reaction involves deprotonation by an active site base Asp-38 and concerted protonation by tyrosine (Tyr)-14 to generate the dienol intermediate. Asp-38 then returns the abstracted proton to the γ-position to complete the isomerization reaction.

There is also a family of aldose–ketose isomerases which catalyse the

Fig. 10.5 Phenylpyruvate tautomerase.

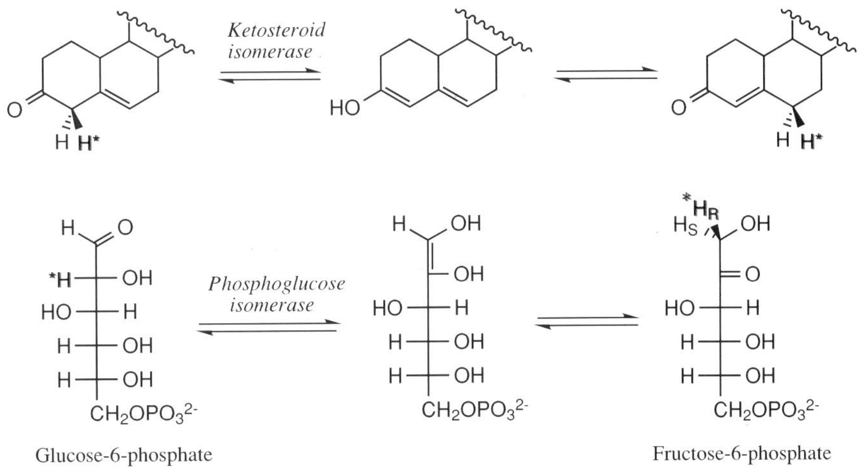

Ketosteroid isomerase

Glucose-6-phosphate

Phosphoglucose isomerase

Fructose-6-phosphate

Fig. 10.6 Ketosteroid isomerase and phosphoglucose isomerase.

interconversion of aldose sugars containing a C-1 aldehyde with ketose sugars containing a C-2 ketone. Phosphoglucose isomerase catalyses the interconversion of glucose-6-phosphate with fructose-6-phosphate, as shown in Fig. 10.6. This reaction proceeds by abstraction of the C-2 proton to generate an enediol intermediate, followed by return of the proton specifically at the *proR* position of C-1. Intramolecular transfer of H* is observed in both directions, indicating that a single active site base is responsible.

10.4   Allylic isomerases

We have already met a few examples of allylic (or 1,3-) migrations: the ketosteroid isomerase reaction illustrated in the previous section is effectively a 1,3-hydrogen migration; and we have encountered 1,3-migrations of allylic pyrophosphates in the terpene cyclase reactions of Chapter 7, Section 7.3. There are many allylic isomerases which operate in many different areas of biological chemistry: I will briefly mention two examples which are important for the cellular assembly of fatty acids and terpenoid natural products respectively.

In Chapter 8, Section 8.2 we met the enzyme β-hydroxydecanoyl thioester dehydratase, which catalyses the elimination of β-hydroxy-thioesters in fatty acid assembly. This enzyme also catalyses the isomerization of the *trans*-2,3-alkenyl thioester to the *cis*-3,4-alkenyl thioester, as shown in Fig. 10.7. The dehydration reaction catalysed by this enzyme utilizes an active site histidine base, His-70 (see Chapter 8, Section 8.2). This base is also thought to be involved in the isomerization reaction. The reaction involves the transfer of the C-4 *proR* hydrogen to the C-2 *proS* position. This

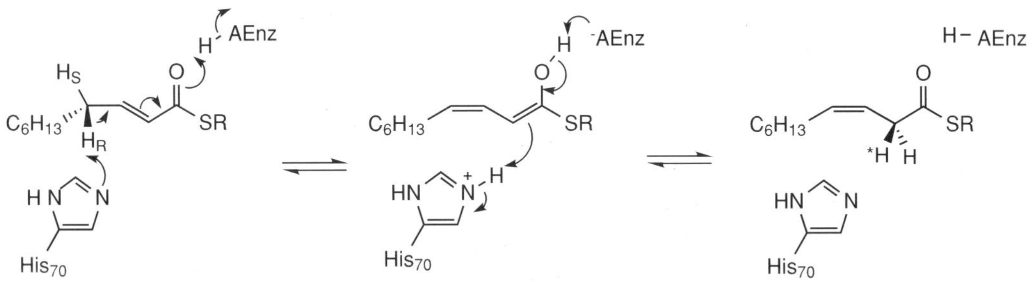

Fig. 10.7 Mechanism of β-decanoyl thioester isomerase.

is defined as a 1,3-suprafacial hydrogen shift, since the hydrogen is transferred across the same face of the molecule. It is thought that His-70 mediates this transfer via an enediol intermediate, as shown in Fig. 10.7. However, it is worthy of note that no intramolecular hydrogen transfer is observed in this case, unlike examples such as ketosteroid isomerase, suggesting that the protonated histidine exchanges readily with solvent water.

Isopentenyl pyrophosphate (IPP) isomerase catalyses the interconversion of IPP and dimethylallyl pyrophosphate (DMAPP), the two five-carbon building blocks for the assembly of terpene natural products (also see Chapter 7, Section 7.3). The enzymatic reaction involves stereospecific removal of the *proR* hydrogen at C-2, and delivery of a hydrogen onto the *re*-face of the 4,5-double bond. The stereochemistry was established by conversion of a stereospecifically labelled substrate in ^2H_2O (as shown in Fig. 10.8), generating a chiral methyl group of *R* configuration at C-4. This reaction is defined as a 1,3-antarafacial hydrogen shift, since the hydrogen is re-inserted on the opposite face of the molecule.

The antarafacial stereochemistry of hydrogen transfer suggests the involvement of two active site groups in the mechanism. This was confirmed by subsequent chemical modification studies. Site-directed mutagenesis experiments have implicated cysteine-(Cys)-139 as an acidic group which protonates the 4,5-double bond, generating a tertiary carbonium ion intermediate. Stereospecific deprotonation is then carried out by Glu-207, as shown in Fig. 10.9. The existence of the carbonium ion intermediate is supported by the potent inhibition of this enzyme (K_i 1.4×10^{-11} M) by a tertiary amine substrate analogue. At neutral pH this amine is protonated,

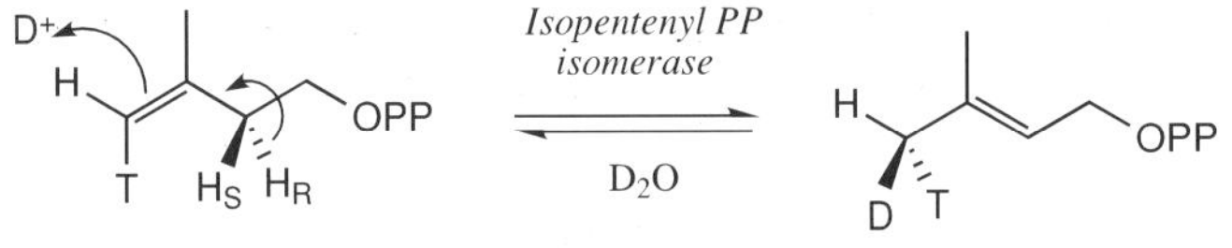

Fig. 10.8 Stereochemistry of IPP isomerase.

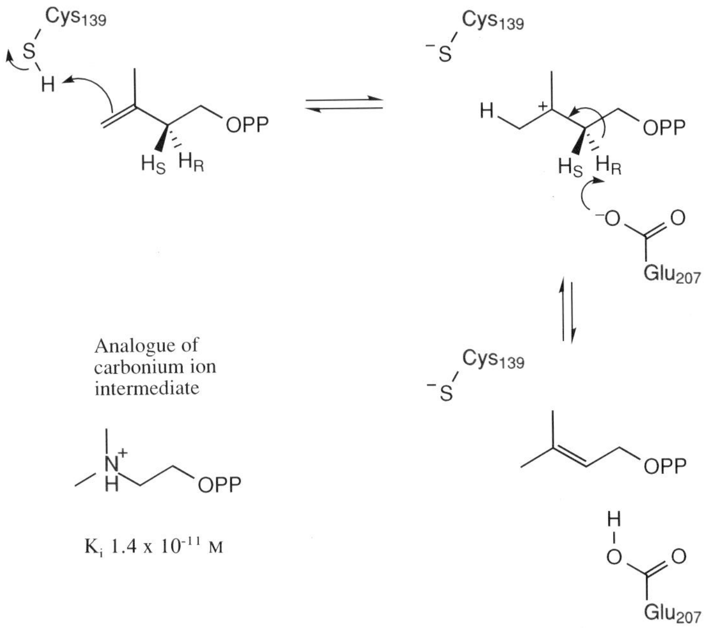

Fig. 10.9 Mechanism for IPP isomerase.

and the ammonium cation mimics the carbonium ion intermediate which is bound tightly by the enzyme.

10.5 Case study: chorismate mutase

Pericyclic reactions are commonly used in organic synthesis; the Diels–Alder reaction being an important synthetic reaction for the stereoselective synthesis of cyclohexane rings. However, only one enzyme-catalysed pericyclic reaction has been well characterized, which is the reaction catalysed by chorismate mutase.

Chorismate mutase catalyses the Claisen rearrangement of chorismic acid into prephenic acid. This reaction is important for plants and microorganisms, which utilize prephenic acid as a precursor to the amino acids L-phenylalanine and L-tyrosine (Fig. 10.10). This reaction occurs non-enzymatically at a significant rate, contributing to the singular instability of the important metabolite chorismic acid. However, the reaction is accelerated 10^6-fold by the enzyme chorismate mutase.

The stereochemistry of the enzyme-catalysed and uncatalysed reactions has been investigated, revealing that both reactions proceed through a chair-like transition state involving the *trans*-diaxial conformer of the substrate, as shown in Fig. 10.11. Tritium substitution at C-5 or C-9 gives no secondary kinetic isotope effect on the enzyme-catalysed reaction, whereas

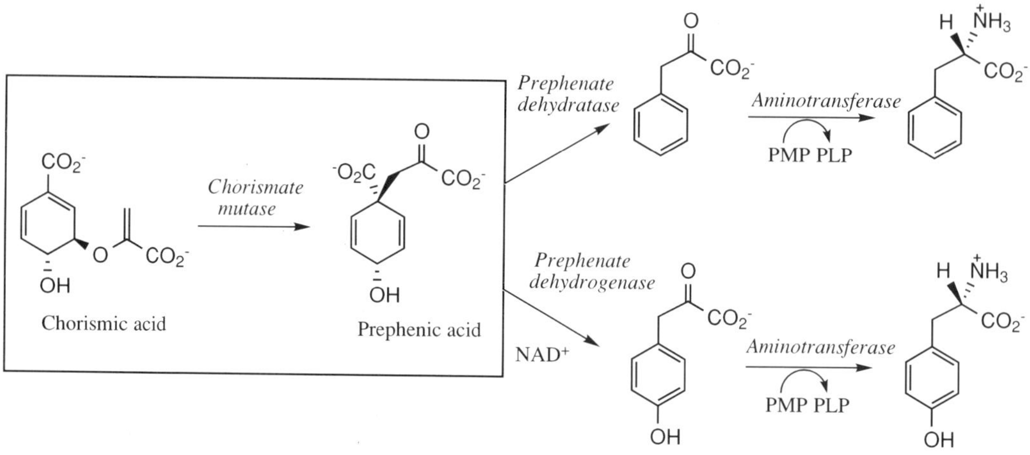

Fig. 10.10 Reaction catalysed by chorismate mutase. NAD, nicotinamide adenine dinucleotide; PMP, pyridoxamine-5'-phosphate.

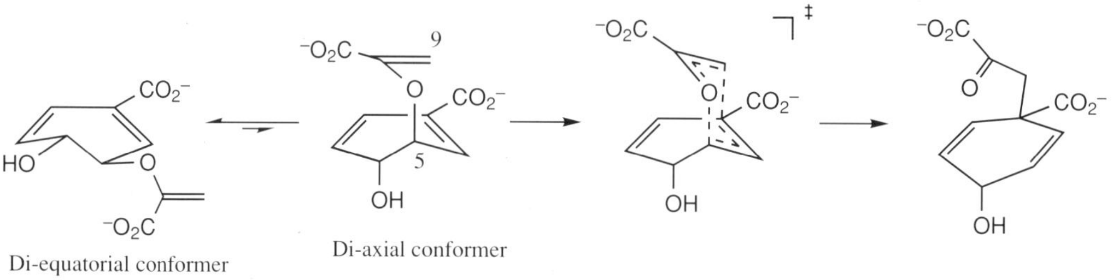

Fig. 10.11 Mechanism for chorismate mutase.

such isotope effects are observed on the uncatalysed reaction. This suggests that the rate-determining step of the enzyme-catalysed reaction is substrate binding, hence isotope effects on subsequent steps would not be observed. This situation in which there is a high preference for catalytic processing of a bound substrate, rather than dissociation, is termed a 'high commitment to catalysis'.

How does the enzyme achieve the 10^6-fold rate acceleration? If the mechanisms of the uncatalysed and enzyme-catalysed reactions are similar, does the rate acceleration come from transition state stabilization? A bicyclic transition state analogue has been synthesized for the chorismate mutase reaction, as shown in Fig. 10.12. This analogue inhibits the enzymatic reaction strongly (K_i 3 μM), suggesting that the enzyme does selectively bind the transition state of the reaction (see section 3.4). This analogue has been used to prepare catalytic antibodies (see Chapter 11, Section 11.3) capable of catalysing the chorismate mutase reaction to the extent of 10^4-fold over the uncatalysed reaction.

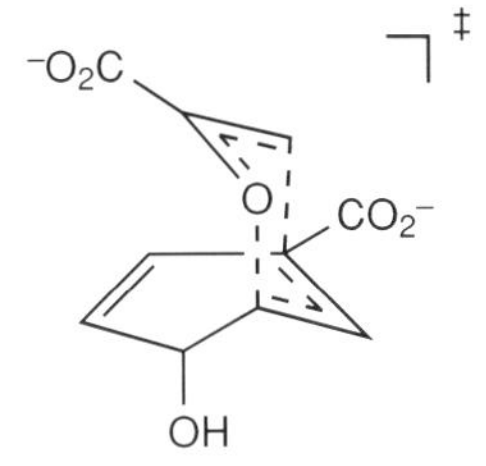

Chair-like transition state

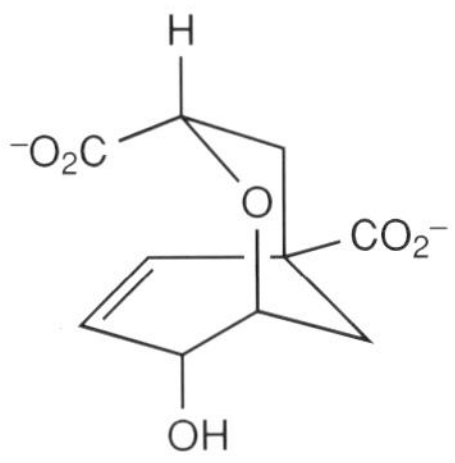

Transition state analogue

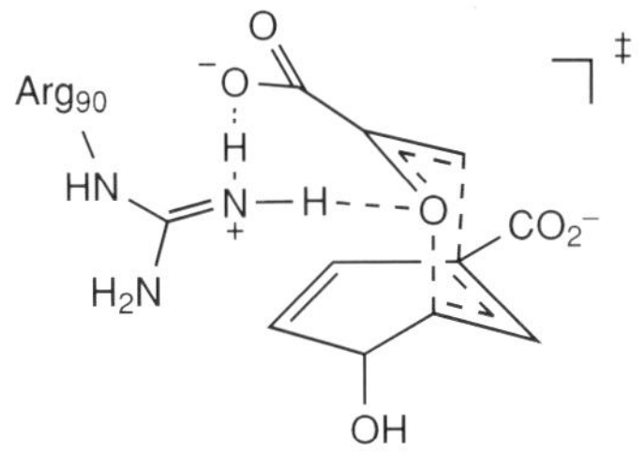

Transition state stabilization
by arginine-90

Fig. 10.12 Transition state analogue for chorismate mutase.

The chorismate mutase enzyme from *Bacillus subtilis* has been crystallized recently in the presence of the transition state analogue, and the X-ray crystal structure solved (Plate 10.1, facing p. 152). This enzyme is a trimer of identical 127-amino acid subunits. Each subunit contains five strands of β-sheet and two α-helices. The β-strands of each subunit pack together to form the trimer structure, with active sites at the interface of each pair of subunits.

Examination of the active site structure has revealed that the ether oxygen of the analogue lies within hydrogen-bonding distance of the guanidinium side chain of arginine (Arg)-90, as shown in Fig. 10.12. Molecular modelling studies on the enzyme–substrate and enzyme–transition state complexes have revealed that Arg-90 can form a favourable electrostatic/hydrogen-bonding interaction with the transition state. This interaction and other transition state binding appear to provide sufficient stabilization to account for the 10^6-fold rate acceleration. This type of electrophilic catalysis is precedented in synthetic organic chemistry in the form of Lewis acid catalysis of pericyclic reactions, such as the Claisen re-arrangement.

10.6 Vitamin B_{12}-dependent re-arrangements

Vitamin B_{12} has the most complex structure of all of the vitamins. The X-ray crystal structure of vitamin B_{12} was solved in 1961 by D.C. Hodgkin. The

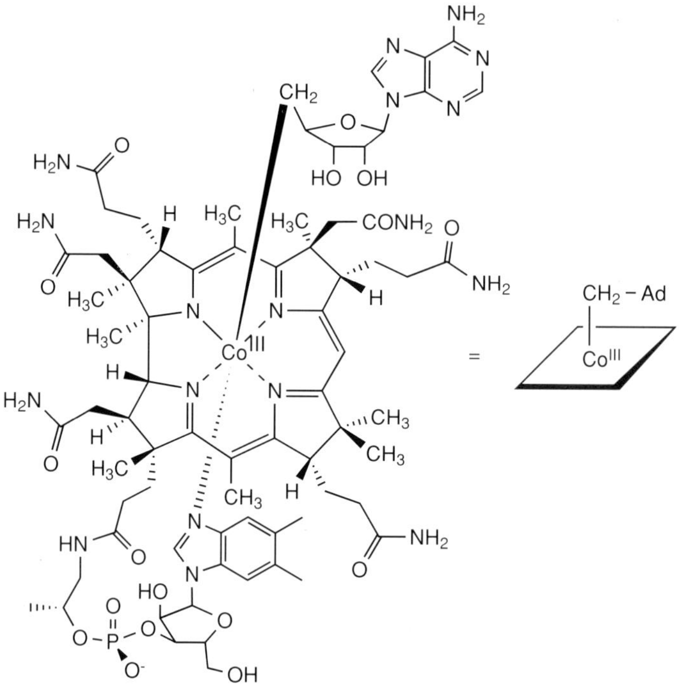

Fig. 10.13 Structure of vitamin B_{12} cofactor.

structure, shown in Fig. 10.13, consists of an extensively modified porphyrin ring system, containing a central cobalt (Co^{3+}) ion. The two axial ligands are a benzimidazole nucleotide and an adenosyl group. The cobalt–carbon bond formed with the adenosyl ligand is weak and susceptible to homolysis, and this is the initiation step for the vitamin B_{12}-dependent reactions.

We shall consider two vitamin B_{12}-dependent re-arrangements: propane-diol dehydrase and methylmalonyl coenzyme A (CoA) mutase. Both reactions involve the 1,2-migration of a hydrogen atom, and the corresponding 1,2-migration of another substituent, either –OH or –CO_2H, as shown in Fig. 10.14.

Propanediol dehydrase catalyses the re-arrangement of propane-1,2-diol to propionaldehyde. There is no incorporation of solvent hydrogens during the reaction, implying that there is an intramolecular hydrogen transfer. Stereospecific labelling studies have shown that the reaction involves the removal of the *proS* hydrogen at C-1. This hydrogen atom is transferred specifically to the *proS* position at C-2, giving an inversion of configuration at C-2.

Tritium labelling of the C-1 *proS* hydrogen gives rise to exchange of ^3H into the adenosyl 5′-position. This implies that there is an adenosyl 5′-CH_3 intermediate in the enzyme mechanism formed by homolysis of the

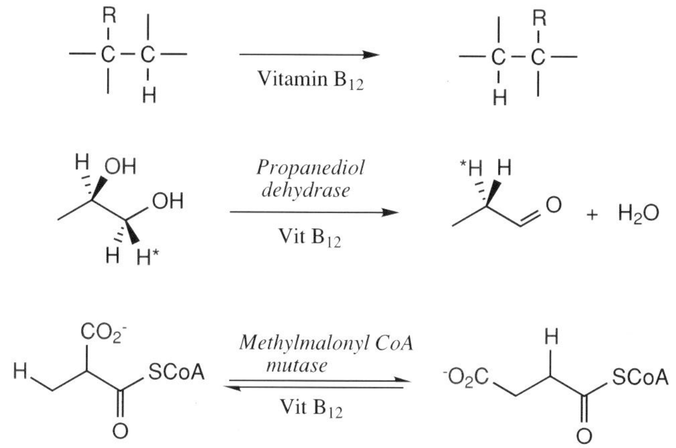

Fig. 10.14 Vitamin B_{12}-dependent reactions.

adenosyl–cobalt bond and hydrogen atom transfer from the substrate. Homolysis of the adenosyl–cobalt bond is further supported by the detection of Co^{2+} intermediates in the reaction by stopped flow electron spin resonance spectroscopy studies.

The proposed mechanism for propanediol dehydrase is shown in Fig. 10.15. Initiation of the reaction by homolysis of the carbon–cobalt bond generates an adenosyl radical, which abstracts the C-1 *proS* hydrogen. Recombination of the substrate radical with Co^{2+} gives an intermediate containing a substrate–cobalt σ-bond. There is some uncertainty over the next isomerization step; however, it has been proposed that hydroxide ion is displaced by cobalt to give a cobalt π-complex, which is attacked by hydroxide at C-1 to generate a second σ-complex. Homolysis to give a secondary substrate radical is followed by delivery of H˙ from the adenosyl group and regeneration of the B_{12} cofactor. The hydrate product is then dehydrated stereospecifically, losing the migrating hydroxyl group to give the product aldehyde.

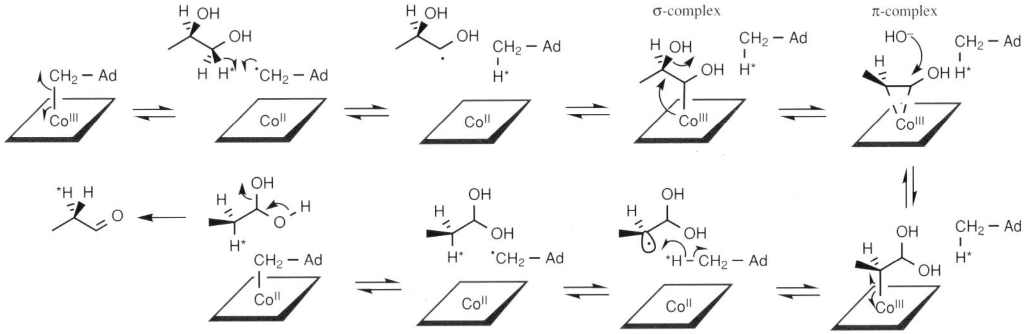

Fig. 10.15 Mechanism for propanediol dehydrase.

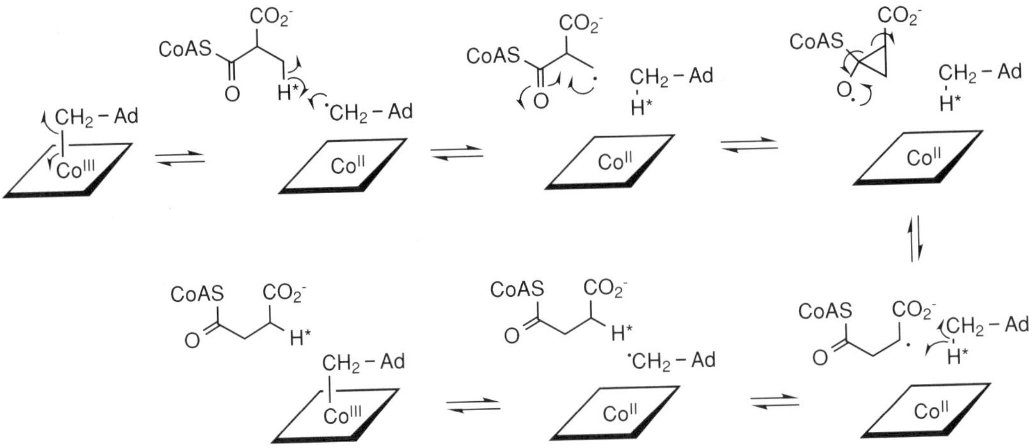

Fig. 10.16 Mechanism for methylmalonyl CoA mutase.

The methylmalonyl CoA mutase reaction is also initiated by formation of the adenosyl radical. This abstracts a hydrogen atom from the substrate methyl group, as shown in Fig. 10.16. Intramolecular ring closure of this primary radical onto the thioester carbonyl forms a cyclopropyl–oxy radical intermediate. Fragmentation of this strained intermediate by cleavage of a different carbon–carbon bond generates a more stable secondary radical adjacent to the carboxylate group. Attachment of the abstracted hydrogen atom gives the succinyl CoA product, and regenerates the vitamin B_{12} cofactor. There is convincing precedent for this mechanism from organic reactions which perform this type of re-arrangement via free radical intermediates.

There are several other vitamin B_{12}-dependent re-arrangements known, each of which involves the kind of migration shown in Fig. 10.14, and each of which can be rationalized by radical intermediates. We have seen in Chapter 6 how the use of radical intermediates in enzyme-catalysed redox reactions allows the catalysis of remarkable reactions. Nature uses both vitamin B_{12} and molecular oxygen to bring about some very interesting reactions involving free radical chemistry.

Problems

1 2,6-Diaminopimelic acid is a naturally occurring amino acid involved in the biosynthesis of L-lysine. *SS*-Diaminopimelate is converted to *RS*-diaminopimelate by a cofactor-independent epimerase enzyme, and *RS*-diaminopimelate is then decarboxylated to L-lysine by a PLP-dependent decarboxylase. What product would you expect if these two reactions were carried out consecutively in ^2H_2O?

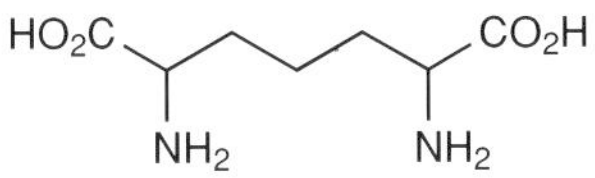

2,6-Diaminopimelic acid

2 The following reactions are part of a pathway for degradation of phenol in *Pseudomonas putida*. Suggest a mechanism for the isomerase reaction. This organism is found to degrade 2-chlorophenol quite readily. What do you think the fate of the additional chlorine atom is?

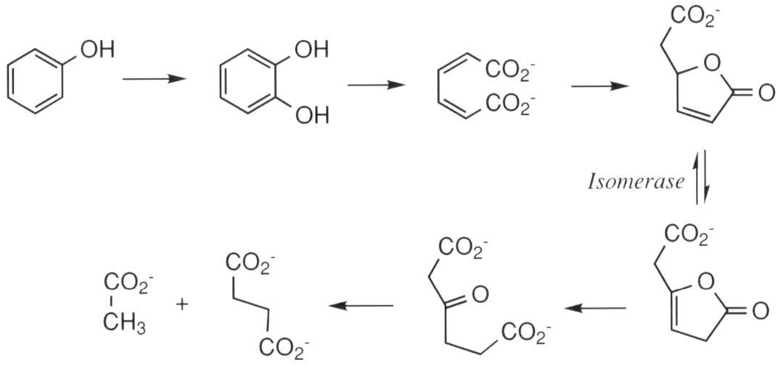

Isomerase

3 Suggest a mechanism for the hydrolase enzyme below, which is involved in an aromatic degradation pathway in *E. coli*.

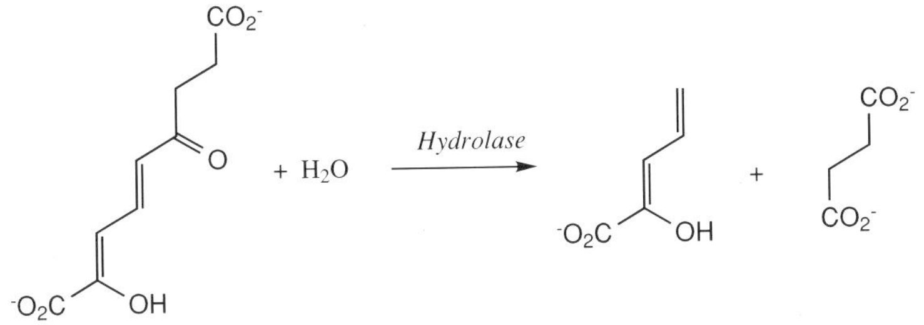

Hydrolase

+ H$_2$O

4 Peptidyl–proline amide bonds are sometimes found in a kinetically stable *cis*-conformation rather than the thermodynamically more favourable *trans*-conformation. An enzyme activity has been found which is capable of catalysing the *cis–trans* isomerization of such peptidyl–proline amide bonds, as shown below. Note: this enzyme is strongly inhibited by the immunosuppressant cyclosporin A, shown in Chapter 1, Fig. 1.3. Suggest possible mechanisms for this isomerase reaction.

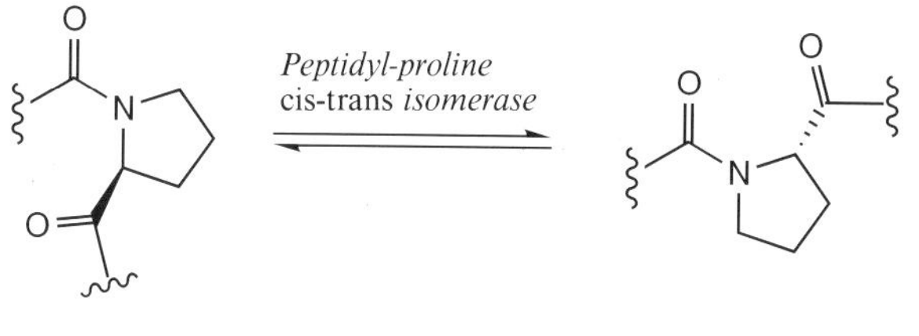

Peptidyl-proline cis-trans isomerase

5 Suggest stepwise and concerted mechanisms for the phosphoenol pyruvate mutase reaction shown below. How might you distinguish between these mechanisms?

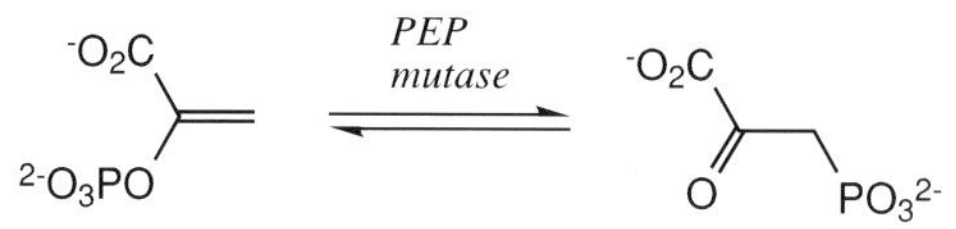

6 The alkaloid natural product hyoscyamine is generated by an enzyme-catalysed intramolecular re-arrangement of littorine, as shown below. The stereochemistry of the conversion is also shown. Suggest a mechanism for this reaction involving vitamin B_{12} as a cofactor.

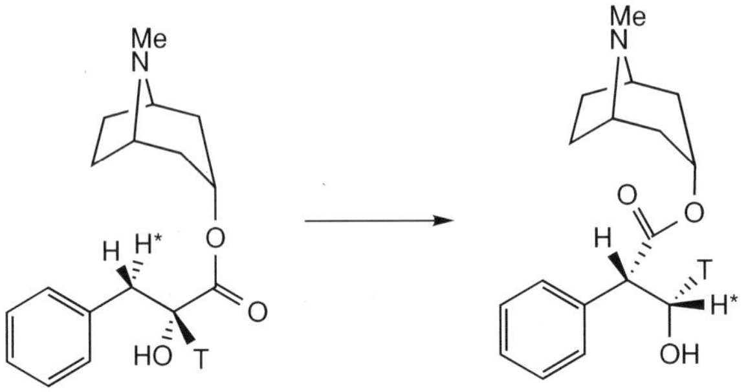

Littorine   Hyoscyamine

Further reading

General

Walsh, C.T. (1979) *Enzymatic Reaction Mechanisms*. Freeman, San Francisco.

Racemases

Adams, E. (1976) *Adv Enzymol*, **44**, 69–138.

Cofactor-independent racemases

Gallo, K.A., Tanner, M.E. & Knowles, J.R. (1993) *Biochemistry*, **32**, 3991–7.
Wiseman, J.S. & Nichols, J.S. (1984) *J Biol Chem*, **259**, 8907–14.

Mandelate racemase

Kenyon, G.L., Gerlt, J.A., Petsko, G.A. & Kozarich, J.W. (1995) *Acc Chem Res*, **28**, 178–86.

Allylic isomerases

Schwab, J.M. & Henderson, B.S. (1990) *Chem Rev*, **90**, 1203–45.

Ketosteroid isomerase

Kuliopulos, A., Mildvan, A.S., Shortle, D. & Talalay, P. (1989) *Biochemistry*, **28**, 149–59.
Xue, L., Talalay, P. & Mildvan, A.S. (1990) *Biochemistry*, **29**, 7491–500.

IPP/DMAPP isomerase

Reardon, J.E. & Abeles, R.H. (1985) *J Am Chem Soc*, **107**, 4078–9.
Reardon, J.E. & Abeles, R.H. (1986) *Biochemistry*, **25**, 5609–16.
Street, I.P., Coffman, H.R., Baker, J.A. & Poulter, C.D. (1994) *Biochemistry*, **33**, 4212–17.

Aldose–ketose isomerases

Rose, I.A. (1975) *Adv Enzymol*, **43**, 491–518.

Chorismate mutase

Chook, Y.M., Ke, H. & Lipscomb, W.N. (1993) *Proc Natl Acad Sci USA*, **90**, 8600–3.
Copley, S.D. & Knowles, J.R. (1985) *J Am Chem Soc*, **107**, 5306–8.
Davidson, M.N., Gould, I.R. & Hillier, I.H. (1995) *J Chem Soc Chem Commun*, 63–4.
Sogo, S.G., Widlanski, T.S., Hoare, J.H., Grimshaw, C.E., Berchtold, G.A. & Knowles, J.R. (1984) *J Am Chem Soc*, **106**, 2701–3.

Vitamin B_{12}-dependent enzymes

Abeles, R.H. & Dolphin, D. (1976) *Acc Chem Res*, **9**, 114–20.
Frey, P.A. (1990) *Chem Rev*, **90**, 1343–57.
Lenhert, P.G. & Hodgkin, D.C. (1961) *Nature*, **192**, 937–8.

11   Non-enzymatic Biological Catalysis

11.1   Introduction

In the final chapter I wish to address the question: are enzymes unique in their ability to catalyse biochemical reactions? The discovery of enzymes and the elucidation of their function provides a basis for understanding how cellular biological chemistry can be catalysed at rates sufficient to sustain life. Many of the reactions catalysed by enzymes are highly complex, yet they can be rationalized in terms of selective enzyme–substrate interactions and well-precedented chemical reactions. Could such selective catalysis be carried out by other biological macromolecules? The answer is yes: we shall meet examples of naturally occurring ribonucleic acid (RNA) molecules capable of catalysing selective self-splicing reactions, and we shall see how the immune system can be used to produce catalytic antibodies.

Finally, a challenging problem for the biological chemist is: if you think you understand how enzymes work, can you design synthetic molecules capable of enzyme-like catalytic properties? This is an exciting area of current research, and we shall see a few ingenious examples of how this problem has been addressed.

11.2   Catalytic RNA

RNA is an important cellular material involved in various aspects of protein biosynthesis. There are three types of RNA which are found in all cells: (i) messenger RNA (mRNA) into which the deoxyribonucleic acid (DNA) sequence of a gene is transcribed before being translated into protein; (ii) transfer RNA (tRNA) which is used to activate amino acids for protein biosynthesis; and (iii) ribosomal RNA (rRNA) which comprises the ribosome protein assembly apparatus. Thus, information transfer and cellular structures were thought in the 1970s to be the only functions of RNA.

In the early 1980s the groups of T.R. Cech and S. Altman reported the startling discovery that certain RNA molecules were capable of catalysing chemical reactions without the assistance of proteins. The first catalytic RNA (or 'ribozyme') to be identified was found in the large subunit rRNA of a ciliated protozoan *Tetrahymena thermophila*. This RNA molecule has the remarkable ability to cut itself out of a large piece of RNA, leaving a shortened RNA template which is subsequently used for protein biosynthesis.

This self-splicing reaction requires no protein-based molecules, but requires two cofactors: a divalent metal ion such as magnesium (Mg^{2+}); and a guanosine (G) monomer, which can either be phosphorylated or bear a free 5'-hydroxyl group.

The mechanism of this self-splicing reaction has been elucidated, and is shown in Fig. 11.1. The RNA precursor is able to bind the guanosine cofactor in the presence of Mg^{2+} ions. The 3'-hydroxyl group of the guanosine

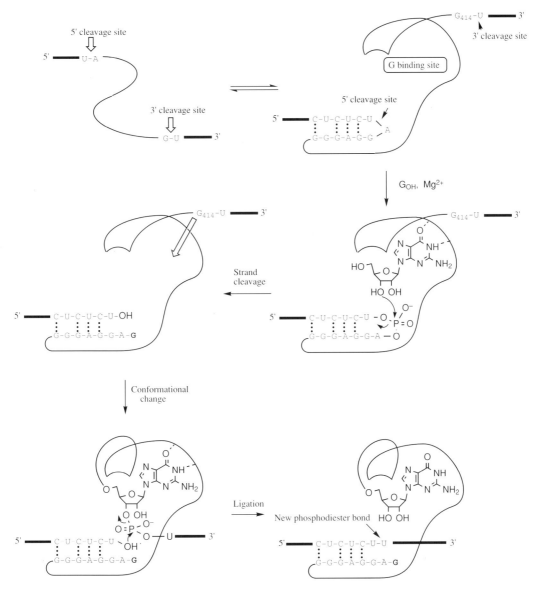

Fig. 11.1 Mechanism of self-splicing reaction of *Tetrahymena* ribozyme.

cofactor then attacks the phosphodiester linkage at the 5'-cleavage site, displacing a free 3'-OH and becoming covalently attached to the RNA molecules. It is thought that the Mg^{2+} acts as a Lewis acid to assist the leaving group properties of the departing 3'-OH. The cleaved RNA strand remains non-covalently bound to the ribozyme through a six-base complementary 'guide sequence', whilst a conformational change brings G_{414} into the guanosine binding site. The free uridine (U) 3'-OH then attacks the G_{414}–U phosphodiester bond, forming the new phosphodiester bond of the shortened RNA, and releasing the ribozyme which terminates in G_{414}.

How effective is this 'ribozyme' as a catalyst? The site of cleavage and sequence selectively of the *Tetrahymena* ribozyme is very high, and the rate constant calculated for the transesterification step is an impressive 350 min^{-1}, some 10^{11}-fold greater than the rate of the corresponding uncatalysed reaction. The rate of reaction is in fact limited by formation of the starting complex rather than the phosphotransfer step, so the *Tetrahymena* ribozyme is highly efficient at catalysing its own self-splicing. However, it is debatable whether the *Tetrahymena* ribozyme can be classified as a true catalyst, since it is modified in structure by the reaction.

There are now a number of examples of such catalytic RNA species, but many questions remain to be resolved. What sort of secondary/tertiary structures are present in catalytic RNA? How does catalytic RNA bind its substrate and cofactors? Do all catalytic RNA use metal ion cofactors, or can they make use of functional groups present in the bases for catalysis? What other types of chemical reactions can they catalyse?

These questions are important since RNA may have been the precursor to DNA as genetic material in the evolution of life on Earth. Whilst DNA is a superb carrier of genetic information, it requires protein in order for the information to be expressed. Proteins, on the other hand, can catalyse lots of important chemical reactions, but require DNA to encode their sequences. Which came first? The answer to this chicken-and-egg problem is probably neither: the fact that RNA can catalyse chemical reactions *and* carry genetic information offers the possibility that it might have been the information storage system of primitive pre-cellular life. In this respect it is interesting to note that many of the coenzymes that are used today in enzymatic reactions contain ribonucleotides: adenosine triphosphate, nicotinamide adenine dinucleotide, flavin adenine dinucleotide, s-adenosyl methionine and vitamin B_{12}. Maybe these molecules (or their ancestors) were key players in prebiotic chemistry.

11.3   Catalytic antibodies

The immune system acts as a major line of defence against foreign substances, be they toxins, proteins or invading micro-organisms. Upon detection of a foreign antigen the immune system generates a 'library' of up to 10^9 antibodies, some of which bind tightly and specifically to the antigen, allowing it to be targeted and destroyed by the body's 'killer' T-cells. The structure of antibodies is illustrated in Fig. 11.2. They are protein molecules consisting of four polypeptide chains, two heavy (H) chains and two light (L) chains, linked together by disulphide bridges. At the two ends of each Y-shaped antibody are the 'variable' regions of the antibody in which variation in sequence is found between antibodies. Most variation is found in the 'hypervariable' regions which make up the antigen combining sites.

Antibodies bind their antigen target very tightly, typical K_d values being 10^{-9}–10^{-11} M, and very selectively. This selectivity of binding is reminiscent of the selectivity of enzyme active sites for their substrates. So, they satisfy one of the criteria for enzyme catalysis: selective substrate recognition. Could antibodies also catalyse chemical reactions? This question was answered by the groups of R.A. Lerner and P.G. Schultz in 1986. They recognized that the major factor underlying enzymatic catalysis is transition state stabilization. If an antibody could be generated which specifically recognizes the transition state of a chemical reaction, and also binds less tightly the substrate and product of the reaction, then it should catalyse the reaction.

Insight into the mechanisms of chemical reactions has provided us with good models for the structures of their transition states. In many cases this has permitted the synthesis of transition state analogues. We have seen earlier how such molecules can act as potent inhibitors of enzyme-catalysed reactions, precisely because enzymes bind the transition state more tightly than either the substrate or product of the reaction. Using a synthetic

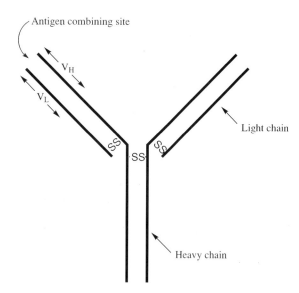

Fig. 11.2 Generalized structure of antibodies. SS, disulphide bridge; V_H, V_L, variable regions of heavy and light chains.

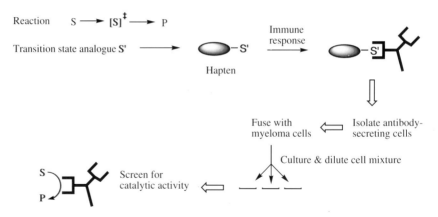

Fig. 11.3 Isolation of catalytic antibodies. P, product; S, substrate.

transition state analogue for a chemical reaction, a protocol was devised for the generation of monoclonal (i.e. single, homogeneous) catalytic antibodies, as shown in Fig. 11.3.

First, the transition state analogue is attached to a protein such as bovine serum albumen in order to generate a sizeable immune response. The 'hapten' thus formed is injected into a mouse or rabbit and the immune response triggered. The antibody secreting cells are then isolated, and these cells are immortalized by fusing them with myeloma (cancer) cells. The immortalized antibody secreting cells are then diluted and screened for catalytic activity, usually by a colorimetric assay. The most active cell lines are then selected, allowing the monoclonal antibodies to be purified and analysed kinetically.

The first catalytic antibodies were generated using phosphonate ester transition state analogues for the tetrahedral intermediate involved in ester hydrolysis reactions. For example, the phosphonate ester shown in Fig. 11.4 acts as a transition state analogue for the hydrolysis of the corresponding ester. Haptens based on this transition state analogue elicited antibodies capable of catalysing the hydrolysis of this ester at rates 10^3–10^5-fold greater then the rate of uncatalysed ester hydrolysis.

Subsequently, haptens have been designed for the more challenging amide hydrolysis reaction. Using a phosphonamidate analogue shown in Fig. 11.5, a catalytic antibody 43C9 has been elicited which is capable of catalysing the amide hydrolysis reaction shown. Antibody 43C9 accelerates the hydrolysis of this amide by a factor of 10^6 at pH 9.0, and the Michaelis–Menten kinetic parameters K_M and k_{cat} can be measured as 0.56 mM and 0.08 min^{-1}, respectively. These catalytic properties are slow by the standards of enzyme-catalysed reactions, but they validate the idea that catalytic antibodies can, in principle, be designed and generated.

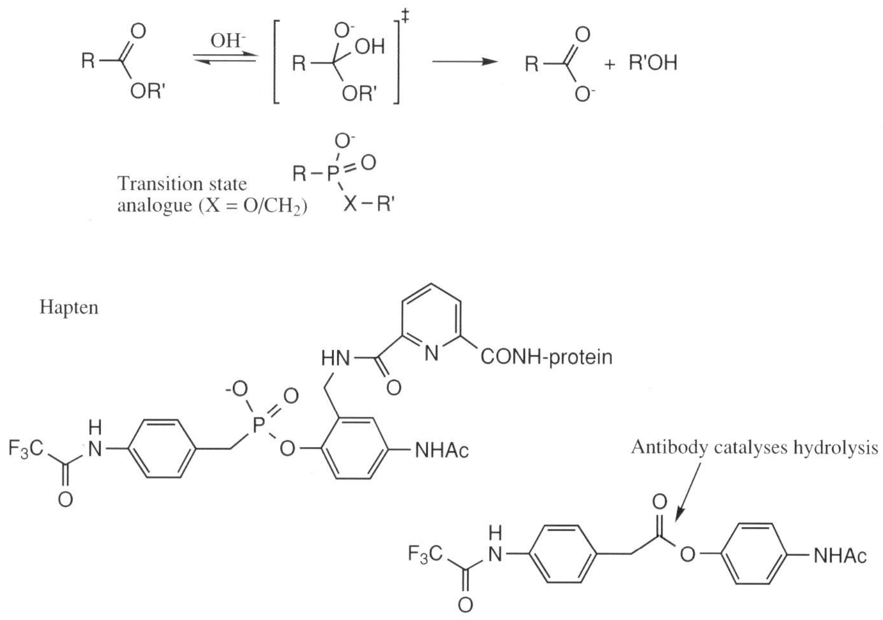

Fig. 11.4 Catalytic antibodies for ester hydrolysis.

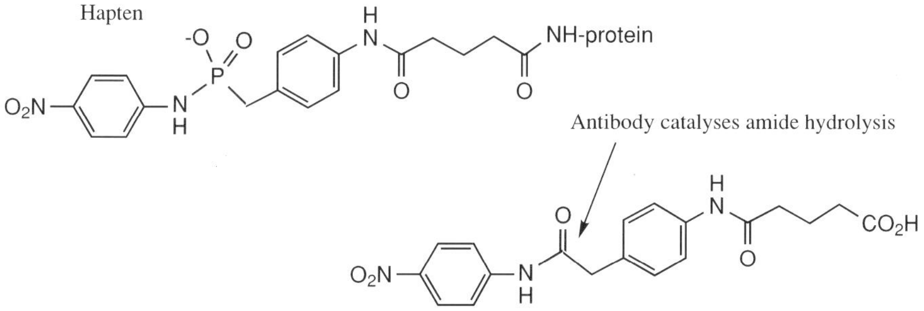

Fig. 11.5 A catalytic antibody for amide hydrolysis.

How are hydrolysis reactions catalysed at the antigen combining sites of these antibodies? The amide and ester hydrolysis reactions catalysed by antibody 43C9 have been analysed in some detail, and evidence has been obtained for the existence of a covalent intermediate in the reaction formed by nucleophilic attack of a histidine (His) side chain. Determination of the amino acid sequence of the antibody, together with molecular modelling studies, suggested that His-L91 and arginine (Arg)-L96 might be involved in antibody catalysis. Mutation of either residue to glutamine gave catalytically inactive antibodies, confirming this suggestion. A mechanism proposed for the ester hydrolysis reaction is shown in Fig. 11.6. In this mechanism His-L91 acts as a nucleophile and Arg-L96 stabilizes the tetrahedral

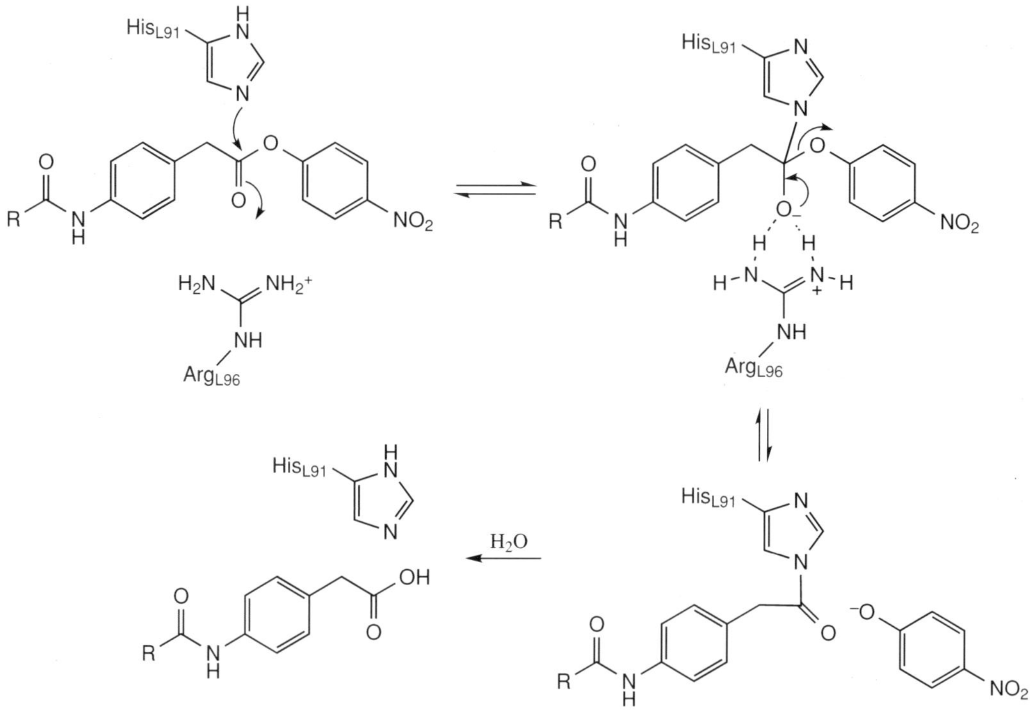

Fig. 11.6 Proposed mechanism for the antibody 43C9-catalysed ester hydrolysis reaction.

transition state in the reaction. Not surprisingly, this antibody is strongly inhibited by the original phosphonamidate analogue.

Catalytic antibodies have now been generated for many types of reactions, and the interested reader is referred to several reviews at the end of the chapter. The final example was mentioned in Chapter 10, Section 10.5. Transition state analogues have been synthesized for the chorismate mutase-catalysed re-arrangement of chorismate to prephenate, shown in Fig. 11.7. One such analogue has been used to generate catalytic antibodies capable of accelerating the re-arrangement by 10^4-fold (as compared to 10^6-fold for the enzyme-catalysed reaction). The kinetic parameters for this catalytic antibody are a K_M of 260 μM and a k_{cat} of 2.7 min^{-1}.

This area of research is of fundamental interest for the understanding of biological catalysis, as much for its limitations as for its successes. The fact that catalytic antibodies can be generated supports the idea that enzymes achieve much of their rate acceleration through transition state stabilization. However, the observation that most catalytic antibodies are much slower than their enzyme counterparts raises the question: what other factors do enzymes utilize to achieve their additional rate acceleration? Perhaps the answer is that enzymes have had millions of years to perfect their catalytic

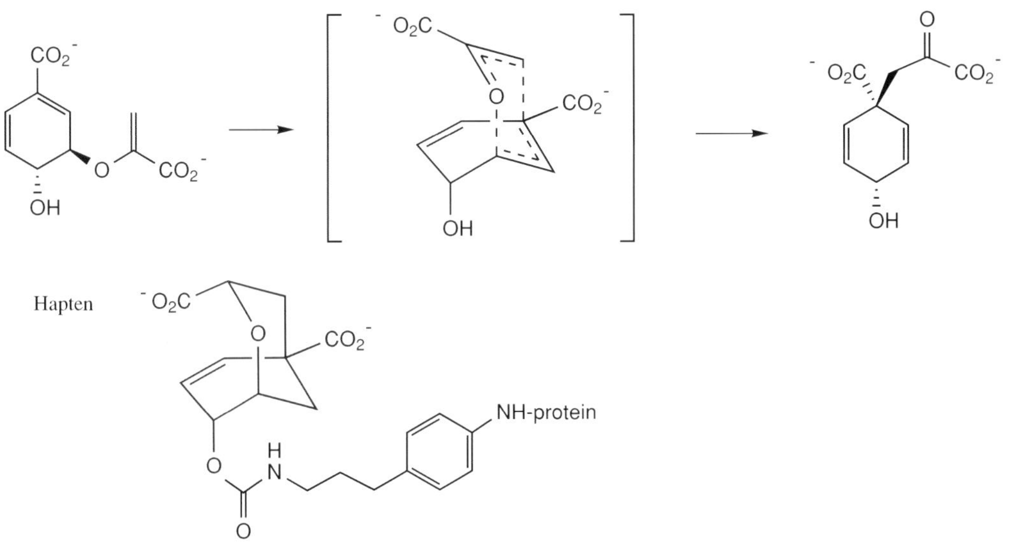

Fig. 11.7 A catalytic antibody for the chorismate mutase reactions.

abilities through such subtle ploys as use of strain, or bifunctional catalysis. The other common limitation of catalytic antibodies is that they often suffer from product inhibition. As discussed in Chapter 3, Section 3.4 it is thermodynamically unfavourable for enzymes to bind their substrates strongly, so perhaps catalytic antibodies are less able to discriminate between transition state and product(s).

11.4   Synthetic enzyme models

One of the final frontiers in biological chemistry is the design of synthetic models for biological catalysis. Can we design and synthesize small or medium-sized molecules which mimic enzymes?

First of all we need to define what requirements we need of such a model, as follows.

1 *Selective substrate binding.* The catalyst must be able to bind its substrate effectively and specifically via a combination of electrostatic, hydrogen-bonding and hydrophobic interactions.

2 *Pre-organization.* The catalyst must have a rigid, well-defined three-dimensional structure, so that when it binds its substrate there is little loss of entropy in going to the transition state of the reaction.

3 *Catalytic groups.* There must be suitably positioned catalytic groups arranged convergently (i.e. pointing inwards towards the substrate), implying that quite a large cavity is required.

4 *Physical properties.* If the catalyst is to be useful in aqueous solution, it should be water soluble, be able to bind its substrate in water, and be active at close to pH 7.

These requirements turn out to be extremely demanding in practice, so there are only a fairly small number of synthetic enzyme-like catalysts which have been developed to date. However, this is a rapidly emerging area of research, so I will give a few examples of current models and advise the interested reader to watch this space!

Historically, the first type of enzyme models developed by D.J. ·Cram used cationic binding sites provided by 'crown ethers', a family of cyclic polyether molecules with a high affinity for metal ions. As well as binding metal ions, crown ethers can bind substituted ammonium cations, which was exploited in the design of the model shown in Fig. 11.8. This model contains pendant thiol groups which can act as catalytic groups for ester hydrolysis, analogous to the cysteine proteases. This catalyst was found to accelerate the hydrolysis of amino acid p-nitrophenyl esters in ethanol by 10^2–10^3-fold compared with an acyclic version, showing the importance of pre-organization of the catalyst. This system also showed some enantioselectivity, being selective for D-amino acid esters by 5–10-fold.

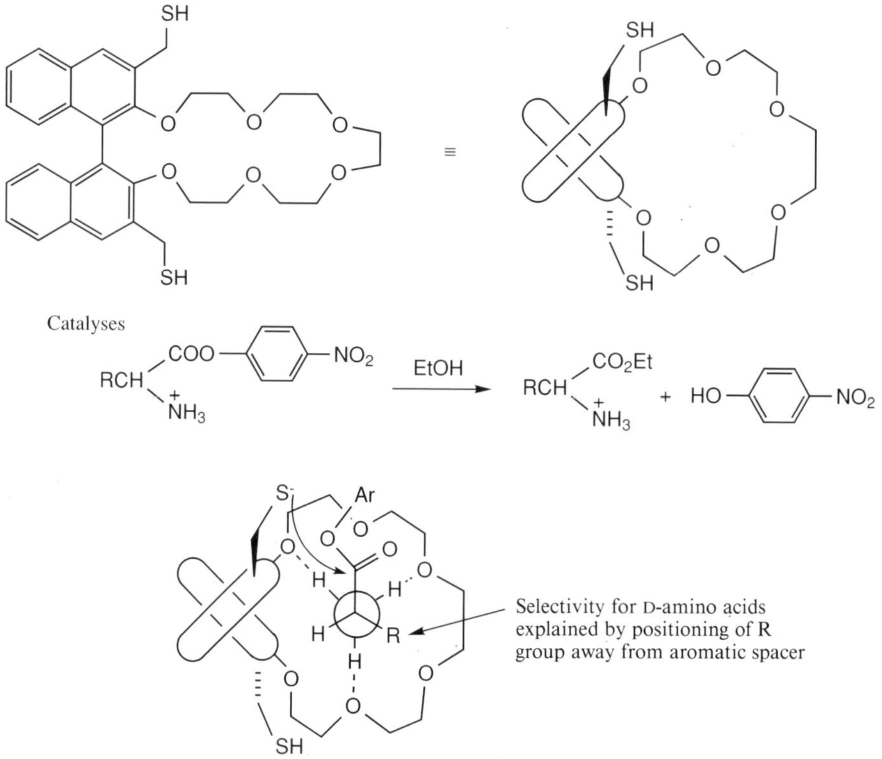

Fig. 11.8 Crown ether catalytic binding site. EtOH, ethanol.

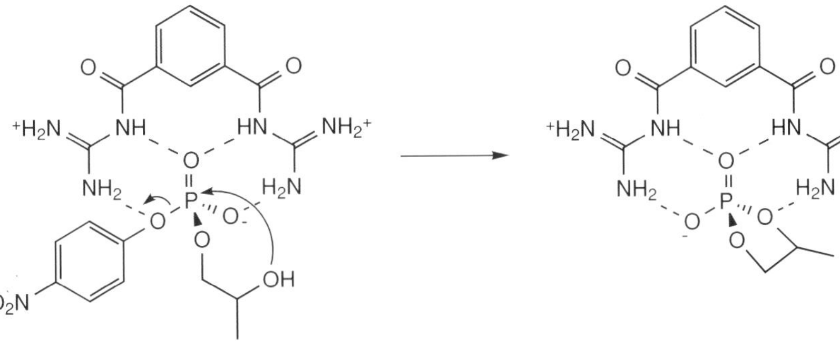

Fig. 11.9 Anionic binding site.

More recently, binding sites have been developed for anionic substrates. One example developed by A.D. Hamilton is a small molecule containing two gaunidinium side chains which is able to bind phosphodiesters via electrostatic and hydrogen-bonding interactions. This system was able to accelerate the rate of an intramolecular phosphodiester hydrolysis reaction, shown in Fig. 11.9, by 700-fold, presumably by a combination of transition state stabilization and protonation of the leaving group.

The most versatile family of enzyme models in current use are the cyclodextrins developed by R. Breslow. These are a family of cyclic oligosaccharide molecules which form a bucket-shaped cavity capable of forming tight complexes with aromatic molecules. These systems offer the advantages that the size of the central cavity is quite large, the binding is effective, they are water soluble and they contain pendant hydroxyl groups around the rim of the 'bucket' which can be functionalized with catalytic groups.

By attaching a blocking agent to a hydroxyl group on one face of the bucket, a more hydrophobic cavity is obtained which is more effective at binding aromatic molecules. One such blocked cavity accelerated the hydrolysis of a *meta*-substituted phenyl acetate, shown in Fig. 11.10, by 3300-fold over the rate of uncatalysed hydrolysis. This can be rationalized, as shown in Fig. 11.10, by the participation of the free hydroxyl groups on the rim of the cavity, which come into close proximity with the *meta*-substituent.

Functionalization of the free hydroxyl groups on the rim of the cavity opens up new possibilities for the design of enzyme models. One example is the cyclodextrin shown in Fig. 11.11, functionalized with two imidazole side chains. This derivative was found to catalyse the regioselective hydrolysis of a cyclic phosphodiester substrate, with a rate constant of 3×10^{-4} s^{-1}. The pH/rate profile of this catalyst showed that maximum activity was obtained at pH 6.3, at which point one of the imidazole groups is protonated and the other deprotonated. The proposed mechanism shown in Fig. 11.11 involves bifunctional acid/base catalysis by the two imidazole groups. The

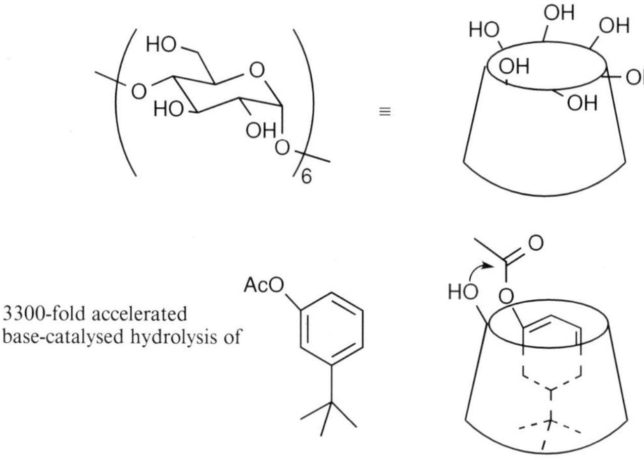

Fig. 11.10 Cyclodextrin-catalysed ester hydrolysis.

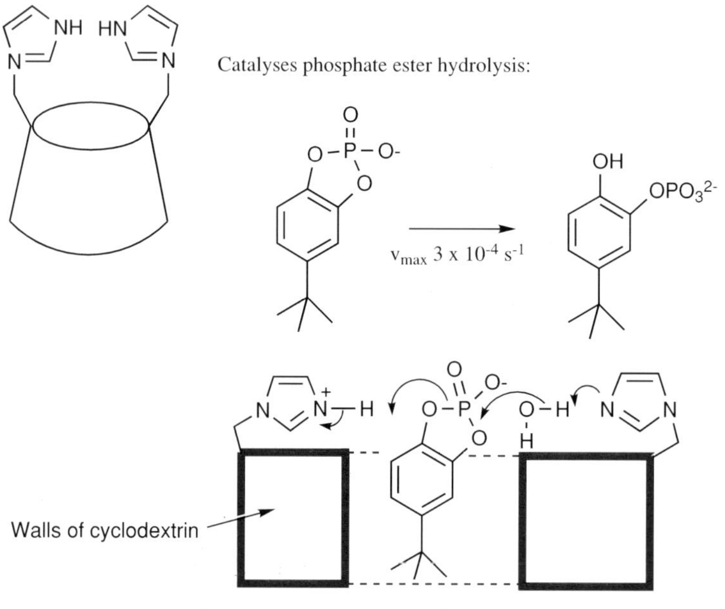

3300-fold accelerated
base-catalysed hydrolysis of

Catalyses phosphate ester hydrolysis:

v_{max} 3 x 10^{-4} s^{-1}

Walls of cyclodextrin

Fig. 11.11 Functionalized cyclodextrin model of ribonuclease A.

regiospecificity of phosphodiester hydrolysis can be rationalized by the orientation of the aromatic substrate in the activity, as shown. This system is a very elegant model for the mechanism of action of ribonuclease A (see Chapter 5, Section 5.5).

Many other types of host–guest systems have been developed in recent years, for example using metal ions to co-ordinate substrate and catalyst functional groups. One alternative approach with which to finish is the use of polymers to generate chiral cavities containing catalytic groups. The idea of

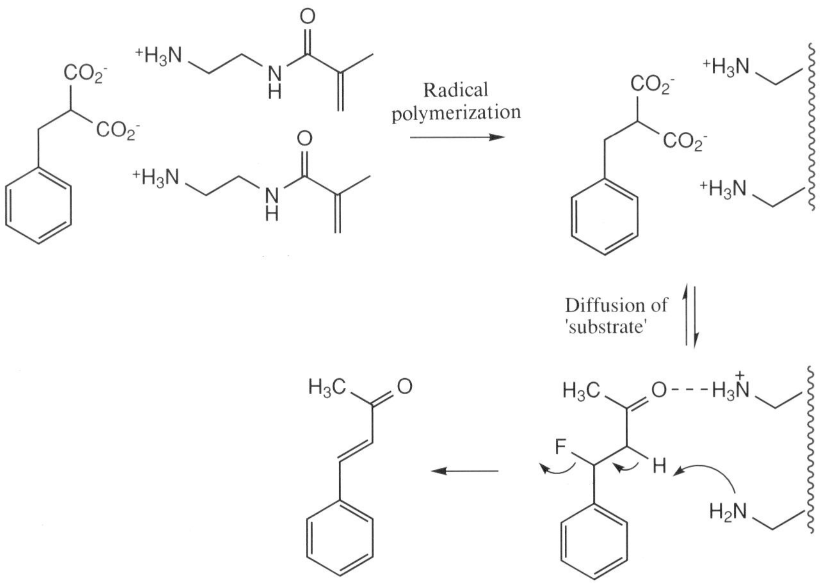

Fig. 11.12 Catalytic polymers.

this approach is to use a small molecule as a template for the co-polymerization of regular cross-linked polymer. After polymerization the template is removed, leaving a complementary cavity inside the polymer which can bind molecules of related structure. For example, co-polymerization of an aromatic dicarboxylic acid with an amino-acrylamide derivative gives a polymer containing pendant amino groups contained within cavities inside the polymer, as shown in Fig. 11.12. After removal of the template, this polymer was found to catalyse the β-elimination reaction of a related substrate with enzyme-like Michaelis–Menten kinetics (K_M 27 mM, k_{cat} 1.1×10^{-2} min^{-1}).

In summary, chemists are using rational design and the ability to synthesize unnatural three-dimensional structures to try to mimic the remarkable catalytic properties of enzymes, and perhaps to generate new types of catalysts. Such catalysts might, for example, be able to catalyse new reactions or operate in organic solvents or at high temperatures. Our current efforts seem crude compared with the biological counterparts, but the future holds great promise.

Problems

1 How reversible do you think the *Tetrahymena* ribozyme-catalysed self-splicing reaction is? What is the potential significance of the reverse reaction?

2 Suggest a hapten that could be used to induce catalytic antibodies for the lactonization reaction below.

3 Explain why the hapten below elicited antibodies capable of catalysing the elimination reaction shown.

4 What type of catalytic properties would you expect from a cyclodextrin covalently modified with: (i) a thiol group; (ii) a riboflavin group? (Note: reduced riboflavin is rapidly oxidized by dioxygen in aerobic solutions.)

Further reading

Catalytic RNA

Cech, T.R. & Bass, B.L. (1986) *Annu Rev Biochem*, **55**, 599–630.
Cech, T.R., Herschlag, D., Piccirilli, J.A. & Pyle, A.M. (1992) *J Biol Chem*, **267**, 17 479–82.

Catalytic antibodies

Benkovic, S.J. (1992) *Annu Rev Biochem*, **61**, 29–54.
Schultz, P.G. (1989) *Acc Chem Res*, **22**, 287–94.
Schultz, P.G. & Lerner, R.A. (1993) *Acc Chem Res*, **26**, 391–5.
Stewart, J.D., Liotta, L.J. & Benkovic, S.J. (1993) *Acc Chem Res*, **26**, 396–404.

Enzyme models

Breslow, R. (1986) *Adv Enzymol*, **58**, 1–60.
Breslow, R. (1995) *Acc Chem Res*, **28**, 146–53.
Dugas, H. (1996) *Bioorganic Chemistry—a Chemical Approach to Enzyme Action*, 3rd edn. Springer-Verlag, New York.
Jubian, V., Dixon, R.P. & Hamilton, A.D. (1992) *J Am Chem Soc*, **114**, 1120–1.
Beach, J.V. & Shea, K.J. (1994) *J Am Chem Soc*, **116**, 379–80.
Cram, D.J. & Cram, J.M. (1978) *Acc Chem Res*, **11**, 8–14.

Appendix 1: Rules for Stereochemical Nomenclature

In order to establish the configuration of a chiral centre containing four different substituents, the four substituents are ranked according to their atomic mass. The substituent with highest atomic mass is labelled '1', the next highest '2', etc. (i.e. O > N > C > H). The molecule is then drawn with substituent '4' pointing away from you, and the substituents 1, 2 and 3 connected in the direction $1\rightarrow2\rightarrow3$. If the arrows form a clockwise right-handed screw, then the centre has the R configuration. If the arrows form an anticlockwise left-handed screw, then the centre has the S configuration. For example, the specifically deuterated sample of ethanol below has four different substituents (O > C > D > H), and the C-1 centre is designated as R. This molecule is then written as $1R$-[1-^{2}H]-ethanol.

1R-[1-^{2}H]-ethanol

Commonly, two or more of the α-substituents have the same atomic mass. In order to prioritize two such α-substituents the atomic mass of their respective β-substituents is analysed in the same way, and the one with higher-ranking β-substituents is given the higher ranking. For example, L-alanine below has two carbon substituents, but the carboxyl group has oxygen β-substituents whereas the methyl group has hydrogen β-substituents. If an α-substituent is attached by a double bond to an atom X, then for the purposes of this analysis the α-substituent has two X β-substituents. L-Alanine is therefore written as 2S-alanine.

L-alanine = 2S-alanine

Prochiral configuration is obtained by replacing the substituent of interest with the next highest available isotope. Commonly, hydrogens attached to prochiral centres are of interest in enzyme-catalysed reactions, in which case the hydrogen is replaced with a deuterium atom, and the configuration of the resulting chiral centre analysed as above. Thus, in the labelled ethanol molecule above the substituent replaced with deuterium is the *proR* hydrogen.

Finally, the direction of attack onto a double bond can be defined in a similar fashion, by designating the three substituents around the sp^2-carbon of interest as 1, 2 and 3 as above. Look down onto one face of the double bond, and if the substituents are arranged in clockwise fashion then you are looking at the *re*-face. If they are arranged in anticlockwise fashion then you are looking at the *si*-face. For example, the two faces of the carbonyl group of glyceraldehyde-3-phosphate are indicated below.

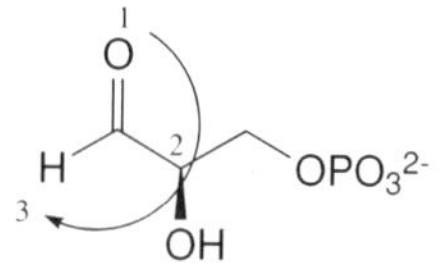

Looking at the *re*- face
of the C-1 carbonyl of
glyceraldehyde-3-phosphate

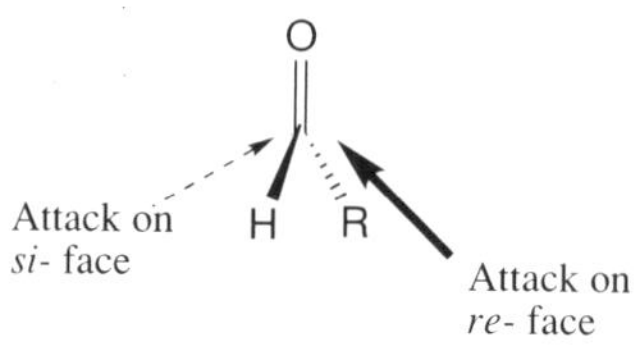

Appendix 2: Amino Acid Abbreviations

Abbreviation		Amino acid	Side chain
A	Ala	Alanine	$-CH_3$
C	Cys	Cysteine	$-CH_2SH$
D	Asp	Aspartic acid	$-CH_2CO_2H$
E	Glu	Glutamic acid	$-(CH_2)_2CO_2H$
F	Phe	Phenylalanine	$-CH_2Ph$
G	Gly	Glycine	$-H$
H	His	Histidine	$-CH_2-imidazole$
I	Ile	Isoleucine	$-CH(CH_3)CH_2CH_3$
K	Lys	Lysine	$-(CH_2)_4NH_2$
L	Leu	Leucine	$-CH_2CH(CH_3)_2$
M	Met	Methionine	$-(CH_2)_2SCH_3$
N	Asn	Asparagine	$-CH_2CONH_2$
P	Pro	Proline	$-(CH_3)_3-N_\alpha$ (cyclic)
Q	Gln	Glutamine	$-(CH_2)_2CONH_2$
R	Arg	Arginine	$-(CH_2)_3NHC(NH_2)^+$
S	Ser	Serine	$-CH_2OH$
T	Thr	Threonine	$-CH(CH_3)OH$
V	Val	Valine	$-CH(CH_3)_2$
W	Trp	Tryptophan	$-CH_2-indole$
Y	Tyr	Tyrosine	$-CH_2-C_6H_4-OH$

Appendix 3: A Simple Demonstration of Enzyme Catalysis

An important (and enjoyable) part of chemistry is to be able to demonstrate chemical principles by experiment. Many of the experiments described in this book are highly technical, but in principle enzymes can be isolated easily from a variety of natural sources. For readers who would like to observe enzyme catalysis at first hand, I include a brief procedure for a second year experiment I have set up at Southampton, which involves an esterase activity easily isolated from orange peel!

The experiment involves the regioselective enzymatic hydrolysis of hydroxybenzoic acid diester derivatives to either the hydroxy-ester or the acyl acid product (Fig. A3.1). The *para*-substituted diester methyl-4-acetoxy-benzoate is commercially available (Aldrich Chemical Co.). Alternatively, diester derivatives of *para-, meta-* or *ortho*-hydroxybenzoic acid can be readily synthesized by acid-catalysed esterification of the carboxyl group, followed by pyridine-catalysed acylation of the phenolic hydroxyl group (Fig. A3.2; see Vogel for standard procedures). The enzymatic hydrolysis of each diester derivative is then assayed against orange peel esterase (as a crude extract) and commercially available (Sigma Chemical Co.) pig liver esterase (PLE) and porcine pancreatic lipase (PPL), using thin-layer chroma-tography to monitor the appearance of one or other hydrolysis product and to measure the rate of hydrolysis.

Preparation of orange peel extract

Peel the outer layer (the 'zest') of one or two oranges using a knife. Add the peel to 50–60 ml of 50 mM sodium citrate buffer (pH 5.5) containing 2.3% sodium chloride, and homogenize in a blender for 2 min until homogeneous. If you have access to a centrifuge, then spin at 12 000 g for 10 min, and decant the clear orange supernatant into a beaker for use in the enzyme assays. The extract is stable for at least 24 h if kept on ice.

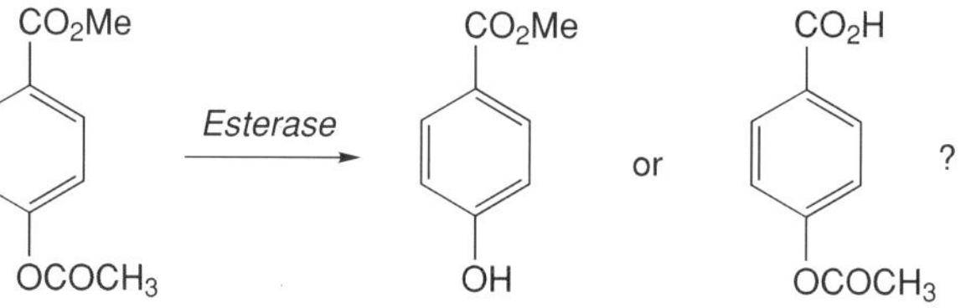

Fig. A3.1.

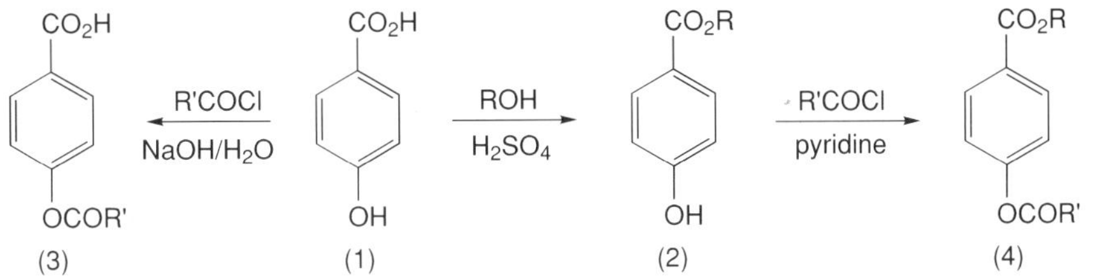

Fig. A3.2.

Assays for enzyme-catalysed hydrolysis

Devise a suitable thin-layer chromatography system to separate the diester (4) from hydroxy-ester (2) and acylated acid (3). Typically, 1 : 1 ethyl acetate/ petroleum ether (60–80° fraction); dichloromethane; or dichloromethane/ 10% methanol are useful eluents. Dissolve 0.1 g of your diester (4) in 5 ml acetone, and set up the following incubations in screw-topped vials:

1   0.1 ml diester (4) solution +0.9 ml 50 mM sodium citrate buffer (pH 5.5);

2   0.1 ml diester (4) solution +0.9 ml orange peel extract;

3   0.1 ml diester (4) solution +0.9 ml PLE stock (1.0 unit ml^{-1} in 50 mM potassium phosphate buffer (pH 7.0));

4   0.1 ml diester (4) solution +0.9 ml PPL stock (1.0 unit ml^{-1} in 50 mM potassium phosphate buffer (pH 7.0)).

Analyse the four incubations by thin-layer chromatography at 30 min, 1 h, 2 h and 24 h time points, and you should see enzymatic hydrolysis to (2) or (3). In all cases that we have examined some enzymatic hydrolysis was observed with at least one of the three enzymes, and in nearly all cases the enzymatic hydrolysis was specific for production of either hydroxy-ester (2) or acyl acid (3).

Further reading

Bugg, T.D.H., Lewin, A.M. & Catlin, E.R. (1997) *J Chem Ed*, **74**, 105–7.
Vogel, A.I. (1989) *Vogel's Textbook of Practical Organic Chemistry*, 5th edn. Longman Harlow.

Appendix 4: Answers to problems

Chapter 2

1 pH 4. (a) Arginine, lysine, histidine, (b) None.

pH 7. (a) Arginine, lysine, histidine (partially). (b) Aspartate, glutamate.

pH 10. (a) Arginine. (b) Aspartate, glutamate, cysteine, tyrosine (partially).

2 Donation of nitrogen lone pair into $C=O$ bond requires nitrogen lone pair to be parallel with π bond. In fact there is some double bond character in the $C=N$ bond, so the amide bond is planar and rigid. *Trans*-conformation more favourable due to steric repulsions between α-carbons in *cis*-conformation. Proline forms a secondary amide linkage in which there is a much smaller difference in energy between *cis*- and *trans*-conformations.

3 First reading frame: Thr–Ala–Glu–Asn–Phe–Ala–Pro–Ser–Agr–Stop.

Second reading frame: Arg–Leu–Lys–Thr–Ser–His–Gln–Val–Asp–.

Third reading frame: Gly–Stop (–Lys–Leu–Arg–Thr–Lys–Ser–Ile).

Stop codon is impossible in the middle of a gene, so second reading frame appears to be the right one.

4 1(Met) × 4(Ala) × 6(Leu) × 6(Ser) × 2(His) × 2(Asp) × 1(Trp) × 2(Phe) × 6(Arg) × 4(Val) = 27 648. A primer based on the His–Asp–Trp–Phe sequence would have a one in eight chance of being correct.

5 (a) This α-helix has six leucines on one face, forming a very hydrophobic surface. This leads to self-aggregation in water to form a four-helix bundle with hydrophobic side chains on the inside and hydrophilic side chains on the surface. There are also favourable Glu_4–Lys_8 and Glu_5–Lys_9 electrostatic interactions which stabilize the helix.

(b) This α-helix has five lysines on one face, designed to mimic the active site of acetoacetate decarboxylase, where the proximity of a second lysine residue leads to a lower pK_a value and a more nucleophilic lysine. This helix showed some catalytic activity as an oxaloacetate decarboxylase.

Chapter 3

1 Intramolecular base catalysis, with the internal tertiary amine acting as a base to deprotonate an attacking water molecule. Tertiary amines are good bases and poor nucleophiles, so nucleophilic catalysis is not feasible.

2 Intramolecular acid catalysis (carboxylic acid will be protonated at pH 4).

3 (a) Phenoxide ion attacks to form a five-membered ring. Effective concentration 7.3×10^4 M. Large rate acceleration due to intramolecular nucleophilic attack, five-membered ring.
(b) Either: (i) general base-catalysed attack of water; or (ii) nucleophilic attack by active site aspartate to give covalent ester intermediate (similar to haloalkane dehalogenase). To examine mechanism (ii) could try to detect covalent intermediate (e.g. by stopped flow methods). Rate enhancement through transition state stabilization, bifunctional catalysis, etc.

4 In the first catalytic cycle, the oxygen atom introduced into the product comes from the aspartate nucleophile, via hydrolysis of the ester intermediate. In subsequent cycles ^{18}O label becomes incorporated into the active site aspartate and is transferred to product.

Chapter 4

1 787 units mg^{-1} ÷ 28 mg μmol^{-1} = 28 μmol product min^{-1} μmol^{-1} enzyme. So, $k_{cat} = 0.47$ s^{-1}

2 Use Lineweaver–Burk or Eadie–Hofstee plot. $v_{max} = 6.0$ nmol min^{-1}, $K_M = 0.71$ mM. $k_{cat} = 2.0$ s^{-1}, $k_{cat}/K_M = 2800$ M^{-1} s^{-1}.

4 Retention of stereochemistry. Not consistent with an $S_N 2$-type displacement.

5 Imine linkage formed between aldehyde group and ε-amino group of an active site lysine residue. Enzymatic reaction goes with retention of configuration at phosphorus, whereas non-enzymatic reaction goes with inversion. Suggests that non-enzymatic reaction is a single displacement, whereas enzymatic reaction is probably a double displacement reaction proceeding via a phosphoenzyme intermediate.

6 Could try to detect phosphoenzyme intermediate using ^{32}P-labelled substrate. In D_2O should see ^2H incorporation into acetaldehyde.

Chapter 5

1 Hydrolysis of acyl enzyme intermediate is rate-limiting step in this case. Rapid formation of acyl enzyme intermediate, releasing a stoichiometric amount of p-nitrophenol, followed by a slower hydrolysis step.

2 Mechanism as for chymotrypsin, via acyl enzyme intermediate. Phosphorylation of active site serine by organophosphorous inhibitors gives a tetrahedral adduct resembling the tetrahedral intermediate in the mechanism, which is hydrolysed very slowly. Differences in toxicity due to: (i) presence of sulphur on parathion, which de-activates the phosphate ester (in insect this is rapidly oxidized to the phosphate oxyester, which then kills the insect!); (ii) differences in active site structure between human and insect enzymes.

Antidote binds to choline site through positively charged pyridinium group. Hydroxylamine group is a potent nucleophile which attacks the neighbouring tetrahedral phosphate ester. Thus, the rate of hydrolysis of the tetrahedral phosphate adduct is rapidly accelerated.

3 Aldehyde is attacked by active site cysteine, generating a thio-hemiacetal intermediate which mimics the tetrahedral intermediate of the normal enzymatic reaction, and is hence bound tightly by the enzyme.

4 Acetate kinase gives acyl phosphate intermediate, acetate thiokinase involves acyl adenylate (RCO.AMP) intermediate.

5 Glycogen phosphorylase cleaves glucose units successively from end of chain. Reaction proceeds with retention of configuration at the anomeric position, so a covalent intermediate is probably formed (cf. lysozyme) by attack of an active site carboxylate. Displacement by phosphate gives α-D-glucose-1-phosphate. Phosphoglucose isomerase contains a phosphorylated enzyme species which transfers phosphate to C-6 to give 1,6-diphosphoglucose. Dephosphorylation at C-1 re-generates phosphoenzyme species. Glucose-6-phosphatase is straightforward phosphate monoester hydrolysis. Defect in glycogen phosphorylase leads to inability to utilize glycogen, so unable to maintain periods of physical exercise.

Chapter 6

1 Alcohol dehydrogenase: $(-0.16(CH_3CHO))-(-0.32(NAD^+)) = +0.16$ V.
Enoyl reductase: $(+0.19(\text{enoyl CoA})) - (-0.32(NAD^+)) = +0.49$ V.
Acyl CoA dehydrogenase: $(+0.25(\text{cytc}_{ox})) - (+0.19(\text{enoyl CoA})) = +0.06$ V.

In acyl CoA dehydrogenase the redox potential for the intermediate FAD must be close to $+0.19$ V if electron transfer is to be thermodynamically favourable. This is right at the top end of the redox potential range for flavin.

2 Enzyme transfers *pro*R hydrogen of NADPH onto C-3 position of substrate. Overall, *syn*-addition of hydrogens from NADPH and water.

3 Transfer of H* onto enzyme-bound NAD$^+$. Resulting C-4 ketone assists the E1cb elimination of C-6 hydroxyl group to give unsaturated ketone intermediate. Transfer of H* from cofactor to C-6 gives product.

4 Transfer of *proS* hydrogen of NADPH to FAD. Then, reverse of acyl CoA dehydrogenase mechanism: transfer of H˙ onto β-position to give α-radical; electron transfer from flavin semiquinone to give α-carbanion; protonation from water at α-position. Note that the same hydrogen transferred from NADPH to FAD is then transferred to substrate. Transfer of H$^-$ is also possible.

5 Formation of flavin hydroperoxide intermediate from FADH$_2$ and O$_2$. Attack on flavin hydroperoxide *para* to phenolic hydroxyl group, followed by elimination of nitrite to give quinone. Quinone then reduced to hydroquinone by second equivalent of NADH.

6 Mechanism of hydroxylation as for general mechanism, via iron(IV)–oxo species. Triple bond of inhibitor is epoxidized to give reactive alkene epoxide intermediate, which re-arranges with 1,2-shift of H* to give a ketene intermediate. This is attacked either by water, giving the by-product, or by an active site nucleophile, leading to covalent modification.

Chapter 7

1 Opening of monosaccharide at C-1 reveals aldehyde substrate. Since enzyme requires no cofactors it presumably proceeds through imine linkage at C-2 of pyruvate, followed by deprotonation at C-3 to give enamine intermediate. Carbon–carbon bond formation between enamine and aldehyde, followed by hydrolysis of resulting imine linkage.

2 Sequential addition of three malonyl CoA units as for fatty acid synthase gives a tetraketide intermediate. Formation of carbanion between first and second ketone groups, followed by reaction with thioester terminus, leads to formation of chalcone. Formation of carbanion adjacent to thioester terminus, followed by reaction with first ketone group, leads after decarboxylation (β-keto acid) to resveratrol. Very similar reactions, so similar

active sites, but differences in position of carbanion formation and carbon–carbon bond formation.

3 Attack of bicarbonate onto phosphate monoester gives enol intermediate and carboxyphosphate. Attack of enol at C-3 onto carboxyphosphate gives carboxylated product.

4 Reverse of normal biotin mechanism. Attack of N-1 of biotin cofactor onto oxaloacetate gives pyruvate and carboxy-biotin intermediate, which carboxylates propionyl CoA to give methylmalonyl CoA.

5 Attack of TPP anion onto keto group gives tetrahedral adduct. Cleavage of α,β-bond using TPP as electron sink gives enamine intermediate. This reacts with aldehyde of second substrate to give another tetrahedral adduct. Detachment from cofactor re-generates TPP anion.

6 Attack of TPP anion onto ketone gives tetrahedral adduct. Decarboxylation gives enamine intermediate as in normal mechanism. At this point the enamine intermediate is oxidized by FAD to give acetyl adduct (probably via H˙ transfer followed by single electron transfer). Hydrolysis by attack of water gives acetate product, and re-generates TPP anion. $FADH_2$ re-oxidized to FAD by O_2.

7 1,3-Migration of pyrophosphate to give linalyl PP. Attack at C-1 to form six-membered ring gives tertiary carbonium ion. Formation of second ring with concerted attack of pyrophosphate gives product. No positional isotope exchange in PP means that PP is bound very tightly (to Mg^{2+}) by enzyme, and is not even free to rotate.

8 Reaction to copalyl PP commenced by protonation of terminal alkene to give tertiary carbonium ion. Two ring closures followed by loss of proton to form C=C. Loss of PP followed by formation of six-membered ring. Closure to form final five-membered ring followed by 1,2-alkyl shift and elimination.

9 Aromatic precursor made from tetraketide intemediate which is methylated by S-adenosyl methionine: carbanion formation between first and second ketones is followed by attack on terminal thioester carbonyl. Formation of two phenoxy radicals *para* to methyl group. Carbon–carbon bond formation via radical coupling. Re-aromatization of left-hand ring followed by attack of phenoxide onto α,β-unsaturated ketone of right-hand ring.

Chapter 8

1 Treat with substrate and NaB^3H_4 (or ^{14}C-substrate $+ NaBH_4$), degrade labelled enzyme by protease digestion, purify labelled peptide and sequence.

2 Overall *syn*-addition. Attack of carboxylate onto α,β-unsaturated carboxylic acid to give enolate intermediate, which protonates on same face.

3 Possibilities are: (i) elimination of water to give enol intermediate, which protonates in β-position; (ii) 1,2-hydride shift from α- to β-position (cf. pinacol re-arrangement). In mechanism (ii) an α-^2H-substituent would be transferred to β-position, in mechanism (i) probably not (unless single base responsible for proton transfer). No intramolecular proton transfer observed in practice, so probably mechanism (i).

4 Oxidation of C-3′-hydroxyl group by enzyme-bound NAD^+. Formation of C-3′-ketone assists elimination of methionine by E1cb mechanism, giving α,β-unsaturated ketone intermediate. Addition of water at C-5′, followed by reduction of C-3′-ketone, re-generating enzyme-bound NAD^+.

5 Isochorismate synthase could be 1,5-addition of water, but all three reactions can be rationalized by attack of an enzyme active site nucleophile at C-2 and allylic displacement of water, giving a covalent intermediate. Attack of water at C-6 gives isochorismate. Attack of ammonia at C-6 or C-4, followed by elimination of pyruvate, gives anthranilate or *p*-aminobenzoate, respectively.

Chapter 9

1 Use of PLP as a four-electron sink. Formation of threonine–PLP adduct, followed by deprotonation in α-position, gives ketimine intermediate. Deprotonation at β-position possible using imine as electron sink. Either one-or two-base mechanisms possible for epimerization process.

2 Formation of aldimine adduct followed by α-deprotonation gives α,β-unsaturated ketimine intermediate. This can be attacked by an active site nucleophile at γ-position to covalently modify enzyme.

3 Formation of aldimine adduct with inhibitor followed by α-deprotonation give a ketimine intermediate which is a tautomeric form of an aromatic amine. Rapid aromatization gives a covalently modified PLP adduct.

4 Use of PLP as a four-electron sink. Formation of amino acid–PLP adduct followed by elimination of phosphate (removal of *proS* hydrogen) gives β,γ-unsaturated intermediate. Protonation at γ-position, followed by attack of water at β-position, and re-protonation at α-position.

5 Attachment of PLP onto α-amino group followed by α-deprotonation gives ketimine intermediate. Hydration of γ-ketone group is followed by cleavage of β,γ-bond, using imine as an electron sink. C–C cleavage and re-protonation proceeds with overall retention of stereochemistry.

Chapter 10

1 One S centre is epimerized by the epimerase enzyme with introduction of an α-^{2}H, but if left to equilibrate both S centres would undergo enzyme-catalysed exchange with ^2H_2O. R centre is decarboxylated, with replacement by ^{2}H, so in principle three atoms of ^{2}H would be found in the L-lysine product.

2 Deprotonation at γ-position to form dienol intermediate is followed by re-protonation at α-position. 2-Chlorophenol is processed to give δ-chloro intermediate. Upon deprotonation at the γ-position loss of chloride ion gives a γ,δ-unsaturated lactone. This is processed by opening of the lactone, and reduction of the α,β-double bond by an NADH-dependent reductase.

3 Protonation of the dienol at C-5 is followed by attack of water (or an active site nucleophile) at the C-6 ketone. Cleavage of the C-5–C-6 bond is then facilitated by the presence of an α,β-unsaturated ketone group which can act as an electron sink.

4 Possibilities are: (i) reversible attack of an active site nucleophile at the amide carbonyl, allowing free rotation of the tetrahedral intermediate; (ii) a 'strain' mechanism in which the enzyme binds the substrate in a strained conformation close to the transition state for rotation of the amide bond. Available evidence points to mechanism (ii), and it is thought that cyclosporin A acts as a mimic of this strained intermediate. Note: the immunosuppressant activity is due to the complex formed beween cyclosporin A and this protein, which is called cyclophilin.

5 Concerted four-electron pericyclic reaction is a disfavoured process. Stepwise reaction possible by transfer of phosphate group to active site group, followed by reaction of enol intermediate with phosphoenzyme species.

6 Abstraction of H* by adenosyl radical, generating a radical intermediate which attacks carbonyl. Cleavage of cyclopropyl intermediate and attachment of H* from adenosine completes reaction.

Chapter 11

1 In principle the reaction is reversible, which would allow the ribozyme to integrate itself into a piece of RNA (reminiscent of the life cycle of some viruses).

2 A cyclic phosphonate ester would be a good transition state analogue. Such an analogue was used to elicit antibodies capable of catalysing this lactonization reaction with an enantiomeric excess of 94%.

3 At neutral pH the tertiary amine will be protonated. When exposed to the immune response, this will elicit a complementary antibody containing a negatively charged carboxylate group in close proximity. This group functions as a base for the elimination reaction. This 'bait and switch' trick has been used in other cases also.

4 (a) Might expect to mimic cysteine proteases, hydrolysing aromatic ester (and possibly amide) substrates. (b) Might expect to mimic FAD-containing oxidases, oxidizing aromatic amines and aromatic thioethers (to sulphoxides).

Index